Very High Speed Integrated Circuits: Heterostructure

SEMICONDUCTORS
AND SEMIMETALS
Volume 30

Semiconductors and Semimetals

A Treatise

Edited by **R. K. Willardson**
ENIMONT AMERICA, INC.
PHOENIX, ARIZONA

Albert C. Beer
BATTELLE COLUMBUS LABORATORIES
COLUMBUS, OHIO

Very High Speed Integrated Circuits: Heterostructure

SEMICONDUCTORS
AND SEMIMETALS

Volume 30

Volume Editor

TOSHIAKI IKOMA

CENTER FOR FUNCTION-ORIENTED ELECTRONICS
INSTITUTE OF INDUSTRIAL SCIENCE
UNIVERSITY OF TOKYO
TOKYO, JAPAN

ACADEMIC PRESS, INC.
Harcourt Brace Jovanovich, Publishers

Boston   San Diego   New York
London   Sydney   Tokyo   Toronto

This book is printed on acid-free paper. ∞

COPYRIGHT © 1990 BY ACADEMIC PRESS, INC.
ALL RIGHTS RESERVED.
NO PART OF THIS PUBLICATION MAY BE REPRODUCED OR
TRANSMITTED IN ANY FORM OR BY ANY MEANS, ELECTRONIC
OR MECHANICAL, INCLUDING PHOTOCOPY, RECORDING, OR
ANY INFORMATION STORAGE AND RETRIEVAL SYSTEM, WITHOUT
PERMISSION IN WRITING FROM THE PUBLISHER.

ACADEMIC PRESS, INC.
1250 Sixth Avenue, San Diego, CA 92101

United Kingdom Edition published by
ACADEMIC PRESS LIMITED.
24-28 Oval Road, London NW1 7DX

The Library of Congress has cataloged this serial title as follows:

Semiconductors and semimetals.—Vol. 1—New York: Academic Press, 1966-

v.: ill.; 24 cm.

Irregular.
Each vol. has also a distinctive title.
Edited by R. K. Willardson and Albert C. Beer.
ISSN 0080-8784=Semiconductors and semimetals

1. Semiconductors—Collected works. 2. Semimetals—Collected works.
I. Willardson, Robert K. II. Beer, Albert C.
QC610.9.S48   621.3815′2—dc19   85-642319
                                          AACR 2   MARC-S

Library of Congress   [8709]
ISBN 0-12-752130-5 (v. 30)

Printed in the United States of America
90 91 92 93   9 8 7 6 5 4 3 2 1

Contents

LIST OF CONTRIBUTORS vii
PREFACE ix

Chapter 1 Fundamentals of Epitaxial Growth and Atomic Layer Epitaxy

H. Watanabe, T. Mizutani, and A. Usui

I. Introduction 1
II. Kinetic and Thermodynamic Comparison of LPE, VPE, MOCVD and MBE . 4
III. Fundamental Growth Parameters in MBE 23
IV. Atomic Layer Epitaxy of GaAs 32
References 50

Chapter 2 Characteristics of Two-Dimensional Electron Gas in III–V Compound Heterostructures Grown by MBE

S. Hiyamizu

I. Introduction 53
II. Modulation Doping in GaAs/AlGaAs Superlattices 55
III. Selectively Doped GaAs/n-AlGaAs Heterostructures 58
IV. Improved Selectively Doped Heterostructures 82
V. Summary 100
References 101

Chapter 3 Metalorganic Vapor Phase Epitaxy for High-Quality Active Layers

T. Nakanisi

I. Introduction 105
II. Growth and Characterization of High-Quality GaAs and GaAlAs . . . 106
III. Fabrication of Device Structures 134
IV. Summary and Future Problems 150
References 152

Chapter 4 High Electron Mobility Transistor and LSI Applications
T. Mimura

I. Introduction 157
II. HEMT Structures and Fabrication Process 158
III. Electrical Characteristics 160
IV. LSI Technology 173
V. Summary 183
    Appendix A 184
    Appendix B 186
    Appendix C 187
    Appendix D 192
    References 192

Chapter 5 Hetero-Bipolar Transistor and Its LSI Application
T. Sugeta and T. Ishibashi

I. Introduction 195
II. Basic Device Characteristics 197
III. Fabrication of AlGaAs/GaAs HBTs 208
IV. High-Frequency Characteristics 216
V. High-Speed Integrated Circuits 221
VI. Conclusions 226
    References 227

Chapter 6 Optoelectronic Integrated Circuits
H. Matsueda, T. Tanaka, and M. Nakamura

I. Introduction 231
II. Design Considerations 234
III. Fabrication Processes 264
IV. High-Speed Transmitter and Receiver 269
V. Conclusions 281
    References 282

INDEX 285
CONTENTS OF PREVIOUS VOLUMES 291

List of Contributors

Numbers in parentheses indicate the pages on which the authors' contributions begin.

S. HIYAMIZU, *Faculty of Engineering Science, Osaka University, 1-1, Machikaneyama-machi, Toyonaka 560, Japan (53)*

T. ISHIBASHI, *Nippon Telegraph and Telephone Company, Electrical Communication Laboratory, 3-1, Morinosato Wakamiya, Atsugi, Kanagawa, 243-01, Japan (195)*

H. MATSUEDA, *Central Research Laboratory, Hitachi Ltd., 1-280, Higashikoigakubo, Kokubunji, Tokyo, 185 Japan (231)*

T. MIMURA, *Fujitsu Laboratories, Ltd., Compound Semiconductor Devices Laboratory, 10-1, Morinosato Wakamiya, Atsugi, Kanagawa 243-01 Japan (157)*

T. MIZUTANI, *Nippon Electric Company, Ltd., 4-1-1, Miyazaki, Miyamae-ku, Kawasaki, Kanagawa 213, Japan (1)*

M. NAKAMURA, *Central Research Laboratory, Hitachi Ltd., 1-280, Higashikoigakubo, Kokubunji, Tokyo, 185 Japan (231)*

T. NAKANISI, *Electron Devices Laboratory, Research and Development Center, Toshiba Company, 1, Komukaitoshiba-cho, Saiwai-ku, Kawasaki, Kanagawa, 210 Japan (105)*

T. SUGETA, *Atsugi Electrical Communication laboratory, Nippon Telegraph and Telephone Company, 3-1, Morinosato Wakamiya, Atsugi, Kanagawa, 243-01 Japan (195)*

T. TANAKA, *Central Research Laboratory, Hitachi Ltd., 1-28, Higashikoigakumbo, Kobunji, Tokyo, 185 Japan (231)*

A. USUI, *Nippon Electric Company, Ltd., Fundamental Research Laboratories, 4-1-1, Miyazki, Miyamae-ku, Kawasaki, Kanagawa 214, Japan (1)*

H. WATANABE, *Microelectronics Research Laboratory, Nippon Electric Company, Ltd., 4-1-1, Miyazaki, Miyamae-ku, Kawasaki, Kanagawa 213, Japan (1)*

Preface

Very high speed integrated circuits are important for supercomputers and for digital communications such as integrated digital networks and land mobile communications. Although silicon integrated circuit technologies are progressing into the submicron range of the scaling rule and the nanosecond range of operational speed, silicon itself has a limit of low mobility as compared with gallium arsenide and some of the other III-V semiconductors. Besides the high electron mobility, III-V semiconductors have the advantage that different compounds, or their alloys, can form a high quality hetero-interface, where a sharp potential change confines electrons into two-dimensional electron gas. The mobility can be made very large in such a two-dimensional electron system, and a field effect transistor that utilizes the two-dimensional electron gas as a channel has the advantages of high speed and low noise. This transistor is referred to using such acronyms as HEMT (high electron mobility transistor), MODFET (modulation doped field effect transistor), and selectively-doped FET, and it is important for very high speed LSIs.

The hetero-junction can also improve a bipolar transistor performance when it is used in the emitter-base junction. This is called HBT (hetero-bipolar transistor), which was proposed long ago and recently has been proved to be excellent. The technologies for using hetero-junction transistors and their integrated circuits include thin film growth of alloy semiconductors on compound substrates, which have been advanced very much in recent years.

Another advantage of hetero structures of III-V semiconductors is the integration of electron devices with photonic devices, thereby yielding optoelectronic integrated circuits. Because optical information processing and optical communication technology play a vital role in the present technical world, optoelectronic integrated circuits have become increasingly important.

This volume is about heterostructures and their circuit applications. Chapter 1 describes the fundamental aspects of epitaxial growth, which forms the basis for molecular beam epitaxy and metalorganic epitaxy; these are the subjects of Chapters 2 and 3. Chapter 4 covers HEMT and its LSI applications and Chapter 5 is dedicated to the hetero-bipolar transistor and its LSI applications. Finally, Chapter 5 describes the technology of optoelectronic circuits.

Because it presents state-of-the-art technologies of III-V heterostructure devices and LSIs, this volume will be very valuable to researchers and engineers working directly in this field as well as to engineers working in telecommunications and high speed computers.

I am indebted to the contributors and editors of this series for making this treatise possible. This book is dedicated especially to Dr. T. Nakanishi, who unfortunately passed away before the publication of this volume.

Toshiaki Ikoma

CHAPTER 1

Fundamentals of Epitaxial Growth and Atomic Layer Epitaxy

Hisatsune Watanabe
Takashi Mizutani
Akira Usui

NEC CORPORATION
FUNDAMENTAL RESEARCH LABORATORIES
TSUKUBA, JAPAN

I. INTRODUCTION	1
II. KINETIC AND THERMODYNAMIC COMPARISON OF LPE, VPE, MOCVD AND MBE	4
1. *Degree of Nonequilibrium State*	4
2. *Chemical Potential Difference: Driving Force of Growth* .	9
3. *V/III Concentration Ratio in Growth Environment* . . .	19
4. *Arsenic Pressure in Growth Environment*	21
III. FUNDAMENTAL GROWTH PARAMETERS IN MBE	23
5. *Controllability in MBE*	23
6. *Mean Free Path and Impinging Rate*	25
7. *Flux Intensity*	27
8. *Real Substrate Temperature in Vacuum*	29
IV. ATOMIC LAYER EPITAXY OF GaAs	32
9. *Growth Principle and Apparatus*	32
10. *Grown Thickness in a Single ALE Cycle*	36
11. *Surface Coverage of Adsorbed Species*	40
12. *Cycle Speed*	44
13. *Advantages of ALE*	44
REFERENCES	50

I. Introduction

GaAs high-speed or microwave devices such as Gunn or impact-avalanche transit-time (IMPATT) diodes and field effect transistors (FETs) have been developed using epitaxial wafers. Now, four epitaxial methods, LPE (liquid-phase epitaxy), halogen transport VPE (vapor-phase epitaxy), MOCVD (metal organic chemical vapor deposition) and MBE (molecular-beam epitaxy), are most widely used in the fields of research and development, and also in production plants. Besides these four epitaxial techniques, various methods have been proposed. There are conceptually new methods and modifications of previous ones, some of them will be used in plants as techniques are improved in the future, but here the discussion is concentrated on the four methods.

Several names such as MOCVD, MOVPE, and OMVPE, have been given to a single method, because of the lack of a world-accepted rule of naming. In this chapter, the terms LPE, VPE, MOCVD, and MBE are used because they are the most popular at present. Their definitions for GaAs growth are as follows:

LPE Growth from arsenic-saturated liquid gallium.
VPE Growth from gas mixture of gallium chloride (GaCl) and arsenic gas. Chloride VPE: $Ga/AsCl_3/H_2$. Hydride VPE: $Ga/HCl/AsH_3/H_2$.
MOCVD Growth from organic gallium gas such as trimethyl gallium (TMG) or triethyl gallium (TEG) and arsine.
MBE Growth from neutral gallium and arsenic molecular beams in vacuum.

LPE was developed for making microwave devices such as Gunn or IMPATT diodes, but now it is exclusively used to make optical devices such as light-emitting diodes, lasers and detectors. Halogen transport vapor-phase epitaxy has been studied by many researchers using various source-gas combinations. At present, two methods are used in production lines. GaAs FETs are fabricated from epitaxial wafers grown by the $Ga/AsCl_3/H_2$ system, which is called chloride VPE because arsenic chloride is used as a starting material, whereas GaAsP LEDs are made by the $Ga/HCl/AsH_3/PH_3/H_2$ system called hydride VPE. MOCVD and MBE were developed for use in production plants of advanced microwave and optical devices that need very sharp interface profiles.

In Part II, a comparative study of these four epitaxial methods is attempted with kinetical and thermodynamical calculations. Questions of what is equilibrium growth, what is nonequilibrium, and how far it is from an equilibrium state are discussed not only qualitatively, but also quantitatively. It will be concluded with calculated values that LPE is nearest to an equilibrium state, whereas MOCVD and MBE are in extreme nonequilibrium. VPE is located between them. We can divide these four methods depending on arsenic-to-gallium concentration ratio (V/III) in their growth environments. Point defects have been frequently discussed from the stoichiometric conditions described by the V/III ratio. However, it will be shown that the GaAs main point defect called EL2 is strongly dependent on the arsenic partial pressure during epitaxial growth. From this point of view, the four methods should be divided into two groups: the LPE–MBE group and the VPE–MOCVD group. The difference of arsenic partial pressures between the two groups is five orders of magnitude. An interesting suggestion is made that an optimum arsenic pressure lies between them.

MBE is carried out under extremely nonequilibrium conditions. It is difficult to determine temperatures and pressures in a vacuum chamber, because these quantities can be defined only under the equilibrium state. Therefore, it is important to treat them as hypothetical quantities when a quantitative study is made. In Part III, MBE growth parameters will be studied in detail from this point of view. MBE has an extremely high controllability for designing an epitaxial multilayer for heterojunction devices, but the growth parameters reported in the literature are surprisingly obscure. This may be because of the lack of a standard measurement method for such nonequilibrium parameters. Among them, mean free path, impinging rate, absolute flux intensity and real substrate temperature will be examined in detail. These fundamental growth parameters should be given in the literature with their measurement methods, or described as estimated or indirect values. This is an essential requirement for a kinetic study of MBE growth and also is very useful for comparing results reported by different research organizations.

In Part IV, a recently developed epitaxial method, atomic layer epitaxy (ALE), is introduced. The ALE method has a monolayer-level controllability with an extremely high uniformity of grown-layer thickness. It will become a powerful technique for obtaining epitaxial layers for ultrathin-film devices and inevitably will be used as a technique to produce epitaxial wafers for integration circuits consisting of monolayer-level controlled devices. The ALE growth condition is completely out of the equilibrium state. For example, no arsenic overpressure is supplied during the gallium adsorption interval, and no gallium is supplied when arsenic gas is introduced to react with the adsorbed species. Therefore, it is impossible to describe the growth system by conventional thermodynamic equations. ALE is only applicable to crystals having a low dissociation pressure, or it can be carried out under the condition that growth reaction is in a completely kinetic limit—namely, when the dissociation velocity can be described to be zero. Kinetic study of the ALE method has just begun, and no detailed analysis has been reported. In Part IV, after a description of the ALE principle and results reported so far, an elementary study of surface reaction kinetics will be tried. Surface coverage calculations will be compared with experimental results, and a kinetic discussion based on impinging rates will be made to improve the slow cycle time of the ALE method. Finally, advantages of the ALE method will be stressed as a highly advanced fabrication technology for monolayer-level controlled future LSIs.

II. Kinetic and Thermodynamic Comparison of LPE, VPE, MOCVD and MBE

It frequently has been said that LPE and VPE are near-equilibrium growths, while MOCVD and MBE are nonequilibrium ones. However, as far as crystal growth is epitaxial, at least the growing surface must be in equilibrium because impinged and adsorbed molecules are given enough time to migrate to epitaxial lattice sites, which are minimum-energy positions of a crystal surface. Otherwise, a polycrystalline or amorphous film would be formed. So, we need an explanation of why we can say near-equilibrium or nonequilibrium, and also how nonequilibrium the growth is.

1. Degree of Nonequilibrium State

An epitaxial growth process is divided into two fundamental processes: an interaction process between the environment and the growing surface, and a surface process. The former includes adsorption, chemisorption, desorption and gas-phase diffusion. The latter includes surface phenomena such as migration of adsorbed species and growth reaction. Whenever growth is epitaxial, the second process must be much faster than the first one. Therefore, whether the state during growth is equilibrium or nonequilibrium can be expressed by quantitatively studying the interaction process between the environment and the growing surface. A nonequilibrium state is determined by the balance of impinging and escaping rates of rate-determining species.

a. Impinging Rate

The frequency of the interaction between the environment and the growing surface is proportional to the impinging and escaping rates of the rate-determining species. J_i, the impinging rate of species i, is a product of its concentration N_i and mean velocity v_i.

$$J_i = \tfrac{1}{4} N_i \times v_i, \tag{1}$$

where the factor $\tfrac{1}{4}$ comes from the averaging integration over the incident angle and the molecular velocity.

Since the rate-determining species in LPE is arsenic in liquid gallium, N_i means arsenic concentration in gallium or arsenic solubility in gallium, C_i:

$$N_i = C_i : \text{LPE}. \tag{2}$$

For VPE, N_i is directly calculated from the partial pressure P_i of the rate-determining species and the substrate temperature T_{sub}:

$$N_i = P_i/RT_{\text{sub}} : \text{VPE}. \tag{3}$$

For MOCVD, it is difficult to determine N_i from the growth environment. The MOCVD rate-determining species has been believed to be the metal organic gallium source, but the molecular form of the rate-determining species is not known. It has not been determined whether it is an original form such as trimethyl gallium, or a radical that is partially decomposed from trimethyl gallium.

The rate-determining species in MBE is well known to be atomic gallium, but its flux intensity at the substrate surface has not been measured as an absolute value.

Therefore, for MOCVD and MBE, concentrations of rate-determining species are estimated from their actual growth rates by Eq. (1);

$$N_i = 4J_i/v_i, \qquad (4)$$

where we assume that J_i is equal to the incorporation rate of gallium, which is identical with the growth rate expressed in unit of atoms/cm² sec. For example, a typical growth rate of 1 μm/hr on a ⟨100⟩ surface corresponds to 6×10^{14} atoms/cm² sec.

The mean velocity v_i in MOCVD is calculated by the gas temperature and the mass M_i of the rate-determining species. We assume that the gas temperature near the substrate surface is equal to the substrate temperature, so that

$$v_i = (8kT_{\text{sub}}/\pi M_i)^{1/2}: \text{MOCVD}. \qquad (5)$$

In MBE, molecular velocity is uniquely determined by cell temperature T_{cell}. Here we have to assume that the cell is Knudsen-type, which enables us to obtain the molecular velocity of the ejected species from the equilibrium pressure in the cell:

$$v_i = (8kT_{\text{cell}}/\pi M_i)^{1/2}: \text{MBE}, \qquad (6)$$

M_i and k are molecular weight of species i and the Boltzmann constant, respectively.

Figure 1 shows concentration profiles of rate-determining species in LPE, VPE, MOCVD, and MBE as a function of distance from the surface. The concentrations at the surface are calculated by Eq. (2) for LPE, by Eq. (3) for VPE, by Eqs. (4) and (5) for MOCVD, and by Eqs. (4) and (6) for MBE. The concentration profiles are schematically shown (not to scale). The concentration of rate-determining species increases from the value at the substrate surface to that in the source environment. Concentrations at the substrate surface in LPE and VPE are assumed to be almost equal to equilibrium values at the substrate temperature. In LPE, for example, the arsenic concentration near the substrate is about 10^{21} cm⁻³ at a typical growth temperature of 800°C, since the temperature decrease for LPE

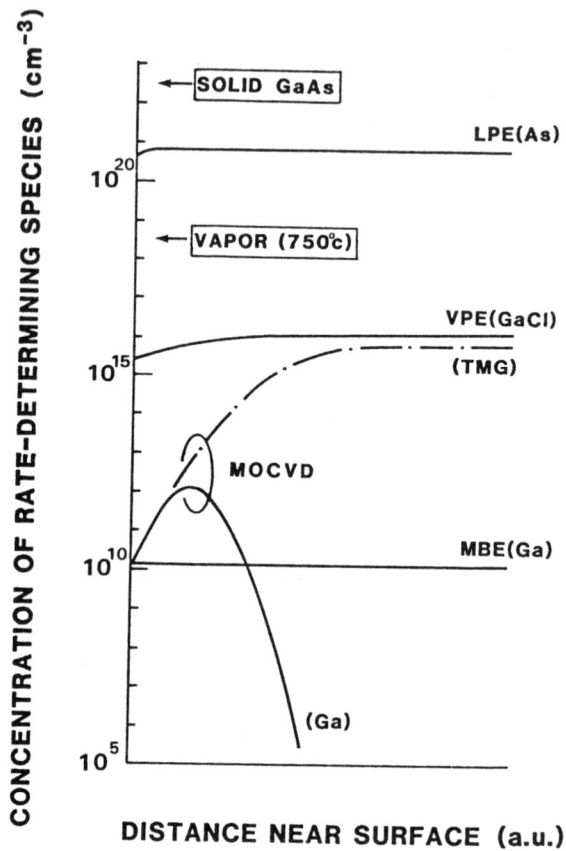

FIG. 1. Concentration profile of rate-determining species in LPE, VPE, MOCVD, and MBE.

growth is usually 2–10°C, the saturated concentration is only a few percent larger than the equilibrium one. The transition distance in LPE is the arsenic diffusion boundary layer, which is fairly small (on the order of a millimeter) as compared with those in other methods because of the low liquid-phase diffusion constant of arsenic.

Molecular density at a typical VPE growth temperature of 750°C is about 7×10^{18} molecules/cm^3 at 1 atm. In VPE, the major reactive species are GaCl, As$_4$ and HCl together with the carrier gas, hydrogen. Among them, GaCl is regarded as the rate-determining species because the GaAs growth rate is most sensitive to GaCl partial pressure (Shaw, 1975). GaCl partial pressure is controlled to about 1×10^{-3} atm in both chloride and hydride VPE so as to obtain high-quality epitaxial layers with mirror surfaces. This pressure corresponds to 7×10^{15} GaCl molecules/cm^3.

GaCl molecules diffuse through the boundary layer that develops above the substrate surface. A concentration gradient exists in the boundary layer.

The corresponding situation in MOCVD is relatively complicated. The growth rate in MOCVD is proportional to the partial pressure of the group III metal organic source. It has been recognized that the group III source gas is the rate-determining species. Since the metal organic source is very unstable at high temperatures, it must be decomposed to atomic gallium and byproduct hydrocarbons before it reaches the substrate, or partially decomposed to some radicals such as $Ga(CH_3)_x$ ($x = 1$ or 2). It has not been identified what kinds of molecules reach the substrate and determine the growth rate. Here, it is assumed that the real rate-determining species is not the original TMG, but atomic gallium or some radicals, and its concentration profile must be similar to the one shown in Fig. 1. The concentration of the atomic gallium or radicals increases as the substrate is approached because of the TMG decomposition, but it decreases to the surface concentration with a profile determined by the diffusion in the boundary layer.

In MBE, the growth rate is exactly proportional to the flux intensity of the gallium molecular beam. Atomic gallium evaporated from a gallium cell is the rate-determining species. Since mutual atomic collision in the molecular beam does not occur in the high-vacuum MBE chamber, the concentration of gallium in the molecular beam is decreased by a factor of $1/r^2$, where r is the distance from the gallium cell. The concentration gradient caused by this is very small in the millimeter range of the distance, so an almost horizontal line is drawn in Fig. 1.

b. Escape Rate

The escape rate J_e is the number of atoms or molecules escaping from the growing surface by reflection, desorption or sublimation.

In LPE, an extremely small percentage of the impinging arsenic is consumed by the epitaxial layer ($J_e \fallingdotseq J_i$), and the small difference gives the net growth rate. In VPE, epitaxial growth takes place with reduction of GaCl vapor by hydrogen, resulting in a decrease in GaCl partial pressure. According to a thermodynamic equilibrium calculation for a growth temperature of 750°C, the GaCl equilibrium partial pressure at the substrate surface is at most 10% lower than the initial GaCl partial pressure in the source gas. This means $J_e \fallingdotseq J_i$ for GaCl in VPE.

In MOCVD and MBE at substrate temperatures around 650°C or lower, the etching rate by sublimation is negligibly small as compared with the growth rate, so $J_e \ll J_i$.

c. *Nonequilibrium State Parameter* λ

Even if a small part of the environment near the substrate surface is in thermodynamic equilibrium, the whole growth system does not reach an equilibrium state unless growth is stopped. In order to express how the system is nonequilibrium as a whole, a nonequilibrium parameter λ is introduced as defined by the following equation:

$$\lambda = (J_i - J_e)/(J_i + J_e). \tag{7}$$

The parameter λ means the net difference between the impinging rate and the escaping rate, being normalized by the total frequency of surface phenomena $(J_i + J_e)$. Since the growth rate is the difference of the impinging rate and the escaping rate,

$$\lambda = (\text{growth rate})/(J_i + J_e). \tag{8}$$

According to this definition, $-1 \leq \lambda \leq 1$. The nonequilibrium parameter λ is zero when the system is in equilibrium, i.e., at zero growth rate $(J_i = J_e)$. λ becomes nearly unity when the growth takes place under extreme nonequilibrium conditions, where the impinging rate is much larger than the escaping rate. A small positive value means growth under near-equilibrium conditions. Negative values mean substrate etching, and $\lambda = -1$ corresponds to an etching reaction with the nonequilibrium extreme $(J_i \ll J_e)$.

Calculated values of λ for typical growth conditions of LPE, VPE, MOCVD, and MBE are summarized in Table I. The nonequilibrium parameter λ is almost zero for LPE and VPE, which means that these methods are near-equilibrium systems. The most striking feature is the presence of a large difference in the nonequilibrium parameter between LPE and VPE. We have frequently approximated for LPE and VPE that the growth environment very near the substrate reaches an equilibrium state, and reactant concentrations at the substrate surface are calculated as perfect-equilibrium ones. This is true as long as the concentration difference between points near to and far from the surface is in question, because the resultant difference is actually small as compared with those for MOCVD and MBE. However, as show in Table I, when we discuss the system by using the nonequilibrium parameter, LPE and VPE should be said to be near equilibrium. Comparing LPE and VPE, LPE is much closer to zero than VPE. Since we assume that the etching reaction can be neglected in MOCVD and MBE, the nonequilibrium parameter becomes unity. This means a completely nonequilibrium growth. Exactly speaking, we cannot ignore etching or sublimation of the GaAs substrate at growth temperatures around 650°C. At this substrate temperature, arsenic escapes with a relatively high partial pressure, and we need arsenic overpressure to suppress thermal decomposition of the substrate

TABLE I

Concentration of Rate-Determining Species at Surface C_i, Impinging Rates R_i, Typical Growth Rate R_g, Mean Velocity of Impinging Species $\langle v \rangle$ and Nonequilibrium Factor λ

Growth method	C_i^a (cm^{-3})	R_i (1/cm² sec)	R_g (1/cm² sec)	$\langle v \rangle^b$ (cm/sec)	λ
LPE	1×10^{21}	1×10^{25}	4×10^{17}	4×10^4	4×10^{-8}
VPE	7×10^{15}	8×10^{19}	1×10^{16}	3×10^4	1×10^{-4}
MOCVD	4×10^{16}	6×10^{14}	6×10^{14}	4×10^4	1
MBE	4×10^{10}	6×10^{14}	6×10^{14}	4×10^4	1

^{a} Values of C_i in MOVCD and MBE are calculated from values of R_i, which are assumed to be equal to growth rates R_g, and from values of mean velocity $\langle v \rangle$.

^{b} Mean velocity $\langle v \rangle$ is calculated as that of the impinging species. In the cases of LPE, VPE and also MOCVD, the mean velocity is determined by the environment temperature near the substrate surface. It is assumed that the environment temperature is equal to the substrate temperature of 800°C. In MBE, the mean velocity of the impinging species is uniquely determined by the cell temperature. In calculation, Knudesen-type evaporation at a cell temperature of 950°C is assumed, namely, the mean velocity of molecular beam species is equal to that of molecules with an equilibrium vapor pressure at 950°C.

before epitaxial growth is started. However, the gallium decomposition pressure is fairly low, so we can approximate that the etching rate by thermal decomposition is low enough to assume $J_i \gg J_e$.

The answer to the question of which is more nonequilibrium, MOCVD or MBE, is dependent on growth conditions such as substrate temperature and arsenic pressure. The recent high-temperature-growth MBE seems to need an investigation of the thermodynamic equilibrium state.

A difference in the nonequilibrium parameter of five to seven orders of magnitude is found among these four epitaxial methods. The smallest value, that of LPE, is due to the highest concentration of the rate-determining species. The next smallest, that of VPE, is also very near to zero. Therefore, these two methods can be recognized to be near-equilibrium growth, while both MOCVD and MBE are far from equilibrium and exist at the nonequilibrium limit.

The nonequilibrium parameter λ, defined as a balance of impinging and escape rates, considering that an actual growth rate is a good parameter to express the degree of nonequilibrium state of the growth system. Especially since it can clearly express the small difference between LPE and VPE, both of which have been qualitatively recognized as equilibrium growths.

2. Chemical Potential Difference: Driving Force of Growth

At equilibrium, there is no chemical potential difference between the substrate and the growth environment. To achieve crystal growth, a supersaturated atmosphere is required. Crystal growth continues as long as the chemical potential of the growth environment is higher than that of the substrate.

In this section, the chemical potential difference between the growth environment and the substrate, which is the driving force of growth, is derived for LPE, VPE, MOCVD, and MBE, and a comparative study will be made from a thermodynamic point of view.

a. Chemical Potential of Substrate

The chemical potential of a solid, μ_s, is defined as the Gibbs free energy increase in the solid after adding one mole of an identical solid at constant temperature and pressure.

$$\mu_s = \left.\frac{\partial G}{\partial n}\right|_{T,P}. \tag{9}$$

G and n are the Gibbs free energy and mole number of the solid, respectively. When the Gibbs free energy is expressed as the molar energy, the molar chemical potential is equal to the Gibbs free energy, which is a function of enthalpy H, entropy S and temperature T:

$$\mu_s = G = H - TS = \int_{298}^{T} C_p \, dT - T \int_{298}^{T} \frac{C_p}{T} \, dT. \tag{10}$$

The GaAs specific heat C_p at constant pressure is known as a function of temperature (Cox and Pool, 1967):

$$C_p(\text{GaAs}) = 5.4 + 7.3 \times 10^{-4} T \quad (\text{cal/mol} \cdot \text{K}). \tag{11}$$

The GaAs substrate chemical potential at growth temperature can be calculated from Eqs. (10) and (11).

b. Chemical Potential Difference in LPE

The growth reaction in LPE is expressed as follows:

$$\text{Ga}(l) + \text{As}(\text{in Ga liquid}) = \text{GaAs}(s). \tag{12}$$

In equilibrium at temperature T,

$$\mu_{\text{GaAs}}(T) = \mu_{\text{liquid}}(T)$$
$$= \mu_{\text{Ga}}(T) + \mu_{\text{As}}(T). \tag{13}$$

The chemical potential of component i is defined as

$$\mu_i(T) = \mu_{0,i}(T) + RT \ln A_i, \tag{14}$$

where $\mu_{0,i}(T)$ is the chemical potential of pure liquid of a component i at T, A_i the activity of i in the liquid solution, and R the gas constant. The activity is expressed by the activity coefficient γ_i and the concentration X_i:

$$A_i = \gamma_i X_i. \tag{15}$$

When arsenic-saturated liquid gallium at T_1 is cooled to temperature T_2 without any deposition of solid GaAs, the chemical potential in the supercooled liquid is

$$\mu_i^*(T_2) = \mu_{0,i}(T_2) + RT_2 \ln \gamma_i X_i(T_1). \tag{16}$$

The chemical potential difference between the supercooled liquid and the GaAs substrate, $\Delta\mu_{\text{GaAs}}$, is obtained as follows:

$$\begin{aligned}\Delta\mu_{\text{GaAs}} &= \mu_{\text{liquid}}^*(T_2) - \mu_{\text{GaAs}}(T_2) \\ &= \mu_{\text{Ga}}^*(T_2) + \mu_{\text{As}}^*(T_2) - \mu_{\text{Ga}}(T_2) - \mu_{\text{As}}(T_2) \\ &= \mu_{0,\text{Ga}}(T_2) + RT_2 \ln \gamma_{\text{Ga}} X_{\text{Ga}}(T_1) + \mu_{0,\text{As}}(T_2) + RT_2 \ln \gamma_{\text{As}} X_{\text{As}}(T_1) \\ &\quad - \mu_{0,\text{Ga}}(T_2) - RT_2 \ln \gamma_{\text{Ga}} X_{\text{Ga}}(T_2) - \mu_{0,\text{As}}(T_2) - RT_2 \ln \gamma_{\text{As}} X_{\text{As}}(T_2) \\ &= RT_2 \ln[\gamma_{\text{Ga}} X_{\text{Ga}}(T_1) \gamma_{\text{As}} X_{\text{As}}(T_1)]/[\gamma_{\text{Ga}} X_{\text{Ga}}(T_2) \gamma_{\text{As}} X_{\text{As}}(T_2)]. \end{aligned} \tag{17}$$

Since the activity coefficient is not a strong function of temperature and the gallium concentration difference at T_1 and T_2 is very small, we can assume that $\gamma_i(T_1) \doteq \gamma_i(T_2)$ and $X_{\text{Ga}}(T_1) \doteq X_{\text{Ga}}(T_2)$, and, hence,

$$\Delta\mu_{\text{GaAs}} = RT_2 \ln[X_{\text{As}}(T_1)/X_{\text{As}}(T_2)]. \tag{18}$$

When the equilibrium constant of the reaction in Eq. (12), $K(T)$, is known as a function of temperature, $\Delta\mu_{\text{GaAs}}$ can be directly calculated by substituting $K(T)$ into Eq. (17):

$$\Delta\mu_{\text{GaAs}} = RT_2 \ln[K(T_1)/K(T_2)]. \tag{19}$$

c. Chemical Potential Difference in VPE

In halogen transport vapor-phase epitaxy using the Ga/AsCl$_3$/H$_2$ chloride VPE system or the Ga/HCl/AsH$_3$/H$_2$ hybrid VPE system, the main growth reaction is

$$\text{GaCl} + \tfrac{1}{4}\text{As}_4 + \tfrac{1}{2}\text{H}_2 = \text{GaAs} + \text{HCl}. \tag{20}$$

The chemical potential of a vapor species i is defined by its partial pressure P_i and the reaction coefficient v_i of the reaction in Eq. (20). Here, v_i is $\tfrac{1}{4}$ for As$_4$ and $\tfrac{1}{2}$ for H$_2$:

$$\mu_i(T) = \mu_{0,i}(T) + v_i RT \ln P_i. \tag{21}$$

$\mu_{0,i}(T)$ is the chemical potential of the vapor species i under standard conditions (1 atm and 298 K). The chemical potential difference VPE is the difference between the chemical potential of the source gas that is supercooled to the substrate temperature T_2, $\mu_{\text{gas}}^*(T_2)$, and that of the substrate, $\mu_{\text{GaAs}}(T_2)$.

$$\Delta\mu_{\text{GaAs}} = \mu_{\text{gas}}^*(T_2) - \mu_{\text{GaAs}}(T_2)$$

$$= \mu_{\text{GaCl}}^*(T_2) + \tfrac{1}{4}\mu_{\text{As}_4}^*(T_2) + \tfrac{1}{2}\mu_{\text{H}_2}^*(T_2) - \mu_{\text{HCl}}^*(T_2)$$

$$- \mu_{\text{GaCl}}(T_2) - \tfrac{1}{4}\mu_{\text{As}_4}(T_2) - \tfrac{1}{2}\mu_{\text{H}_2}(T_2) + \mu_{\text{HCl}}(T_2)$$

$$= \mu_{0,\text{GaCl}}(T_2) + RT_2 \ln P_{\text{GaCl}}(T_1) + \tfrac{1}{4}\mu_{0,\text{As}_4}(T_2) + \tfrac{1}{4}RT_2 \ln P_{\text{As}_4}(T_1)$$

$$+ \tfrac{1}{2}\mu_{0,\text{H}_2}(T_2) + \tfrac{1}{2}RT_2 \ln P_{\text{H}_2}(T_1) - \mu_{0,\text{HCl}}(T_2) - RT_2 \ln P_{\text{HCl}}(T_1)$$

$$- [\mu_{0,\text{GaCl}}(T_2) + RT_2 \ln P_{\text{GaCl}}(T_2) + \tfrac{1}{4}\mu_{0,\text{As}_4}(T_2) + \tfrac{1}{4}RT_2 \ln P_{\text{As}_4}(T_2)$$

$$+ \tfrac{1}{2}\mu_{0,\text{H}_2}(T_2) + \tfrac{1}{2}RT_2 \ln P_{\text{H}_2}(T_2) - \mu_{0,\text{HCl}}(T_2) - RT_2 \ln P_{\text{HCl}}(T_2)]$$

$$= RT_2 \ln \frac{P_{\text{GaCl}}(T_1)\{P_{\text{As}_4}(T_1)\}^{1/4}\{P_{\text{H}_2}(T_1)\}^{1/2}\{P_{\text{HCl}}(T_1)\}^{-1}}{P_{\text{GaCl}}(T_2)\{P_{\text{As}_4}(T_2)\}^{1/4}\{P_{\text{H}_2}(T_2)\}^{1/2}\{P_{\text{HCl}}(T_2)\}^{-1}}. \tag{22}$$

Since the equilibrium constant $K(T)$ of the reaction of Eq. (20) is written by

$$K(T) = P_{\text{GaCl}}(T)\{P_{\text{As}_4}(T)\}^{1/4}\{P_{\text{H}_2}(T)\}^{1/2}\{P_{\text{HCl}}(T)\}^{-1}, \tag{23}$$

we obtain the chemical potential difference in VPE with $K(T)$:

$$\Delta\mu_{\text{GaAs}} = RT_2 \ln [K(T_1)/K(T_2)]. \tag{24}$$

It should be noticed that we can use Eq. (24) only for the Ga/AsCl$_3$/H$_2$ VPE system. In this system, chemical reactions in the gallium source region and the substrate region can be expressed as in Eq. (20), where T_1 and T_2 correspond to the gallium source temperature and the substrate temperature, respectively.

On the other hand, in the Ga/HCl/AsH$_3$/H$_2$ hydride VPE system, there is no reaction between the metallic gallium source and arsenic vapor (AsH$_3$) because AsH$_3$ gas bypasses the gallium reservoir, and T_1 does not mean the gallium source temperature. We cannot use Eq. (24) for the hydride VPE system. Instead, the chemical potential of the supercooled atmosphere at the substrate region, $\mu_{\text{gas}}^*(T_2)$, is expressed by the following equation:

$$\mu_{\text{gas}}^*(T_2) = \mu_{\text{GaCl}}^*(T_2) + \tfrac{1}{4}\mu_{\text{As}_4}^*(T_2) + \tfrac{1}{2}\mu_{\text{H}_2}^*(T_2) - \mu_{\text{HCl}}^*(T_2)$$

$$= \mu_{0,\text{GaCl}}(T_2) + RT_2 \ln P_{0,\text{GaCl}} + \tfrac{1}{4}\mu_{0,\text{As}_4}(T_2) + \tfrac{1}{4}RT_2 \ln P_{0,\text{As}_4}$$

$$+ \tfrac{1}{2}\mu_{0,\text{H}_2}(T_2) + \tfrac{1}{2}RT_2 \ln P_{0,\text{H}_2} - \mu_{0,\text{HCl}}(T_2) - RT_2 \ln P_{0,\text{HCl}}$$

$$= \mu_{0,\text{gas}}(T_2) + RT_2 \ln [P_{0,\text{GaCl}}(P_{0,\text{As}_4})^{1/4}(P_{0,\text{H}_2})^{1/2}/P_{0,\text{HCl}}], \tag{25}$$

where

$$\mu_{0,\text{gas}}(T_2) = \mu_{0,\text{GaCl}}(T_2) + \tfrac{1}{4}\mu_{0,\text{As}_4}(T_2) + \tfrac{1}{2}\mu_{0,\text{H}_2}(T_2) - \mu_{0,\text{HCl}}(T_2) \quad (26)$$

is the chemical potential of the gas mixture of GaCl, As_4, H_2, and HCl, which all have one atmospheric pressure at T_2, and $P_{0,i}$ is the partial pressure of the species i in the supercooled vapor phase at the substrate region before the growth reaction proceeds. When all the specific heats of GaCl, As_4, H_2, and HCl are available, we can calculate $\mu_{0,\text{gas}}(T_2)$. However, this is not the usual case. Instead of Eq. (26), it is better to use the following equation:

$$\Delta\mu_{\text{GaAs}} = RT_2 \ln \frac{P_{0,\text{GaCl}}\{P_{0,\text{As}_4}\}^{1/4}\{P_{0,\text{H}_2}\}^{1/2}\{P_{0,\text{HCl}}\}^{-1}}{P_{\text{GaCl}}(T_2)\{P_{\text{As}_4}(T_2)\}^{1/4}\{P_{\text{H}_2}(T_2)\}^{1/2}\{P_{0,\text{HCl}}(T_2)\}^{-1}}$$

$$= RT_2 \ln [P_{0\,\text{GaCl}}\{P_{0\,\text{As}_4}\}^{1/4}\{P_{0,\text{H}_2}\}^{1/2}\{P_{0\,\text{HCl}}\}^{-1}]/K(T_2). \quad (27)$$

In hydride VPE, HCl and AsH_3 are used as starting gases with controlled partial pressures. The source HCl gas reacts with metallic gallium, forming GaCl gas:

$$HCl + Ga = GaCl + \tfrac{1}{2}H_2. \quad (28)$$

The source arsine gas is decomposed, forming As_4, before the substrate region.

$$4AsH_3 = As_4 + 6H_2. \quad (29)$$

Thus-formed GaCl and As_4 gases are not mixed at the gallium source region but are mixed just before the substrate region, forming the supercooled atmosphere.

$P_{0,\text{GaCl}}$ and P_{0,As_4} are calculated with the equilibrium constants K_1 of Eq. (28) and K_2 of Eq. (29). It is a good approximation to assume

$$P_{0,\text{GaCl}} = P_{i,\text{HCl}} \quad (30)$$

and

$$P_{0,\text{As}_4} = \tfrac{1}{4}P_{i,\text{AsH}_3}, \quad (31)$$

where $P_{i,\text{HCl}}$ and P_{i,AsH_3} are source HCl and AsH_3 pressures that are calculated in terms of those in the substrate region.

In open-tube VPE, we assume $P_{\text{H}_2} \doteqdot 1.0$. In the gallium source region at temperature T_1,

$$K_1(T_1) = P_{\text{HCl}}/P_{\text{GaCl}}(P_{\text{H}_2})^{1/2}$$

$$= P_{\text{HCl}}/P_{\text{GaCl}}. \quad (32)$$

The equilibrium partial pressure of HCl in the metallic gallium source region is very small at gallium temperatures over 850°C:

$$P_{0,\text{HCl}} \ll P_{0,\text{GaCl}}. \quad (33)$$

Thus,
$$P_{0,\text{HCl}} = K_1(T_1)P_{i,\text{HCl}}. \tag{34}$$

By substituting Eqs. (30), (31), and (34) into Eq. (27), we obtain the chemical potential difference in hydride VPE.

d. Chemical Potential Difference in MOCVD

Although the elementary reaction processes of MOCVD are not well known, the chemical potential difference in MOCVD can be calculated from the overall growth reaction and partial pressures of species in the reaction. We need only the chemical potential of the source gas. The overall growth reaction in the $\text{TMG}/\text{AsH}_3/\text{H}_2$ system is expressed as follows:

$$\text{Ga}(\text{CH}_3)_3 + \text{AsH}_3 = \text{GaAs} + 3\text{CH}_4. \tag{35}$$

In an equilibrium state at high temperatures, trimethyl gallium and arsine are completely decomposed:

$$\text{Ga}(\text{CH}_3)_3 + \tfrac{3}{2}\text{H}_2 = \text{Ga} + 3\text{CH}_4; \tag{36}$$

$$\text{AsH}_3 = \tfrac{1}{4}\text{As}_4 + \tfrac{3}{2}\text{H}_2. \tag{37}$$

Therefore, we can use the following imaginary reaction as an overall growth reaction:

$$\text{Ga}(\text{gas}) + \tfrac{1}{4}\text{As}_4(\text{gas}) = \text{GaAs}(\text{solid}), \tag{38}$$

and the imaginary partial pressures of Ga and As_4 are expressed as follows:

$$P_{\text{Ga}} = P_{\text{TMG}}; \quad P_{\text{As}_4} = \tfrac{1}{4}P_{\text{AsH}_3}.$$

From these reactions, the chemical potential difference is expressed as

$$\Delta\mu_{\text{GaAs}} = \mu^*_{\text{gas}} - \mu_{\text{GaAs}}, \tag{39}$$

where

$$\mu^*_{\text{gas}} = \mu_{0,\text{Ga}} + RT_2 \ln P_{\text{Ga}} + \tfrac{1}{4}\mu_{0,\text{As}_4} + \tfrac{1}{4}RT_2 \ln P_{\text{As}_4}$$

$$= \mu_{0,\text{Ga}} + \tfrac{1}{4}\mu_{0,\text{As}_4} + RT_2 \ln P_{\text{TMG}}(\tfrac{1}{4}P_{\text{AsH}_3})^{1/4} \tag{40}$$

and

$$\mu_{\text{GaAs}} = \mu_{\text{eq.Ga}} + \mu_{\text{eq.As}_4}$$

$$= \mu_{0,\text{Ga}} + RT_2 \ln P_{\text{eq.Ga}} + \tfrac{1}{4}\mu_{0,\text{As}_4} + \tfrac{1}{4}RT_2 \ln P_{\text{eq.As}_4}. \tag{41}$$

Hence,

$$\Delta\mu_{\text{GaAs}} = RT_2 \ln \frac{P_{\text{TMG}}(\tfrac{1}{4}P_{\text{AsH}_3})^{1/4}}{P_{\text{eq.Ga}}(P_{\text{eq.As}_4})^{1/4}}, \tag{42}$$

where $P_{eq.\,Ga}$ and $P_{eq.\,As_4}$ are equilibrium vapor pressures of GaAs substrate at temperature T_2, which have been reported in the literature (Arthur, 1967).

e. Chemical Potential Difference in MBE

To calculate the chemical potential we always need concentrations or partial pressures in the growth environment. However, it is difficult to define pressures of gallium and arsenic in their molecular beams. In principle, we cannot define the "pressure" of a molecular beam, because all molecules in a beam fly toward the substrate, which means that the distribution function in a beam deviates far from the equilibrium state. However, in order to compare the chemical potential difference of MBE with those of other epitaxial methods, we define hypothetical pressures of gallium and arsenic for the MBE system. The hypothetical gallium pressure $P_{hyp}(Ga)$ is defined as the pressure that has the same impinging rate of the gallium molecular beam at the substrate surface:

$$P_{hyp}(Ga) = N_{i,\,Ga} R T_{sub}. \tag{43}$$

$N_{i,\,Ga}$ is the hypothetical molecular density in the gallium beam and T_{sub} is the substrate temperature. $N_{i,\,Ga}$ can be calculated from the experimental growth rate G_r and the mean velocity of gallium in the molecular beam v_{Ga}, as in Eq. (4):

$$N_{i,\,Ga} = 4J_{i,\,Ga}/v_{Ga}. \tag{44}$$

It should be noticed that Eq. (4) was obtained for the impinging rate to the substrate from the gas phase in equilibrium. Therefore, $N_{i,\,Ga}$ is the hypothetical molecular density in an equilibrium gas phase that has the same impinging rate as the molecular beam. The gallium impinging rate $J_{i,\,Ga}$ is obtained from the growth rate G_r.

$$J_{i,\,Ga} = G_r/N_s t_m, \tag{45}$$

where N_s and t_m are the number of lattice sites per unit area and the thickness of monolayer along the growth direction, respectively.

The mean velocity of impinging gallium atoms in its molecular beam is expressed by Eq. (6) with the gallium cell temperature T_{cell} and the mass of gallium M_{Ga}:

$$v_{Ga} = (8kT_{cell}/\pi M_{Ga})^{1/2}, \tag{46}$$

where k is the Boltzmann constant. By combining Eqs. (43)–(46),

$$P_{hyp}(Ga) = \frac{2RT_{sub}G_r}{N_s t_m}(2kT_{cell}/\pi M_{Ga})^{-1/2}. \tag{47}$$

The arsenic molecular beam is controlled to be γ times larger than the gallium beam:

$$J_{i,\text{As}_4} = \gamma J_{i,\text{Ga}}; \quad (48)$$

$$P_{\text{hyp}}(\text{As}_4) = \gamma P_{\text{hyp}}(\text{Ga}). \quad (49)$$

From these hypothetical pressures, we can calculate the chemical potential difference in MBE by replacing P_{TMG} and P_{AsH_3} of Eq. (42) by $P_{\text{hyp}}(\text{Ga})$ and $P_{\text{hyp}}(\text{As}_4)$, respectively.

$$\Delta\mu_{\text{GaAs}} = RT_{\text{sub}} \ln \frac{\gamma^{0.25}[P_{\text{hyp}}(\text{Ga})]^{1.25}}{P_{\text{eq.Ga}}(P_{\text{eq.As}_4})^{1/4}}. \quad (50)$$

f. Comparison of Chemical Potential Differences

Figure 2 shows calculated chemical potential differences in LPE, VPE, MOCVD, and MBE, together with those of LEC (liquid-encapsulated Czochkralski) and HB (horizontal Bridgeman) bulk growth methods, which are not calculated but estimated as nearly equilibrium growth.

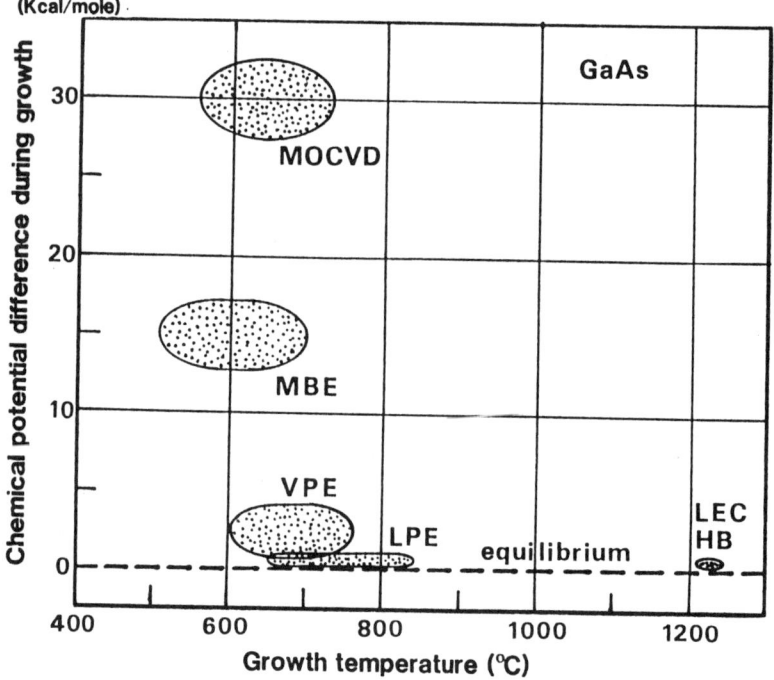

FIG. 2. Chemical potential difference in LPE, VPE, MOCVD, and MBE.

LPE is located at the position nearest to the equilibrium line. This is the other expression of the near-equilibrium state, corresponding to the extremely small value of the nonequilibrium parameter λ shown in Table I. As explained before, the small value of λ in LPE comes from the high impinging rate of arsenic. The calculated chemical potential difference is only several tens of calories per mole for LPE. It should be noticed that the region for LPE is expanded in vertical axis. The regions for LEC and HB are also exaggerated.

The partial pressure of the rate-determining species GaCl in VPE is also very high, as shown in Fig. 1. However, GaCl vapor is stable at high temperatures, and the equilibrium GaCl pressure at the substrate surface is also very high, so the chemical potential difference becomes small. The typical value of the chemical potential difference in VPE is about 2 kcal/mol. This near-equilibrium situation corresponds to the relatively high impinging rate of GaCl shown in Table I. Here the two methods of VPE, Ga/AsCl$_3$/H$_2$ chloride VPE and Ga/HCl/AsH$_3$/H$_2$ hydride VPE, are compared. Suppose that gallium source temperatures in the chloride and hydride systems are controlled to 850 and 800°C, respectively, and the substrate temperatures are the same, 750°C. Although the temperature difference between source gallium and the substrate is larger in the chloride VPE (100°C) than in the hydride VPE (50°C), the chemical potential difference is calculated to be smaller in the chloride VPE than in the hydride VPE system (Mizutani and Watanabe, 1982). This means that the chloride system is more nearly at equilibrium than the hydride system. This is because in the hydride VPE system, the conversion efficiency of source HCl gas to GaCl in the gallium source region is very high, more than 99% even at 800°C, and hence the HCl pressure in the gallium source region $P_{0,\text{HCl}}$ is very small, while the equilibrium HCl partial pressure at the substrate surface $P_{\text{HCl}}(T_2)$ is not so small. Therefore, as shown by Eq. (27), the chemical potential difference becomes large. On the other hand, in chloride VPE, the HCl partial pressure difference between the arsenic-saturated gallium source and the substrate regions is relatively small and those for GaCl and As$_4$ are also small, resulting in a small chemical potential difference.

The chemical potential difference in MBE is between those of VPE and MOCVD, nearly half of that for MOCVD. It should be noticed that the chemical potential difference in MBE is calculated assuming a hypothetical partial pressure of gallium and arsenic. The meaning of the hypothetical pressure must be discussed in more detail. In MBE, every gallium atom hits a growing surface only once. This is true as long as the gallium sticking coefficient is unity. Furthermore, their impinging direction is always normal to the surface. (Strictly speaking, the impinging direction is not normal to the surface, because of the arrangement of the effusion cell and the substrate holder, which is rotated to obtain uniform grown thickness.) The situation seems quite different from normal gas behavior, where gaseous molecules interact with a solid surface many times and the impinging angle is randomly distributed from 0 to 90 degrees. However, as long as growth is epitaxial, the impinging direction is not a problem. Even if gallium is assumed to act as an ideal gas and interact with the substrate surface with various incident angles, an essentially important factor is how many atoms hit the surface per unit time. This is because the sticking coefficient is always unity, being almost independent of the incident angle. From the viewpoint of the impinging frequency, the question of how large the driving force of growth is is equivalent to the question of how intense the molecular beam is. On the other hand, the higher the temperature, the higher the chemical potential of the substrate. The decomposition pressure increases exponentially and the chemical potential difference becomes small, as in Eq. (50). The chemical potential difference in MBE is, therefore, divided into two components. One is between the molecular beam and the surface, and the other is the difference between the adsorbed atoms and the substrate. The large chemical potential difference mainly comes from the former. The latter can be assumed to be very small as long as the growth is epitaxial.

The layer-by-layer growth in MBE confirmed by the observation of Reflected High Energy Electron Diffraction (RHEED) oscillation (Neave *et al.*, 1983) is strong evidence that the adsorbed species migrate a long distance on the surface. The migration length was measured to be 72 Å at 580°C when the Ga flux was 1.2×10^{14} atoms/cm^2 sec and the As$_2$/Ga flux ratio was 3.0 (Neave *et al.*, 1985). Lewis *et al.* (1985) observed that the specular beam intensity gradually recovers the original intensity after the termination of the growth. This observation suggests that the atoms that once formed the crystal lattice have an opportunity to be released onto the surface and migrate far from the embedded place. In other words, the atoms from the embedded surface are not so far from thermal equilibrium, even though the overall growth process of MBE is indeed far from equilibrium.

MOCVD is located at the highest position among the four epitaxial methods. The high chemical potential difference in MOCVD results from the high source-gas pressures of TMG and arsine, ranging from 10^{-4} to 10^{-3} atm, which are very much higher than the equilibrium dissociation pressures of gallium and arsenic. However, as shown in Fig. 1, if the rate-determining species is not TMG but some gallium radical such as Ga(CH$_3$)$_x$ ($x = 1$ or 2) or atomic gallium, the chemical potential difference seems to be remarkably reduced. However, as long as the final product of TMG is atomic gallium in an equilibrium state at the substrate temperature, intermediate byproduct concentrations do not influence the calculation of the chemical potential difference. Contrary to this equilibrium consideration, the nonequilibrium parameter is directly affected by impinging and escaping rates.

Therefore, we conclude that the chemical potential difference expresses the degree of nonequilibrium state of the environment from a thermodynamic viewpoint, while the nonequilibrium parameter λ defined by Eq. (7) or (8) indicates it on the growing surface from a kinetic viewpoint.

3. V/III Concentration Ratio in Growth Environment

Electrical and optical properties of grown layers are strongly dependent on the concentration or beam-flux ratio of arsenic to gallium, written as V/III or [As/Ga]. Point defects such as gallium and arsenic vacancies or their complexes with some interstitial atoms frequently have been discussed using [As/Ga] ratio. For such investigations we need absolute values of [As/Ga] at growth temperatures in these four epitaxial methods.

Figure 3 shows [As/Ga] ratios in growth environments of LPE, VPE, MOCVD, and MBE, together with those of LEC and HB, which are growths from molten GaAs. In the three-phase (solid, liquid and gas) coexisting equilibrium, the [As/Ga] ratio in a liquid phase is uniquely determined as a liquidus line, shown by the dotted line. The [As/Ga] ratio in LPE is almost equal to the liquidus because growth takes place under near-equilibrium conditions with gradually decreasing temperature of the arsenic-saturated

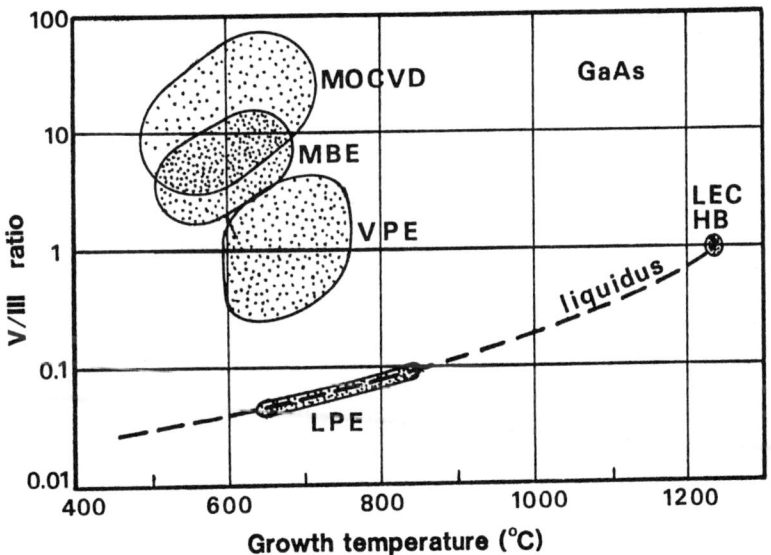

Fig. 3. Arsenic to gallium concentration ratio (V/III) in LPE, VPE, MOCVD, and MBE.

gallium. Since the temperature decrease in the so-called step-cooling LPE method is about 10°C, the [As/Ga] ratio in the supercooled liquid is only 10% larger than the liquidus at any temperatures. When the [As/Ga] ratio in liquid is smaller than the liquidus, a substrate would be etched or would melt back.

In VPE with the hydride system, we can change the [As/Ga] ratio, which is the partial-pressure ratio of AsH_3 and HCl ($=$ GaCl), in a wide range below and above unity. This is a unique characteristic of the hydride VPE method, contrary to other methods. In MBE and MOCVD the [As/Ga] ratio should be always kept two to several tens times higher than unity to avoid deposition of liquid gallium droplets. In the $Ga/AsCl_3/H_2$ chloride method, the [As/Ga] ratio cannot be changed much, being fixed at 0.3–0.7, which is calculated as follows. GaCl partial pressure in the gallium source region is a function of $AsCl_3$ input pressure, gallium source temperature and the degree of perfectness of reaction between flowing $AsCl_3$ and the GaAs crust that covers the liquid gallium source contained in a source boat. We assume that the reaction is completed and the vapor-phase environment attains equilibrium with the gallium source. The GaCl pressure is calculated from Eq. (20), with its equilibrium constant $K(T)$ given by Eq. (23), and the relation

$$P_{GaCl} = 3\eta P_{i, AsCl_3}, \quad (51)$$

where η is the conversion efficiency of HCl, which is converted to GaCl in the gallium source region. η is obtained with the chlorine conservation relation

$$3P_{i, AsCl_3} = P_{GaCl} + P_{HCl}. \quad (52)$$

In chloride VPE,

$$[As/Ga] = \tfrac{1}{4} P_{As_4}/P_{GaCl} = P_{i, AsCl_3}/P_{GaCl} = (3\eta)^{-1} \quad (53)$$

Since η is 0.5 to 0.9 (Watanabe, 1975), being dependent on gallium temperature and the input $AsCl_3$ pressure $P_{i, AsCl_3}$, the [As/Ga] ratio in chloride VPE is 0.67 to 0.37.

In MOCVD, the [As/Ga] ratio is plotted as the input AsH_3 and TMG pressure ratio. In MBE, it is an arsenic-to-gallium flux ratio. However, in both MBE and MOCVD it is not clear which molecule among As_m ($m = 1$, 2 or 4) is most influencial on epitaxial growth. From the viewpoint of equilibrium state, As_4 vapor is dominant rather than As_2 under the normal-growth MOCVD condition, while As_2 is dominant in MBE. The molecular form of gallium-contained species has not been identified in MOCVD; it is atomic gallium in MBE. Decomposition kinetics of TMG and other gallium organic gas were reported by Yoshida et al. (1985) and Nishizawa and Kurabayashi (1983).

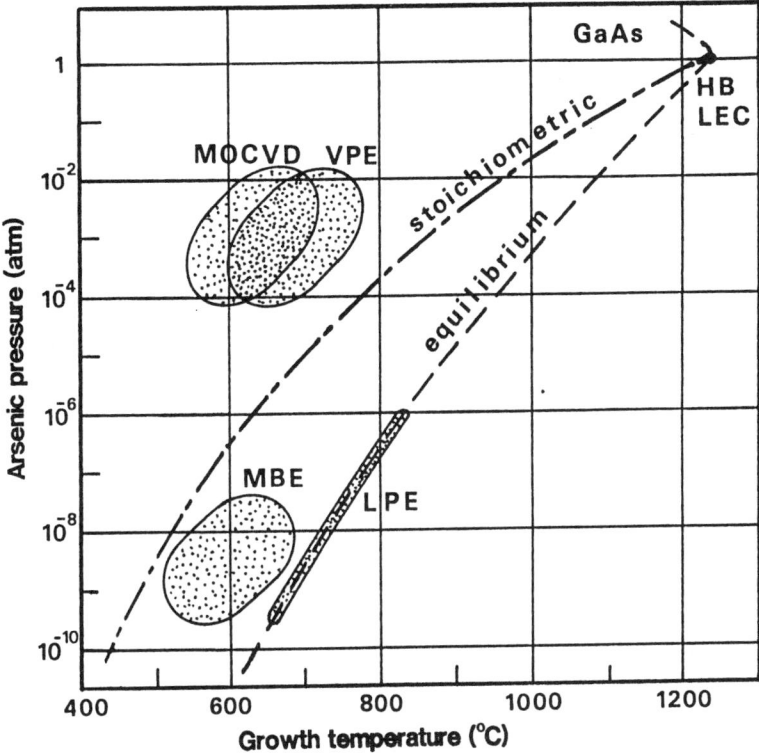

FIG. 4. Arsenic partial pressure in LPE, VPE, MOCVD, and MBE.

4. ARSENIC PRESSURE IN GROWTH ENVIRONMENT

Figure 4 shows arsenic partial pressures during growth in LPE, VPE, MOCVD, and MBE. The arsenic pressures in VPE and MOCVD are plotted as As_4 pressure. Under the usual VPE condition that the growth temperature is about 750°C and total arsenic pressure is greater than 10^{-4} atm, As_4 is the dominant species independent of the starting gas species ($AsCl_3$ or AsH_3). Therefore, the arsenic pressure is one-fourth of the $AsCl_3$ partial pressure in chloride VPE and also one-fourth of the AsH_3 partial pressure in hydride VPE. For MOCVD, the arsenic pressure is plotted also as one-fourth of the input AsH_3 pressure. However, the molecular form of arsenic impinging the substrate surface has not been identified. If the environment near the substrate is equilibrium, the tetramer of arsenic, As_4, must be dominant, but according to a mass spectroscopic study on arsine decomposition (Ban, 1972), source-gas arsine is not completely decomposed to its equilibrium species because of its rather slow decomposition speed. According to

Nishizawa and Kurabayashi (1983), the GaAs surface acts as a catalyst for the decomposition reaction of arsine, and it seems possible to assume that the major species of arsenic gas is the arsenic tetramer near the substrate surface.

The molecular forms of gallium and arsenic molecular beams in MBE have been understood: atomic gallium from a metal gallium source and As_4 from a solid arsenic source. It is impossible to define their partial pressures because the MBE chamber is not in the equilibrium state. However, arsenic pressure in MBE is plotted in Fig. 4 as an apparent pressure, where arsenic pressure is calculated using Eqs. (43)–(49). The apparent pressure measured by a B–A gauge and the arsenic cell temperature are used to calculate the impinging rate.

In LPE, the growing surface exists in a liquid and is not exposed to the vapor phase, so the arsenic pressure during growth seems impossible to define. However, from the viewpoint of equilibrium theory, the chemical potentials must be equal in the three phases (the GaAs substrate, the arsenic-saturated liquid gallium and the vapor phase outside the liquid). If we consider the equilibrium arsenic pressure in LPE, it must be equal to the equilibrium dissociation pressure of the GaAs crystal, which is plotted by the dashed line in Fig. 4. In an actual LPE growth apparatus there is no intentional supply of arsenic gas, so arsenic must be continuously evaporated from the arsenic-saturated gallium source. But arsenic evaporation is suppressed by the GaAs crystal cap that covers the liquid. The arsenic pressure during LPE must be almost equal to the equilibrium pressure.

The broken line in Fig. 4 labeled "stoichiometric" is a curve calculated by the following equation, proposed by Nishizawa *et al.* (1975). They derived it as the optimized arsenic pressure to obtain stoichiometric GaAs epitaxial (LPE) layers and to suppress the crystal quality change in heat treatment with an arsenic overpressure:

$$P_{As}(\text{stoichiometry}) = 2.6 \times 10^6 \exp(-1.05(\text{eV})/kT_{sub}): \text{torr.} \quad (54)$$

This equation was obtained after summarizing several experimental arsenic pressures ($P_{As_4} + P_{As_2}$) that are necessary to keep electrical and crystallographic properties unchanged. Heat treatment at pressures higher and lower than P_{As}(stoichiometry) results in a change in carrier concentration, which is considered to be due to some stoichiometric change such as [As/Ga] change in the solid, or due to generation of interstitials and vacancies. The lattice constants of the heat-treated samples are also changed when arsenic pressure is not adjusted to the value of Eq. (54).

It is clearly seen in Fig. 4 that the four epitaxial methods are divided into two goups. VPE and MOCVD belong to the same group, and LPE and MBE to the other. In VPE and MOCVD the orders of magnitudes of arsenic pressure are almost the same, while in MBE the arsenic pressure is considerably lower but high enough as compared with the equilibrium dissociation pressure. LPE pressure is plotted on the equilibrium line, as explained above.

It is very interesting that in the VPE–MOCVD group the arsenic pressure is higher than the stoichiometric pressure line of Eq. (54), while in the LPE–MBE group it is lower than this line. It has been well recognized that epitaxial layers grown by VPE and MOCVD have the point defects called EL2, but LPE and MBE samples do not. The point defect EL2 is a major electron trap in concentrations ranging from 10^{14} to 10^{16} cm^{-3}. EL2 is regarded as arsenic-rich point defects such as interstitial arsenic-contained antisite defects. The EL2 concentration is a function of the arsenic partial pressure during growth in VPE and MOCVD (Watanabe et al., 1983), where it is increased by increasing arsenic pressure. It seems that the stoichiometric line divides GaAs crystals into two groups: EL2-containing material and no-EL2 material. In other words, EL2 is generated by applying arsenic pressure higher than the stoichiometric line during growth or heat treatment.

III. Fundamental Growth Parameters in MBE

5. Controllability in MBE

The capability of MBE to make an atomically sharp heterojunction has realized various devices that had been expected theoretically to have superior performances, such as HEMTs (high electron mobility transistors) and quantum-well lasers. An advanced MBE growth gave an extremely high mobility over 2×10^6 cm^2/Vsec in an Al$_x$Ga$_{1-x}$As/GaAs modulation doped structure (Hiyamizu et al., 1983), while for optical devices, an extremely low threshold current of 2.5 mA (Tsang et al., 1982) was demonstrated by a double-heterostructure laser using a MBE-grown Al$_x$Ga$_{1-x}$As/GaAs epitaxial layer. High-quality heterointerfaces grown by MBE were also demonstrated by devices using an Al$_x$In$_{1-x}$As/Ga$_y$In$_{1-y}$As system. A transconductance over 700 mS/mm was successfully fabricated as a HEMT structure from an Al$_x$In$_{1-x}$As/Ga$_y$In$_{1-y}$As epitaxial film (Hirose et al., 1986). These results indicate that not only interface but also bulk properties of MBE-grown multilayer structures are very high quality, which have been realized by utilizing several features of MBE controllability. Here, four of them are discussed as inherent advantages of the MBE method.

First, we can expect an extremely fast switching speed in a growth on/off. Start or termination of growth is done by opening or closing the effusion cell shutters. Since gallium molecules fly with a velocity of 10^5 cm/sec, we need not worry about the time lag of growth termination after a shutter is closed, which is estimated to be 0.1 msec when the distance between substrate and effusion cell is 10 cm. During this transient time only 10^{-4} monolayer is grown with a typical MBE growth rate of 3 Å/sec. This is a particularly important character for controlling not only the heterojunction composition profile but also impurity position with an atomic-layer accuracy. A good example of impurity position control is found in the deep-level study as a function of donor position in GaAs/AlAs superlattices (Baba et al., 1983, 1986). A major deep-level center called the DX center in $Ga_xAl_{1-x}As$ alloys, which induces a undesired persistent photoconductivity (PPC), can be eliminated by controlling silicon donor position in a 20 Å-GaAs/20 Å-AlAs short-period superlattice. No PPC behavior is found only when silicon donors are doped in the middle four-monolayer-thick GaAs region, which is sandwiched by two-monolayer-thick undoped inside GaAs layers and 20 Å AlAs outside layers. PPC is strongly enhanced when silicon is doped at the GaAs–AlAs interface.

Second, the MBE growth rate is not influenced by extraneous deposition on the chamber wall. Molecular beams are never trapped midway to the substrate. The growth rate is stable independent of deposition on the substrate holder and the chamber wall behind the substrate. This is a unique advantage of MBE, in contrast to VPE and MOCVD in which extraneous deposition frequently occurs on the reactor inner wall, especially on the tube wall upstream of the substrate. In VPE and MOCVD the extraneous deposition causes growth-rate fluctuation and resultant difficulty in the profile control of alloy composition and impurity concentration. In order to eliminate the extraneous deposition some technique, such as an additional HCl injection method (Mizutani and Watanabe, 1982), is necessary in VPE and MOCVD, while no special trick is needed in MBE.

Third, we can grow epitaxial layers in a wide range of substrate temperature with the same growth rate and also with the same alloy composition. This comes from the fact that the sticking coefficients of group III elements are unity in a wide range of substrate temperatures. This is particularly useful for controlling the growth rate while keeping alloy composition constant at any growth temperature. Growth temperature is frequently changed to obtain optimized properties in devices. For example, for $Al_xGa_{1-x}As$ laser diodes, the growth temperature should be kept relatively high (over 750°C) to obtain an active layer with a high luminescence efficiency. On the other hand, a low growth temperature is recommended to obtain a heavily doped low-resistive contact layer with a doping concentration in the range of 10^{18} cm^{-3}. For the nonalloy ohmic contact of GaAs FETs, a growth temperature as low as 350°C is needed to obtain a silicon-doped contact layer with a carrier concentration over 10^{19} cm^{-3} (Ogawa, 1986). Since the alloy composition is also independent of the growth temperature, grown thickness can be well controlled independently with growth-temperature optimization.

Fourth, the recent discovery of RHEED intensity oscillation during MBE growth is directly connected to a unique advantage apparent only in a vacuum system. The RHEED oscillation was found to be related to the layer-by-layer growth mechanism in the MBE process (Harris et al., 1981; Neave et al., 1983). Connecting this advantage with the real-time surface monitoring technique, we can control grown thickness with monolayer accuracy simply by counting RHEED oscillation peaks (Sakamoto et al., 1984).

From these considerations it seems possible to conclude that MBE growth has been established as a completely controlled technique that gives reproducible results. However, it should be noted that growth parameters in MBE are not always controlled with their absolute values but are only controlled reproducibly with a relative accuracy. For example, the flux intensity of a molecular beam is not measured as an absolute value at the substrate. Furthermore, growth temperatures reported in the literature are not reliable. This is because measurements of these quantities are not easy in a non-equilibrium high-vacuum system. In order to make a critical comparison of reported performances by different MBE groups, it is essential to know absolute values of their growth parameters. In the following section, quantitative analysis is made of fundamental growth parameters in MBE. The background pressure, the impinging rate of residual gas, the flux intensity of the molecular beam, and the real substrate temperature are studied with reference to the importance of their absolute values.

6. MEAN FREE PATH AND IMPINGING RATE

The room-temperature mean free path and impinging rate of nitrogen gas calculated as a function of the pressure are shown in Fig. 5. Mean free path and impinging rates of aluminum, gallium, indium and arsenic molecules evaporated at their actual source temperatures easily can be calculated by multiplying by the individual correction factors listed in Table II. The correction factor for mean free path ranges from 0.26 for As_4 to 1.2 for Ga, and that for impinging rate from 0.43 for As_4 to 2.2 for Al. However, this five-times difference does not affect the conclusion of the following discussion.

MBE growth rate is determined by the impinging rates of group III elements because their sticking coefficients are unity. A growth rate of 1 μm/hr corresponds to a pressure of mid-10^{-7} torr and a mean free path of 100 m. MBE growth usually is carried out at a V/III flux ratio greater than 2.0 to avoid surface morphology degradation caused by liquid gallium condensation. The arsenic mean free path of about 25 m is still much longer than the distance between effusion cells and the substrate, indicating no collision in the molecular beam. This is the most prominent feature of MBE growth and is the basic origin of other characteristics of the MBE method.

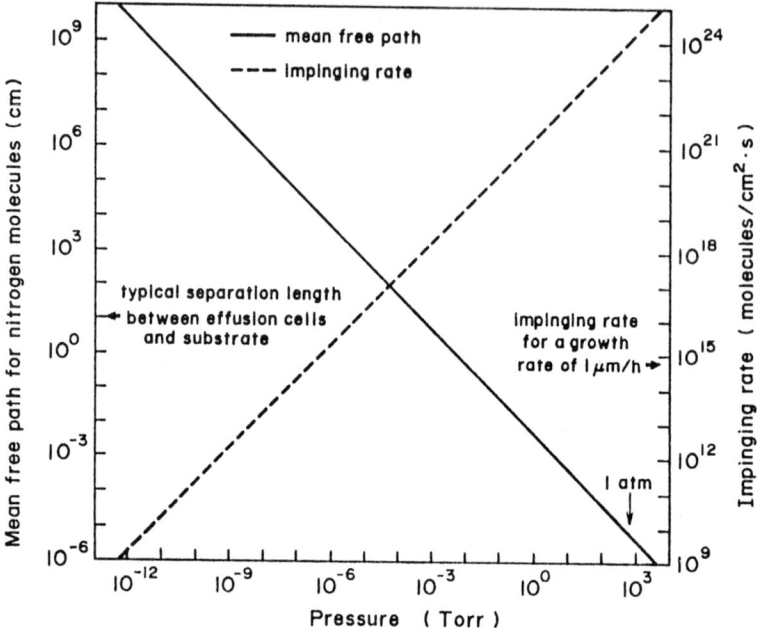

FIG. 5. Mean free path (solid line) and impinging rate (dashed line) of a nitrogen molecule as a function of pressure. Correction factors for other gas molecules are given in Table II.

When the background pressure is 5×10^{-4} torr, the mean free path is 10 cm, which means that at least one molecular collision may occur between source and substrate. The impinging rate of the background species at this pressure is estimated to be 10^{18} molecules/cm² sec. This is 10^3 times larger than that of gallium when the growth rate is assumed to be 1 μm/hr. Therefore, it is almost impossible to expect an epitaxy because there are too many collisions of foreign molecules onto the substrate surface. This is the most

TABLE II

CORRECTION FACTORS FOR THE MEAN FREE PATH AND IMPINGING RATE

Species	Source temperature (K)	Molecular weight (g/mol)	Correction factor	
			Mean free path	Impinging rate
N_2	300	28	1	1
Al	1400	26.9	1.04	2.2
Ga	1250	69.7	1.20	1.29
In	1150	114.8	0.99	0.97
As_4	600	299.7	0.26	0.43

fundamental reason for the necessity of an ultrahigh vacuum system that allows a clean substrate surface to be obtained.

To obtain a residual impurity concentration less than 10^{15} cm^{-3}, we must decrease the impinging rate of the impurity gas to less than 10^8 molecules/cm^2 sec, if the sticking coefficient of the impurity gas is unity. This impinging rate corresponds to a background pressure of 10^{-14} torr. This is again the reason for the necessity of the ultralow background pressure.

In order to reduce the impinging rate of the residual gases, cryo-panels cooled by liquid nitrogen are installed to surround the substrate and source cells as completely as possible. They are very effective for trapping residual gases of CO and H$_2$O and improving the background pressure. This is particularly useful for growing high-quality epitaxial layers containing aluminum, which has large bonding energies for oxygen and carbon as compared with gallium (Phillips, 1981). At present, commercially available MBE systems are guaranteed to achieve an ultimate vacuum better than 5×10^{-11} torr. This background pressure is, however, still 10^3 times higher than the above-calculated pressure. Nevertheless, we can obtain high-purity samples having nondoped carrier concentrations of $2-3 \times 10^{14}$ cm^{-3} (p-type). This is because the main residual gas under ultrahigh vacuum conditions is hydrogen, and the sticking coefficients of impurity gases are much smaller than unity. The vapor pressure of an element that has a large sticking coefficient is necessarily low.

This serendipitous situation arises from the slow velocities of the surface decomposition reactions of carbon-containing gases such as CO and hydrocarbons. Thus it is particularly easy to understand the success of gas-source MBE, which uses metal organic source gas and generates the same order of magnitude of molecular beam intensity of hydrocarbon gases as byproducts. However, these gas species have large sticking coefficients when a substrate surface contains aluminum atoms, resulting in carbon-contaminated growth. The choice of source gas should be carefully studied to reduce the carbon concentration in Al-containing materials.

7. Flux Intensity

Flux intensities of group III molecular beams can be calibrated from an actual growth rate, because group III element sticking coefficients are unity in a wide range of substrate temperatures. Those of group V molecular beams are difficult to determine as absolute values. The sticking coefficient of the group V element gases such as As$_4$, As$_2$, P$_4$, or P$_2$ varies according to the V/III flux intensity ratio (Foxon and Joyce, 1975).

The conventional method of measuring arsenic flux intensity by a flux gauge (B–A gauge) is based on the assumption that the ionization coefficient of As can be estimated from the linear relationship between the ionization coefficient and the electron number of the molecule (Flaim and Ownby, 1971)—an assumption yet to be confirmed. This method also needs correct polar angles of the incident direction of the molecular beams, since the B–A gauge measures only the molecular density where it is located.

The observation of RHEED patterns during growth as a function of V/III ratio gives an alternative method for determining the group V flux intensity. This is based on the following knowledge so far obtained.

(1) (2 × 4) surface reconstruction is obtained when the arsenic coverage is large, namely under so-called As-stabilized conditions, while a (4 × 2) reconstruction pattern is observed under conditions of less arsenic coverage of the Ga-stabilized surface (Drathen et al., 1978; Arthur, 1974).

(2) A Ga-stabilized surface is obtained after arsenic is desorbed to half coverage from an initial As-stabilized surface (Arthur, 1974).

(3) The maximum sticking coefficients of As_4 and As_2 are 0.5 (Foxon and Joyce, 1975) and 1.0 (Foxon and Joyce, 1977), respectively.

This method has, however, some ambiguity because the critical value of the V/III flux intensity ratio where surface reconstruction changes from the As-stabilized surface to the Ga-stabilized one is dependent on the growth temperature (Neave and Joyce, 1978; Chang et al., 1973).

Consequently, the two methods discussed above give relative calibration factors between different research organizations.

Another problem arises from source temperature fluctuation, which takes place after the cell shutter is opened and closed. According to the transient state measurement by Hirose et al. (1986), Ga flux intensity overshoots about 15% of its final intensity after the Ga shutter operation, as demonstrated in Fig. 6. This transient behavior is caused by the cell temperature reduction after the shutter is opened. The shutter is acting as a thermal shield that suppresses temperature lowering due to radiation from the heated cell. When the shutter is opened, the source cell starts to cool down by heat radiation. Inversely, the cell temperature becomes high after the shutter is closed. The source temperature fluctuation was evaluated to be less than $\pm 0.04°C$ from the half-maximum of the X-ray rocking curve of GaInAs layers grown with the transient behavior (Mizutani and Hirose, 1985).

As shown in Fig. 6, after the start of shutter on/off operation with a 30-sec periodicity to grow a GaAs/AlAs superlattice, the initial peak intensity of the gallium beam is gradually increased. This is because the balance between thermal shielding by the shutter and heat escape is not at equilibrium but is tending toward the thermal shielding. The transient behavior makes superlattice property control difficult even if the lattice matching of GaAs/AlAs is not affected by the transient gallium composition fluctuation. It becomes

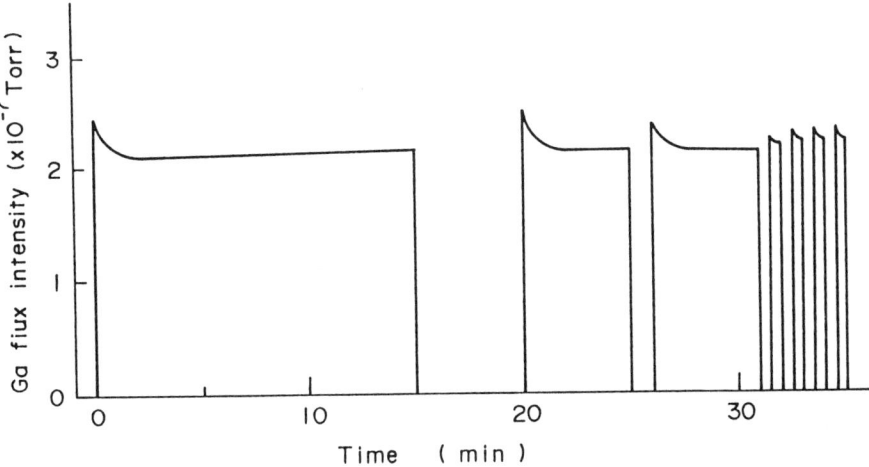

FIG. 6. Gallium flux intensity fluctuation after cell shutter operation in MBE.

serious when accurate composition control is needed, as in the case of AlInAs or GaInAs alloy growth, where the compositional transient in the vicinity of the heterointerface results in local lattice mismatch. When the transient is short, no misfit dislocation would be generated, but electronic properties such as conduction band discontinuity design would be difficult to control. One solution given for avoiding the transient temperature fluctuation was to enlarge the distance between the shutter and the exit of a source cell. It was found that the conventional distance, less than 1 cm, must be increased to more than 5 cm to eliminate the non-steady state effect (Mizutani and Hirose, 1985).

8. Real Substrate Temperature in Vacuum

For critical comparison of a great deal of data reported by different organizations, we need the real or absolute values of growth parameters. Among them, the most ambiguous growth parameter in MBE is substrate temperature. In LPE, VPE, and MOCVD, heat flows effectively by the thermal conduction of the environment gas. In contrast, no gaseous heat conduction is expected in MBE, and heat flow takes place only by radiation and thermal conduction through the substrate holder. Since the growth environment in the MBE chamber does not reach an equilibrium state, it is very difficult to measure the absolute temperature of the substrate surface. Substrate temperature has been measured by a thermocouple that was inserted into a hole drilled parallel to the surface of the GaAs substrate (Foxon et al., 1973) or pressed to the back surface of the substrate by a metal spring (Matysik, 1976). The load-rock system for obtaining high reproducibility makes the temperature measurement difficult. The substrate rotation mechanism is used to obtain high thickness uniformity in the commercial MBE apparatus makes accurate measurement more difficult. Furthermore, there is a large discrepancy, as great as 100°C, between the substrate temperature measured by the thermocouple and that measured by an optical pyrometer, particularly when the substrate is heated up quickly.

Attempts to determine real substrate temperature are summarized in Fig. 7. Two kinds of temperature measurements have been tried as calibration standards for the accurate surface temperature. One is the measurement

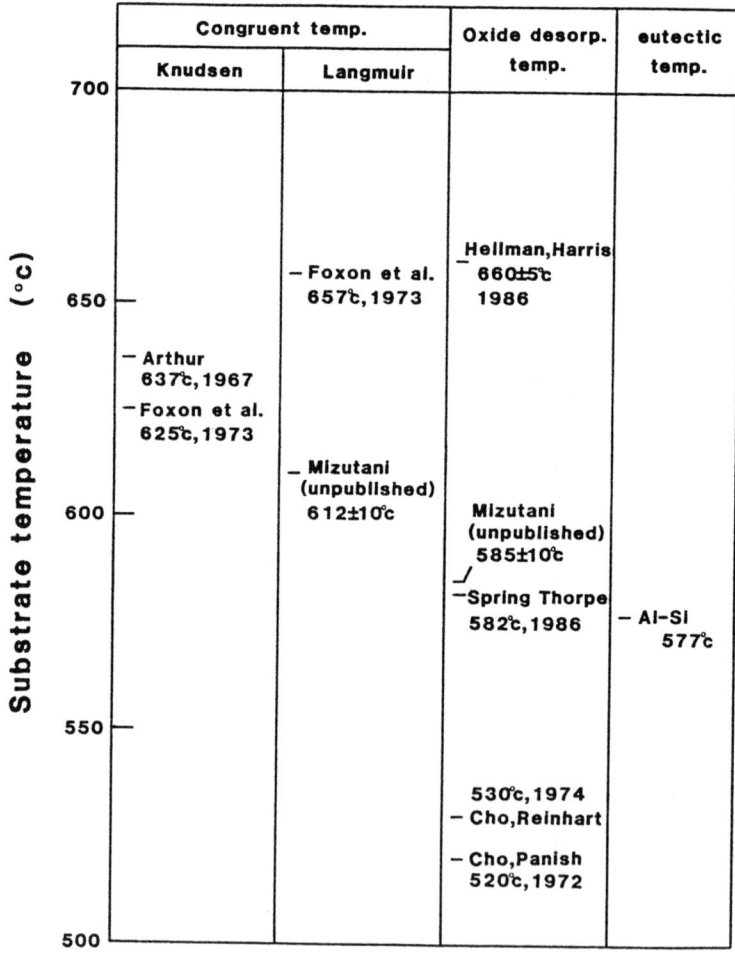

Fig. 7. Comparison of calibration temperatures in vacuum.

1. FUNDAMENTALS OF EPITAXY GROWTH AND ALE 31

of the upper limit of the congruent evaporation temperature, and the other is the temperature at which the surface oxide film of GaAs substrate is evaporated.

Arthur (1967) found that GaAs congruent evaporation from a Knudesen cell occurs at 637°C. Foxon et al. (1973) measured the congruent temperature also under Knudesen cell conditions and found it to be 625°C, which agrees relatively well with Arthur's temperature. However, according to the measurement under Langmuir conditions, congruent evaporation was observed up to 657°C, which is 32°C higher than the temperature measured by the Knudesen cell (Foxon et al., 1973). The temperatures were calibrated with thermocouple measurements. These results indicate that congruent evaporation is easily affected by the environment. The congruent temperature is difficult to use for calibration of the substrate surface temperature.

Cho and Panish (1972) observed a GaAs ⟨100⟩ surface by Auger electron spectroscopy (AES) and found that surface oxygen desorbed at 520°C, which was later corrected to 530°C by Cho and Reinhart (1974). Recently, by the same technique using AES, SpringThorpe and Mandeville (1986) determined this temperature to be 582°C, while by Hellman and Harris (1986) it was claimed to be 655–665°C. These rather large inconsistencies suggest that oxygen desorption measurement is also inadequate as the calibration method.

Very recently, Mizutani (1988) proposed an alternative standard for temperature calibration in which the Si–Al eutectic temperature of 577°C is used. A silicon substrate that is covered with an aluminum thin film is placed on the substrate holder together with a GaAs substrate, and the surface temperatures of the two samples are measured by the pyrometer. The aluminum surface temperature is identified to be 577°C when the emissivity is changed suddenly from that of metal aluminum to that of the Al–Si eutectic. This temperature is used as a standard, and the GaAs surface temperature is determined from a calibration curve obtained from the eutectic temperature. He determined the GaAs congruent temperature and the oxygen desorption temperature to be 612 ± 10°C and 585 ± 10°C, respectively.

A problem in temperature measurement using a pyrometer is that the measured temperature of the substrate surface is dependent on the electrical property of the substrate (Mizutani, 1987). The temperature difference between samples becomes large when the samples are heated. For example, around 500°C, the apparent temperature of n^+-InP substrate is about 60°C higher than that of semi-insulating InP substrate. Similarly, the temperature difference of about 10°C was confirmed for n^+-GaAs as compared with semi-insulating GaAs. These phenomena can be explained as a result of the emissivity difference. The emissivity is a function of the optical absorption intensity (Mizutani, 1987).

IV. Atomic Layer Epitaxy of GaAs

Various kinds of ultrathin-film layered structures that require monolayer controllability of film thickness have been proposed for very high-speed devices. At present, these devices appear only in discrete forms, but they eventually should be integrated as ICs with higher functional performances. In addition, monolayer-level controlled three-dimensional structures also are proposed for obtaining three-dimensionally confined electrons or holes, which will give some unique electronic and optical properties. For instance, in an ultrathin wire structure, which consists of two kinds of multilayers perpendicular to each other, electrons are one-dimensionally confined and an extremely high electron mobility is expected (Sakaki, 1980). To realize such three-dimensionally very thin and complicated structures, a more advanced epitaxial technique is required that has the capability to grow a thin film with monolayer accuracy together with high uniformity on a structured substrate having a v-grooved or mesa structure.

MBE and MOCVD techniques do not meet these next-generation requirements. For instance, run-to-run reproducibility with monolayer accuracy is very difficult to obtain in conventional growth techniques. Furthermore, it is almost impossible to grow an epitaxial film on the side-wall of a mesa-structured substrate with monolayer controllability and reproducibility.

Atom layer epitaxy (ALE), originally proposed by Suntola *et al.* (1980), is very attractive for these purposes because of its unique characteristics. In conventional epitaxial techniques, all elements that compose a grown crystal are simultaneously present in the growth environment. On the other hand, in ALE, these elements are separated and alternately supplied to the growth region, and the substrate surface is alternately exposed to different ambients.

Recently, a general review paper was published by Goodman and Pessa (1986) for ALE of II–VI and III–V compound semiconductors. In the present section we concentrate on (1) the principle and the current status of GaAs ALE with reference to the importance of the degree of surface coverage of adsorbed species; (2) a proposal of a novel idea, "digital epitaxy"; and (3) the unique advantages of ALE over MBE and MOCVD.

9. GROWTH PRINCIPLE AND APPARATUS

a. *Principle of* GaAs *ALE*

Figure 8 shows the sequence of events taking place in GaAs ALE. An essential point in the ALE process is an alternate supply of two source gases. The source gas for gallium is either GaCl, trimethyl gallium (TMG), triethyl gallium (TEG) or diethyl gallium chloride (DEGC), and the source gas for arsenic is AsH_3, As_4 or an organic arsine such as triethyl arsine (TEAs).

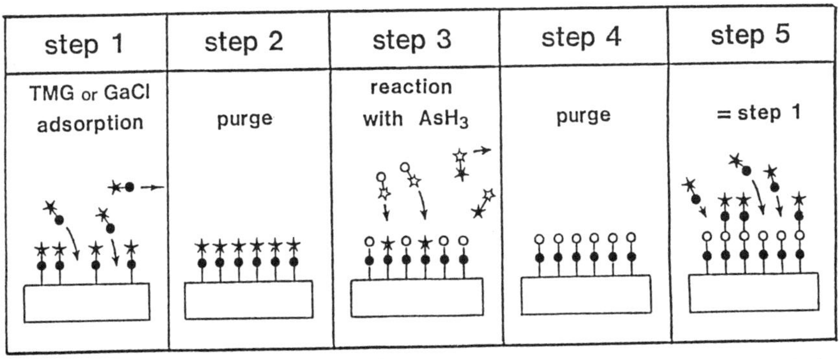

1. FUNDAMENTALS OF EPITAXY GROWTH AND ALE

step 1	step 2	step 3	step 4	step 5
TMG or GaCl adsorption	purge	reaction with AsH₃	purge	= step 1

FIG. 8. Growth sequence is GaAs ALE. (Reprinted with permission from IOP Publishing Ltd., Watanabe, H., and Usui, A., in "Gallium Arsenide and Related Compounds 1986," p. 1.)

The ALE sequence is explained by taking the GaCl-arsine system as an example. In step 1, GaCl gas is supplied to a substrate surface and GaCl molecules are adsorbed on the surface. In step 2, the gas is stopped and purged from the growth region, where GaCl molecules are kept to be adsorbed. In step 3, arsine gas is supplied. Arsine reacts with the adsorbed GaCl molecules, forming a GaAs monolayer and HCl gas. In step 4, arsine gas is purged after completing the reaction with GaCl. Step 5 is the same as step 1. In one ALE cycle from step 1 to step 4, a monolayer is grown.

Thickness can be controlled with a monolayer accuracy by repeating a number of these cycles. However, it should be noticed that grown thickness is a function of the surface coverage in step 2. Only when the surface coverage is exactly unity is the thickness equal to monolayer thickness times the cycle number. Furthermore, unity surface coverage is essentially necessary to obtain an atomically sharp abruptness at the heterostructure interface and a low density of native point defects. Surface coverage will be discussed later in detail.

b. Growth Apparatus

Several experimental results on GaAs ALE have been reported. The gases used for the gallium source are TMG (Nishizawa *et al.*, 1985; Bedair *et al.*, 1985; Doi *et al.*, 1986; Mori *et al.*, 1986); TEG (Nishizawa *et al.*, 1986; Kobayashi *et al.*, 1986); GaCl (Usui and Sunakawa 1986, 1987); and DEGC (Mori *et al.*, 1988). The source gas for arsenic is arsine in all the cases reported.

Usui and Sunakawa (1986) used GaCl and AsH_3 as source gases, where GaCl is obtained by the reaction between HCl and liquid gallium at about 800°C. Their two growth apparatuses are shown in Figs. 9a and b. Both systems are based on dual growth chamber atmospheric pressure VPE. In the

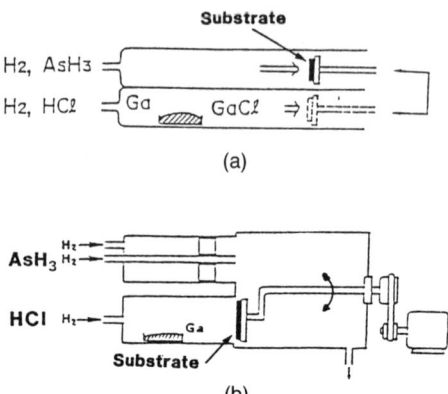

FIG. 9. Growth apparatus of GaAs ALE by GaCl–AsH$_3$ system: (a) dual chamber method at atmospheric pressure (Usui and Sunakawa, 1986); (b) improved dual chamber method at atmospheric pressure (Usui and Sunakawa, 1987).

dual growth chamber method, a substrate is mechanically transferred from one growth chamber to the other to be exposed to GaCl and arsine, alternately. GaCl and arsine are diluted in hydrogen. It takes 45 to 60 sec for a single ALE cycle by the reactor shown in Fig. 9a. This rather long cycle time is due to a mechanical limit of the substrate transfer process, but it is improved to 6 sec by the growth system shown in Fig. 9b, because the substrate is not inserted into the tube but is only rotated at the exits of the two tubes.

Growth apparatuses for GaAs ALE using metal organic source gases are shown in Fig. 10. In ALE experiments by Nishizawa *et al.* (1985), who called it MLE (molecular layer epitaxy) instead of ALE, growth was carried out in the vacuum system shown in Fig. 10a. They used TMG and TEG as a gallium source. It takes about 30 sec for a single cycle, which consists of, for example, 4 sec for TMG introduction, 3 sec for evacuation for the TMG purge, 20 sec for AsH$_3$ introduction for the GaAs monolayer and 3 sec for evacuation for the AsH$_3$ purge. Molecular species in the growth chamber are analyzed by the quadrapole mass analyzer (QMS). Light illumination effects on growth were studied and were found to be effective to improve surface smoothness and electrical properties of grown layers (Nishizawa *et al.*, 1986).

In ALE by Bedair *et al.* (1985), an atmospheric vertical reactor having three independent introduction tubes, shown in Fig. 10b was used. A substrate placed on a horizontal rotating holder is exposed to TMG and AsH$_3$ streams alternately. A single ALE cycle time can be controlled simply by changing the rotation speed, typically 10 sec or shorter for one cycle.

Doi *et al.* used the vertical reactor shown in Fig. 10c, which was the same one originally used to develop their laser–MOVPE technique (Aoyagi *et al.*,

1. FUNDAMENTALS OF EPITAXY GROWTH AND ALE

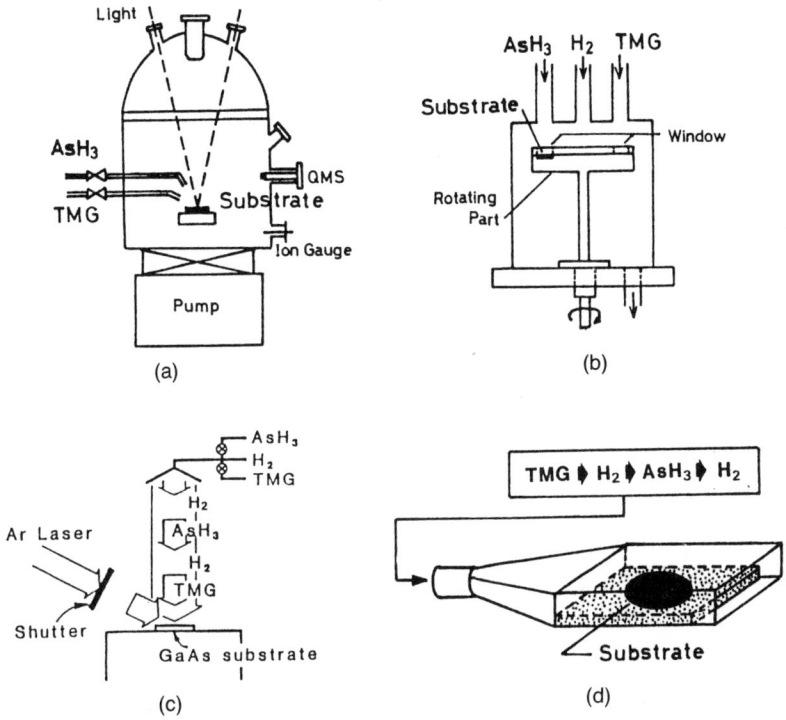

FIG. 10. Growth apparatus of GaAs ALE by TMG-AsH, system: (a) molecular beam method in vacuum, termed molecular layer epitaxy (MLE). (Reprinted by permission of the publisher, The Electrochemical Society, Inc. (1985) Nishizawaa, J., and Abe, H., *J. Electrochem. Soc.* **132**, p. 1197.); (b) rotating substrate method at atmospheric pressure (Bedair *et al.*, 1985); (c) laser-induced adsorption method at 76 torr, termed switched laser metal organic vapor phase epitaxy (SL-MOVPE) (Doi *et al.*, 1986); (d) horizontal narrow chamber method at 100 torr (Mori *et al.*, 1986).

1985). It takes 4 sec for a single ALE-cycle. The periods of TMG introduction, purge, AsH_3 introduction and purge processes are all one second. A laser beam from an argon-gas laser with 5145 Å wavelength was used to irradiate the substrate surface. They studied the illumination effect on the growth rate in each step and found that at temperatures lower than 400°C no growth takes place under no illumination, but the laser light enhances decomposition of TMG adsorbed on the substrate surface. Epitaxial growth occurs selectively only at the irradiated region. No effect of laser illumination is found for the arsine introduction period and the hydrogen purge processes.

Kobayashi *et al.* (1985) proposed a modified ALE method, called flow-rate modulation epitaxy (FME). They used TEG and arsine for gallium and arsenic sources, respectively, which are alternately supplied to the substrate region at an operation pressure of 90 torr. The most significant point in FME is that a small amount of arsine is added during the TEG flow period. Arsine is supplied continuously throughout the process, but its flow rate is changed stepwise. They found that this small amount of arsine addition was very effective to improve the electrical properties of grown layers. They succeeded in obtaining n-type films with fairly high electron mobilities, in contrast to normal TMG- or TEG-ALE, which always results in heavily p-type grown layers. The second unique feature of FME is that there is no hydrogen purge process between TEG and arsine periods. The growth apparatus used was designed so as to obtain rapid exchange of gas composition over the substrate, within 0.1 sec (Kobayashi et al., 1985).

Mori et al. (1986) used the horizontal MOCVD growth apparatus shown in Fig. 10d, which is operated at a total pressure of 100 torr. In their TMG-ALE, the cycle time is 11 sec, which consists of a 3-sec TMG supply, a 2-sec hydrogen purge, a 4-sec arsine supply and a 2-sec hydrogen purge. A highly uniform growth was confirmed by a substrate as large as three inches in diameter. In the same reactor, they tried GaAs ALE by using a diethyl gallium chloride (DEGC) source instead of TMG (Mori et al., 1988). Observed growth features were very similar to the GaCl–AsH$_3$ ALE system, rather than to the TMG–AsH$_3$ one. This is because the DEGC source is decomposed to GaCl and hydrocarbon gases above the substrate surface, and GaCl molecules are adsorbed on the surface.

10. GROWN THICKNESS IN A SINGLE ALE CYCLE

Figure 11 shows thickness grown in a single ALE cycle as a function of the partial pressures of TMG, TEG, DEGC, and GaCl. In all cases, the substrate

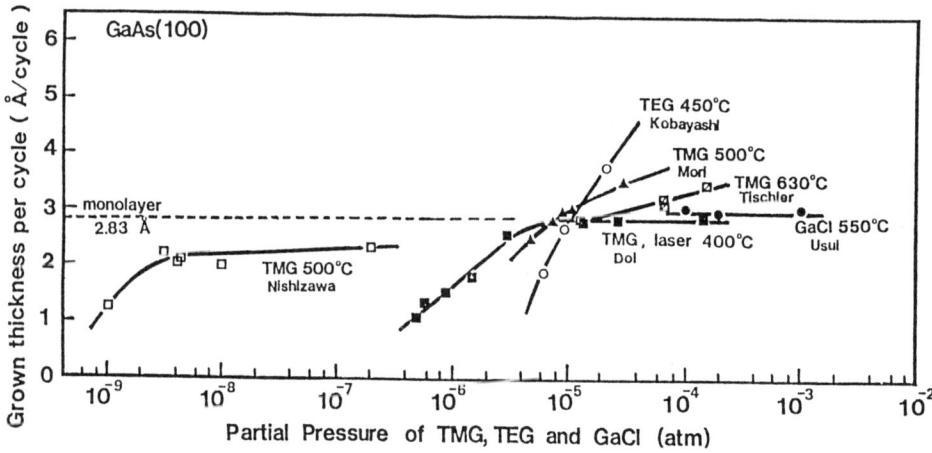

FIG. 11. Dependence of grown thickness in a single ALE cycle upon source gas partial pressure. (Reprinted with permission from IOP Publishing Ltd., Watanabe, H., and Usui, A., in "Gallium Arsenide and Related Compounds 1986," p. 1.)

orientation is ⟨100⟩, which has a monolayer thickness of 2.83 Å. Since it is very difficult to measure directly the thickness grown in a single ALE cycle, the total grown thickness was measured, and the thickness per cycle is calculated by dividing the total thickness by the number of repeated cycles. The total thickness is obtained by measuring the step height at the edge of the SiO_2 mask that covers a part of the substrate surface to prevent deposition.

In TMG-ALE by Nishizawa *et al.* (1985), no carrier gas is used. The source TMG and AsH_3 gases are injected as molecular beams to a substrate surface in a vacuum chamber. Their pressures on the horizontal axis are values measured by the QMS in Fig. 10a. All the other ALE studies were carried out in H_2 carrier gas open-flow systems at total pressures of atmospheric pressure (Tischler, Doi, and Usui), 100 torr (Mori) and 90 torr (Kobayashi). In their ALE experiments, Tischler *et al.* used a type of reactor (Fig. 10b) in which the substrate holder makes it difficult to determine TMG partial pressure over the substrate. Therefore, the partial pressure is assumed to be half of the partial pressure at the TMG inlet.

In the experiment by Nishizawa *et al.*, grown thickness per cycle saturates with increasing TMG pressure. The saturated value is slightly smaller than the theoretical monolayer thickness. This is because of an insufficient supply of AsH_3. The saturation value is increased to 2.83 Å by increasing the AsH_3 pressure (Nishizawa *et al.*, 1986). In Tischler's and Mori's experiments, the thickness per cycle saturates at high TMG pressures, but the saturation seems to be imperfect. As discussed by Nishizawa and Kurabayashi (1986), and by Tischler and Bedair (1986), there must be a self-limiting mechanism in TMG-ALE. But the insufficient saturation observed at high TMG pressures indicates that the surface coverage in step 1 in Fig. 8 is a function of TMG pressure. In addition, the fact that the thickness per cycle exceeds the theoretical monolayer thickness strongly suggests that continuous decomposition of TMG takes place during step 1, which only explains experimental values larger than the theoretical monolayer thickness. TEG-ALE carried out by Kobayashi *et al.* is quite different from TMG-ALE because the saturation is much less sufficient—almost no saturation even at the low temperature of 450°C. As reported by Yoshida *et al.* (1985), TEG thermal decomposition starts in a homogeneous way at temperatures around 300°C or below, which is about 150°C lower than the TMG decomposition temperature. Atomic gallium is produced in the vapor phase and/or the substrate surface, and gallium deposition occurs on previously deposited gallium atoms, which gives a thickness per cycle larger than the theoretical monolayer thickness.

According to Doi *et al.* (1986), no growth was observed in TMG-ALE at 400°C. This is because the TMG thermal decomposition velocity is extremely slow. Laser beam irradiation was found to be very effective to enhance the thermal decomposition of TMG adsorbed on the substrate. Laser irradiation during arsine supply and also hydrogen purge processes have no effect. Saturation of the thickness per cycle at TMG pressures higher than about 10^{-5} atm is perfect. They explained their series of experiments as follows. TMG is partially decomposed to some organic gallium radicals such as mono- or dimethyl gallium, which is adsorbed on arsenic sites but not on gallium sites. This process occurs under no irradiation when the substrate temperature is about 400°C, but because of the high stability of the adsorbed radicals thus formed, no growth reaction proceeds in the next step of arsine supply. In other words, the reaction velocity between the adsorbed species and the impinging arsine is extremely slow at this low temperature. Ar-laser irradiation enhances the decomposition of adsorbed gallium radicals to atomic gallium where the number of generated gallium atoms is equal to the number of adsorbed radicals. Because of the almost zero probability of gallium radical adsorption on thus-generated atomic gallium, no more generation of atomic gallium is expected even if TMG supply is continued during the Ar-laser irradiation period. This is the mechanism for the self-limiting process that assures an exact monolayer deposition in laser-ALE. They call it "SME by SL–MOVPE" (stepwise monolayer epitaxy by switched-laser MOVPE).

In Usui and Sunakawa's GaCl-ALE, the thickness per cycle is completely independent of GaCl partial pressure and equal to the theoretical monolayer thickness in the remarkably wide pressure range investigated. If we assume that the adsorbed species is GaCl, its reduction by hydrogen, forming atomic gallium, does not take place before arsine comes to the adsorbed surface, where the surface coverage of GaCl must be unity to give an exact monolayer thickness, independent of GaCl pressure.

Figure 12 shows a summary of reported temperature dependences of grown thickness per ALE cycle. The maximum temperature of 600°C is a higher limit above which GaAs substrate dissociation is not negligible and only rough surface is obtained.

The results obtained by TMG and TEG sources show a positive dependence. The grown thickness per cycle increases with increasing temperature, and, at high temperatures, it exceeds the theoretical one-monolayer thickness. This fact suggests that during the adsorption step of TMG or TEG, thermal decomposition of TMG and TEG becomes significant with increasing temperature, and the deposition of gallium alone takes place. However, when the thickness per cycle is lower than monolayer thickness, it seems to saturate to monolayer thickness. For this saturation tendency, the existence of a self-limiting mechanism has been proposed (Nishizawa and Kurabayashi, 1986; Tischler and Bedair, 1986).

TMG-ALE by Doi *et al.*, which was carried out under the illumination

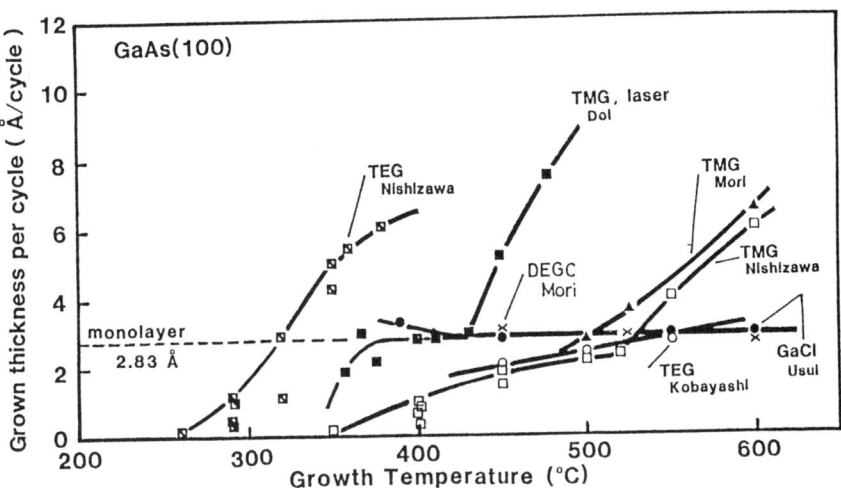

FIG. 12. Dependence of grown thickness in a single ALE cycle upon growth temperature. (Reprinted with permission from IOP Publishing Ltd., Watanabe, H., and Usui, A., in "Gallium Arsenide and Related Compounds 1986," p. 1.)

of argon laser light, has a plateau around 400°C. Without the laser irradiation, no deposition occurs. This fact also suggests the existence of a self-limiting mechanism in the adsorption process under illumination (Doi *et al.*, 1986).

In GaCl- and DEGC-ALE, grown thickness is perfectly independent of temperature. The DEGC-ALE process must be similar to that for GaCl-ALE, because DEGC thermal decomposition results in GaCl and some hydrocarbon gases. The adsorbed species in DEGC-ALE seems to be GaCl, as in GaCl-ALE. The experimentally obtained thickness per cycle agrees well with the theoretical monolayer thickness in the whole temperature range examined. These facts strongly indicate the existence of a perfect self-limiting mechanism. An ideal self-limiting mechanism consists of two things: the surface species are adsorbed with sufficiently high adsorption energy, and the adsorbed species are not adsorbed on the same species—meaning that the adsorption is complete with a monolayer thickness. The former condition is necessary for the purge step after the GaCl or DEGC adsorption step. Otherwise, the adsorbed species is desorbed, and the surface coverage becomes less than unity. The thickness per cycle becomes less than monolayer thickness. The latter condition is also necessary to assure an exact monolayer formation. If the adsorption energy is so strong that the second species can be adsorbed on the first adsorption layer, more than two-monolayer thickness deposition would take place in a single ALE cycle.

These discussions will be developed in more detail in the next section.

11. Surface Coverage of Adsorbed Species

a. Elementary Process in GaAs ALE

The growth reaction in ALE takes place on the substrate surface where an adsorption layer exists. We can consider two possibilities for an adsorption-layer-related chemical reaction. One is the Langmuir–Hinshelwood mechanism, which is a reaction between two adsorbed species. The other is the Rideal mechanism, which is a heterogeneous reaction between a gas-phase species and an adsorbed species. The Langmuir–Hinshelwood mechanism is not likely in ALE, because the surface coverages of the two adsorbed species are less than unity. The total of the coverages is equal to unity or lower. This means that the grown thickness in a single ALE cycle becomes thinner than a monolayer thickness. However, experimental single-ALE-cycle thickness frequently exceeds the crystallographic monolayer thickness. Therefore, we assume that the elementary reaction in ALE obeys the Rideal mechanism.

Suppose that GaAs is formed through an elementary process between an adsorbed species M^{ads} and a gas-phase species N^{gas}:

$$M^{ads} + N^{gas} = GaAs. \tag{55}$$

M^{ads} is either GaX^{ads} or AsY^{ads}, where GaX^{ads} (AsY^{ads}) is the adsorbed species containing gallium (arsenic). In papers reported so far, the adsorbed species M^{ads} has been supposed always to be a gallium-containing species, not an arsenic-containing one. However, it should be noticed that neither experimental direct observation nor theoretical validation was made for the identification of M^{ads}. GaX^{ads} is either the original form of the source gas or that of a partially decomposed molecule. From thermodynamic equilibrium theory, a completely decomposed form, atomic gallium (Ga), should be included as a candidate for GaX^{ads} in TMG-ALE and TEG-ALE systems. Consequently, the following three cases can be considered as elementary processes.

(1) $GaX^{gas} \rightarrow GaX^{ads}$: direct adsorption.
$GaX^{ads} + AsY^{gas} \rightarrow GaAs + X*Y^{gas}$.

(2) $GaX^{gas} \rightarrow GaXr^{ads} + (X-Xr)$: adsorption as a radical.
$GaXr^{ads} + AsY^{gas} \rightarrow GaAs + X*Y^{gas}$.

(3) $GaX^{gas} \rightarrow Ga^{ads} + X^{gas}$: atomic gallium deposition.
$Ga^{ads} + AsY^{gas} \rightarrow GaAs + Y^{gas}$.

GaXr and $X*Y$ are a partially decomposed radical of GaX and a byproduct consisting of X and Y, respectively.

In GaCl-ALE, GaX^{ads} is supposed to be GaCl, the same molecular form as the source gas GaCl, which is produced by the reaction between HCl and a metallic gallium source. This belief is based on growth kinetic studies of GaAs chloride transport VPE (Shaw, 1975; Cadoret and Cadoret, 1975). GaCl adsorption in VPE has been studied as a competitive adsorption with other reactants such as As_4, As_2, and HCl. In their calculations, arsenic adsorption in the form of As_4, As_2, or AsH_3 is negligibly small. The overall growth reaction in GaCl-ALE is assumed to be reaction (1). AsY^{gas} is not identified but is supposed to be As_4 or As_2. A thermodynamic calculation suggests that As_4 must be dominant when AsH_3-containing hydrogen gas attains equilibrium.

In TMG-ALE and TEG-ALE, we assume that the probability of direct adsorption of TMG (TEG) is low and that the adsorbed species is either a radical of TMG (TEG) or atomic gallium. Reactions (2) and (3) are expected for the elementary processes of TMG-ALE and TEG-ALE.

Since TMG and TEG are thermally unstable, atomic gallium is expected as the surface species. The vapor pressure of elemental gallium is low, so that gallium deposition takes place not only on arsenic atoms but also on gallium atoms, resulting in more than one monolayer deposition in a single ALE cycle. Experimental results in Figs. 11 and 12 clearly suggest this at high substrate temperatures, particularly, in TEG-ALE. In TMG-ALE, the thickness in a single cycle exceeds the theoretical monolayer thickness, but it seems to saturate at high source-gas pressures. This fact suggests the existence of a self-limiting mechanism. The self-limiting mechanism is such that no more adsorption can proceed once the bare surface is covered with the adsorption layer, and therefore, the thickness per ALE cycle never exceeds a monolayer thickness. To explain the self-limiting mechanism, the gallium compound radical $GaXr^{ads}$ is assumed to exist as the adsorption layer species (Nishizawa and Kurabayashi, 1986; Tischler and Bedair, 1986; Tischler et al., 1987). Although the gallium compound radical was not identified experimentally, it has been supposed to be $Ga(CH_3)_x$ ($x = 1, 2$) in ALE using trimethyl gallium.

In ALE with a diethyl gallium chloride (DEGC) source, the adsorbed species is supposed to be not organic gallium radicals but GaCl. DEGC is also thermally unstable, but the Ga–Cl bond in DEGC is very strong, and GaCl and hydrocarbon fragments are produced after DEGC thermal decomposition. Therefore the elementary process in DEGC-ALE is expected to be equal to that in GaCl-ALE.

The molecular form of the arsenic-containing species in ALE using AsH_3 as the arsenic source is also unclear. From thermodynamic calculations, As_4 and As_2 molecules are considered as the stable form at growth temperatures. However, since the thermal decomposition speed of arsine is not so fast (Ban, 1972), undecomposed AsH_3 or partially decomposed AsH_2 or AsH must impinge the surface. In the following discussion, a quantitative analysis is made with the assumption that one monolayer is formed when arsenic-containing gas species attack the previously adsorbed gallium-containing adsorption layer. Under this assumption it is not essential to specify the molecular form of the arsenic-containing gas species.

b. Adsorption Equation

The surface coverage θ is defined as a ratio of the molecular density of the adsorbed species, N_{ads}, to the density of the adsorption sites, N_s:

$$\theta = N_{ads}/N_s. \tag{56}$$

N_s corresponds to the surface density of lattice sites. The dependence of the surface coverage θ of the adsorbed species on its partial pressure P has been interpreted by several different adsorption isotherms (T. Keii, 1970). In Henry-type adsorption, the surface coverage is increased proportionally with pressure: $\theta \propto P$. In Freundlich-type adsorption, it increases as $\theta \propto P^{1/n}$ (n: integer). The Brunauer–Emmett–Teller (BET) adsorption isotherm gives a multilayer adsorption, which explains surface coverages of $\theta > 1$. The Langmuir-type isotherm is expressed by

$$\theta = \theta_m \frac{KP}{1 + KP}, \tag{57}$$

where K and θ_m are the adsorption constant and the maximum surface coverage, respectively. The adsorption constant K is a function of temperature T. When the molecular size of the adsorbed species is large and a single adsorbed species occupies more than one adsorption site, the maximum surface coverage θ_m is less than unity.

We assume that the thickness grown in a single ALE cycle, t_c, is related to the surface coverage in step 2 in Fig. 8 and the monolayer thickness d:

$$t_c = \theta d. \tag{58}$$

The experimental curves in Fig. 11 seem to be well explained by one of the isotherms described above. However, all the MO-ALE results in Fig. 12 show increasing θ with T, which is not explained by the normal adsorption types described above. We cannot use the conventional simple adsorption equations for MO-ALE. Although a more detailed study is needed, it seems that in MO-ALE at least two surface reaction processes take place simultaneously. One is a thermal decomposition process of the metal organic source gas that forms adsorbed radicals. The other is a thermal decomposition process of the adsorbed radicals that forms atomic gallium. Both of the processes are enhanced by increasing temperature, but the desorption of the adsorbed species also would be enhanced at higher temperatures. Therefore, the formation of adsorbed species and their desorption are competitive. If the formation speed is higher than that of desorption, atomic gallium is deposited by the thermal decomposition of adsorbed radicals, resulting in gallium-on-gallium deposition, which leads to a surface coverage larger than unity. Since this phenomenon becomes more remarkable with increasing temperature, the results obtained in MO-ALE at high temperatures shown in Fig. 12 are well understood.

GaCl-ALE shows neither partial pressure dependence (Fig. 11) nor temperature dependence (Fig. 12). These results can be explained if we assume that the adsorbed species (GaCl) concentration obeys the Langmuir isotherm expressed by Eq. (57). The surface coverage in this formula approaches unity at high pressures or when the adsorption constant K is large enough. In this limit of $KP \gg 1$, the coverage becomes insensitive not only to P but also to T. The experimental result that the thickness per single ALE cycle is equal to a monolayer thickness indicates $\theta_m = 1.0$. The adsorption layer thickness measurement made by Theeten and Hottier (1976) in chloride VPE and the surface coverage calculation of GaCl in that system by Cadoret (1980) support the Langmuir mechanism in the GaCl–GaAs adsorption system.

c. Coverage Calculation in GaCl-ALE

The adsorption constant K in Eq. (57) is expressed by

$$K = \exp\left(-\frac{\Delta H - T\Delta S}{RT}\right), \tag{59}$$

where ΔH and ΔS are the enthalpy and entropy changes due to adsorption and R is the gas constant. ΔH for GaCl adsorption was found to be 35.2 Kcal/mol by Cadoret (1980) or 55.7 kcal/mol by Korec (1982). Since the discrepancy between these values is not small, we calculate the coverages for the two cases. Since no data on ΔS for GaCl are available, we assume that it is equal to the universal value of entropy change from vapor to liquid, 22 cal/deg. At 450°C, K is calculated to be 6.3×10^5 with Cadoret's ΔH, or to be 1.1×10^{12} with Korec's ΔH. Therefore, we obtain $1 - \theta = 2 \times 10^{-3}$ with Cadoret's ΔH or $1 - \theta = 1 \times 10^{-8}$ with Korec's ΔH when the GaCl partial pressure is 1×10^{-4} atm. Although the assumptions used here must be examined further, it can be said that the GaCl coverage is almost unity, at least more than 99%. This is supported by extrapolating the result of GaCl adsorption measurement in GaAs VPE made by Theeten and Hottier (1976).

TABLE III
Impinging Rates and Degrees of Excess Exposure in Reported GaAs ALE

Gallium source	Growth environment	Impinging rate R_i (molecules/cm² sec)	Exposure time t (sec)	Excess degree $R_i t / N_s$[a]	Reference[b]
TMG	Vacuum	2.0×10^{13}–3.9×10^{15}	4	0.12–25	(1)
TEG	90 torr H_2	3.1×10^{16}–2.2×10^{17}	1	50–350	(2)
TMG	76 torr H_2	5.5×10^{15}–2.5×10^{18}	1	8–4000	(3)
GaCl	760 torr H_2	1.3×10^{18}–1.3×10^{19}	6	12,000–120,000	(4)

[a] Excess degree of exposure of adsorbed species, $R_i t/N_s$, is defined as the ratio of the total molecules supplied during the gallium source adsorption step to the adsorption site density N_s, which is assumed to be 6.4×10^{14} cm^{-2} for GaAs $\langle 100 \rangle$ substrate.

[b] References: (1) Nishizawa et al. (1985); (2) Kobayashi et al. (1986); (3) Doi et al. (1986); (4) Usui and Sunakawa (1986).

12. Cycle Speed

A disadvantage of ALE at present is a slow growth rate or long ALE cycle time. We discuss the cycle time from the viewpoint of the impinging rate of adsorbed species and the exposure time. In Table III, the calculated impinging rates of TMG, TEG and GaCl in reported ALE experiments are shown. The impinging rate R_i is calculated by Eqs. (5) and (6) for the viscous flow systems carried by Kobayashi et al., Doi et al., and Usui et al., and the molecular beam system carried by Nishizawa et al., respectively. The degree of excess supply, $R_i t/N_s$, is found to be fairly large in all the ALE work, where t and N_s are the exposure time of step 2 in Fig. 8 and the density of lattice sites on the GaAs $\langle 100 \rangle$ surface, 6.4×10^{14} cm^{-2}, respectively. The calculated values of the degree of excess supply suggest that we can reduce exposure time by at least one or two orders of magnitude. This means that considerable reduction of the ALE cycle time is possible. However, because of the mechanical limit of gas valve operation or the substrate transfer operation, it is not easy to reduce the cycle time to less that 1 sec. Furthermore, the limits of adsorption speed in step 1 and reaction speed in step 3 in Fig. 8 must be considered. It seems difficult to achieve a cycle time less than 1 sec. The 1-sec ALE cycle corresponds to a growth rate of one monolayer per second, which is equal to the conventional MBE or MOCVD growth rate.

13. Advantages of ALE

a. Easy Control by "Digital Epitaxy"

In conventional epitaxial methods, we must control various growth parameters such as partial pressures and flow rates of source materials, source temperatures and substrate temperature. All of these parameters are analog quantities. In addition, the analog quantity of growth time is adjusted to obtain the desired grown thickness. It is common in LPE, VPE, MOCVD, and MBE. Hereafter, we call these conventional epitaxial methods analog epitaxy. The growth mechanism in analog methods is a continuous sequence of local growth reactions such as step growth or the two-dimensional nucleation growth.

The growth mechanism in ALE is completely different from the analog epitaxial methods. Crystal growth occurs through a surface reaction between impinging species and adsorbed layer species. After sufficient time for the impinging species to completely react with the adsorbed species, no more crystal growth occurs even if more of the impinging species is supplied. To begin the next growth, we supply source gas to form an adsorption layer. By repeating this process we obtain a given thickness. The thickness grown in a single cycle is a function of the surface coverage of the adsorbed layer. If the surface coverage of the adsorbed layer is, for example, half of the lattice site density, i.e., $\theta = 0.5$, the resultant grown thickness is half of the monolayer thickness. The surface coverage of the adsorbed species is usually a function of its partial pressure and the substrate temperature, as discussed before. We need analog control of temperatures, partial pressure and flow rates to obtain the desired thickness of the grown layer. However, under the particular ALE conditions we are free from any analog control, as described below.

As shown in Fig. 8, the thickness per cycle is insensitive to TMG pressures over about 1×10^{-5} atm in Doi et al.'s laser MO-ALE, and to GaCl pressure in the range from 1×10^{-4} to 1×10^{-3} atm in Usui and Sunakawa's GaCl-ALE. Figure 12 shows that the thickness per cycle is independent of growth temperatures, from 370 to 420°C in Doi et al.'s laser TMG-ALE, and from 400 to 600°C in Usui and Sunakawa's GaCl-ALE. The observed thicknesses per cycle agree well with the theoretical monolayer thickness of 2.83 Å. Laser power dependence is also very weak in their temperature range.

Under these particular conditions, the thickness per cycle is a function of neither the source gas pressures not the growth temperature and is equal to the monolayer thickness. Total grown thickness is, therefore, determined only by the number of ALE cycles, and no control of any analog quantities such as temperatures of sources and substrate, partial pressures and flow rates of source gases as well as growth time is required.

Since growth time is irrelevant in determining the grown thickness, now we are free from the requirement to control any analog quantities. The only growth parameter determining the grown thickness is the number of the ALE cycle, which is a digital quantity. Therefore, we can call this particular growth method "digital epitaxy" (Watanabe and Usui, 1987; Usui and Watanabe, 1987).

The advantage of digital epitaxy is its high reproducibility. In principle, there is no run-to-run fluctuation of growth rate. The grown thickness is easily controlled to a designed value with monolayer accuracy by counting the ALE cycle. The feature of the insensitivity of the thickness to the source-gas partial pressure means that no careful design of the reactor, especially the substrate holder and its surroundings, is necessary. This advantage is particularly attractive for designing a multiwafer reactor having a high reproducibility.

So far, digital epitaxy has been confirmed in systems of $GaCl-AsH_3$ (Usui and Sunakawa, 1986), $DEGC-AsH_3$ (Mori et al., 1988) and $TMG-AsH_3$ with laser irradiation (Doi et al., 1986).

Digital epitaxy also has been confirmed for other III-V compound semiconductors. InAs, InP, and GaP ALEs were successfully carried out as digital epitaxy, where the grown thickness per single ALE cycle was found to be insensitive to the partial pressures of InCl and GaCl (Usui and Sunakawa, 1987; Usui et al., 1987).

b. High Uniformity of Grown Thickness

In conventional epitaxy in a gas flow system, thickness nonuniformity has several causes. A boundary layer is formed just above the substrate surface, and the stream line is strongly dependent on the susceptor structure. The thickness of the boundary layer is a function of flow velocity and distance from the leading edge of the substrate. Under conditions in which the growth rate is determined by the diffusion velocity of gaseous reactants in the boundary layer, growth rate nonuniformity appears to be due to nonuniform thickness of the boundary layer. The partial pressures of the source gases are decreased along the flow direction with the growing crystal. This is the so-called depletion effect of source gases, which results in nonuniform thickness. When the growth is controlled by the chemical reaction speed of the adsorbed species, the grown thickness is a function of temperature. If the temperature is not constant over the substrate, the thickness becomes nonuniform.

In MBE, since the flux intensity profile in a molecular beam is not uniform, the substrate must be rotated to assure uniform growth. To obtain good compositional uniformity for an element having a sticking coefficient less than unity, such as arsenic or phosphorus, temperature uniformity is also very important.

On the other hand, ALE growth is not influenced by these parameters. Particularly in digital epitaxy, no control of boundary layer thickness, temperature distribution near a substrate or a substrate rotation is required. In principle, thickness uniformity over a substrate is always assured to be

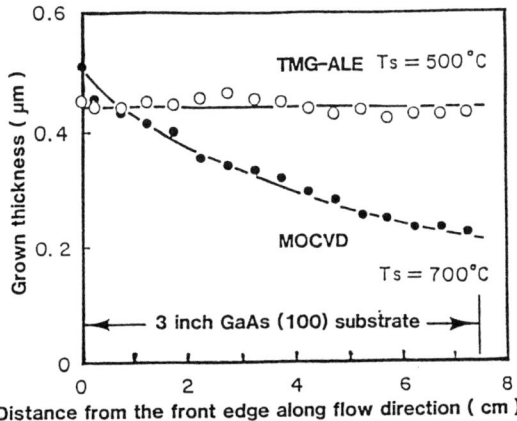

FIG. 13. Thickness uniformity comparison between ALE and MOCVD on a 3-inch substrate in the reactor shown in Fig. 10d. (Reprinted with permission from IOP Publishing Ltd., Watanabe, H., and Usui, A., in "Gallium Arsenide and Related Compounds 1986," p. 1.)

perfect. This is demonstrated in Fig. 13, which is the result of TMG-ALE done by Mori et al. (1986). Exactly speaking, this is not digital epitaxy, because the thickness per cycle is not perfectly independent of TMG pressure and substrate temperature as shown in Figs. 11 and 12. The growth was carried out under the condition that the thickness per cycle becomes equal to the monolayer thickness. Their reactor was rather small, but a substrate as large as three inches in diameter was used for the uniformity experiment. In the MOCVD mode, namely in the case of simultaneous supply of TMG and arsine, grown thickness uniformity is particularly bad. This is because the volume in the substrate region of their reactor is very small. The small height to the ceiling of the substrate holder block strongly induces a depletion effect of the TMG source (Okamoto et al., 1984). In the same flow condition but changing to ALE mode, i.e., from simultaneous to alternate supply of TMG and arsine source gases, the thickness nonuniformity was drastically improved. Thickness variation from edge to edge of the 3-inch substrate is considerably smaller. This result is well explained by the ALE mechanism as an expected advantage. High uniformity of grown thickness was also confirmed for digital epitaxy in the DEGC-AsH_3 system (Usui et al., 1987; Mori et al., 1988).

c. Smooth Surface of Grown Layer

In epitaxial growth various kinds of surface irregularities are generated on the grown surface. Terraces in LPE, hillocks in VPE, and oval defects in MBE or MOCVD are well known to be difficult to completely eliminate. Surface protrusion irregularities are generated by local enhancement of growth that is caused by surface diffusion or migration of adsorbed species to low-potential-energy specular points such as tops and facets of hillocks or contamination microclusters. This is frequently observed as hillock defects on VPE-grown surfaces. To eliminate these kinds of surface defect, slightly inclined orientations such as two to five degrees off from $\langle 100 \rangle$ orientation toward $\langle 100 \rangle$ are used in VPE, where the intentionally introduced surface atomic steps are effective in suppressing undesired surface migrations. Direct supply from the vapor phase to the low-potential-energy points without surface migration is also responsible for the surface irregularities.

A digital ALE produces a high-quality mirror surface. In GaCl-ALE by Usui and Sunakawa (1986), very smooth surfaces without any irregular texture were obtained for the substrate orientations of $\langle 100 \rangle$, 2° off $\langle 100 \rangle$, $\langle 511 \rangle$, $\langle 211 \rangle$ and $\langle 111 \rangle$A,B. This is explained as follows. Since the surface lattice sites are all occupied by the adsorbed species (GaCl) with its surface coverage of unity, the adsorbed species has difficulty migrating and aggregating. No additional supply of the adsorbed species from the vapor phase is possible on a surface with unity surface coverage of the adsorbed species. Second-layer adsorption of GaCl is found to be less possible, as shown in Fig. 11. In ALE, local enhancement of growth does not occur because of these specular characteristics of the ALE process.

However, thermal decomposition occurs at high temperatures when an arsenic overpressure is not applied. In the adsorption process of the group III element, for example, the GaCl adsorption process, there is no arsenic supply. To suppress thermal decomposition of the GaAs substrate during this step, the substrate temperature must be kept low. So, the ALE method must be carried out at low temperatures such as 400 to 500°C, where GaAs thermal decomposition is negligibly small.

d. Uniform Growth in Selective Epitaxy

When growth is restricted to a local area of a substrate, the process is called selective epitaxy. A GaAs layer can be selectively grown through a dielectric film mask, such as SiO_2 in halogen transport VPE. Selective epitaxial growth has been tried to improve GaAs FET performances by reducing the contact resistances of the source and drain electrodes. Heavily doped n^+-GaAs is selectively grown on both sides of a gate electrode. In VPE, the growth rate in the window area is a function of the window area. Figure 14 schematically shows the growth rate dependence of the relative ratio of the window area to masked area. When the window is small, the growth rate is enhanced. This is because reactants impinged on the mask surface are not consumed there, but are supplied into the window area. Species impinged on the mask go to the window area by two ways: migration on the mask surface, and gas-phase

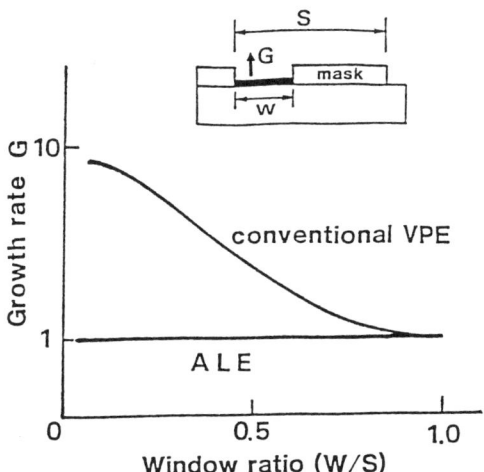

FIG. 14. Growth rate enhancement in selective epitaxy.

diffusion through a desorption-and-adsorption process. When the window is relatively large, the growth rate approaches the value without a mask. However, at the peripheral edge of the window area, the growth rate is larger than that of the inside of the window. This nonuniform edge growth becomes remarkable when the width of the window is larger than the surface migration distance. The local growth enhancement causes faceting of the epitaxial layer, and a flat surface is not obtained in the window area. The faceting appears more frequently at low growth temperatures, where the growth rate is kinetically determined by the surface chemical reaction.

In ALE by the $GaCl-AsH_3$ system, no such problems were found. No surface texture is observed on the grown surface of a selectively grown region. A hillock-free mirror surface with neither edge growth nor facet development is obtained. Thickness in selective epitaxy is readily controlled by adjusting the number of ALE cycles. The grown layer thickness in the selective region is highly uniform, independent of the window size (Watanabe and Usui, 1987).

e. Possibility of Side-wall Epitaxy

In GaCl-ALE, it was found that a single ALE cycle offers monolayer growth on not only $\langle 100 \rangle$ surfaces but also $\langle 111 \rangle A,B$, $\langle 211 \rangle$ and $\langle 511 \rangle$ (Usui and Sunakawa, 1986). This fact suggests good thickness controllability of layers grown on a side-wall of a structured substrate. Various possible side-wall epitaxes are illustrated in Fig. 15. Under the conditions in which digital epitaxy takes place, total grown thickness is determined only by the number

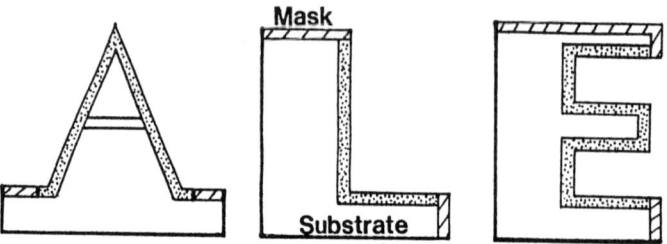

FIG. 15. Possibility of epitaxial growth on side-walls (side-wall epitaxy) by ALE. (Reprinted with permission from IOP Publishing Ltd., Watanabe, H., and Usui, A., in "Gallium Arsenide and Related Compounds 1986," p. 1.)

of ALE cycles, independent of surface orientation. Therefore, if the source gas is given enough time to be adsorbed on the surfaces of such a complicated substrate structure, it would be possible to control the grown thickness at the monolayer level. This is a unique characteristic of the ALE method, and no techniques other than ALE can give such growth controllability. By carrying out such side-wall epitaxy on a substrate that is processed on an epitaxial multilayer wafer, we can obtain novel structures in which electrons or holes are confined to have one- or zero-dimensionally localized properties. Since uniformity and thickness controllability are expected to be excellent wth a high reproducibility, we can expect integrated circuits using such complicated low-dimensionally confined electronic properties in future.

ACKNOWLEDGMENTS

The authors wish to express their deep appreciation to H. Sunakawa, K. Mori, and M. Yoshida for their kind supply of ALE data and stimulating discussions on the ALE mechanism. They also wish to thank M. Ogawa, T. Baba, and K. Hirose for their useful discussions on the characteristics of MBE-grown films and MBE growth kinetics. They should like to thank F. Saito and D. Shinoda for their kind interest in this work and for continuous support. The authors are greatly indebted to Professor T. Ikoma for his support to complete the present chapter.

REFERENCES

Aoyagi, Y., Masuda, S., and Namba, S. (1985). *Appl. Phys. Lett.* **47**(?), 95.
Arthur, J. R. (1967). *J. Phys. Chem. Solids* **28**, 2257.
Arthur, J. R. (1974). *Surf. Sci.* **43**, 449.
Baba, T., Mizutani, T., and Ogawa, M. (1983). *Japan. J. Appl. Phys.* **22**, L627.
Baba, T., Ogawa, M., and Mizutani, T. (1986). *Surf. Sci.* **174**, 408.
Ban, V. S. (1972). *J. Cryst. Growth* **17**, 19.
Bedair, S. M., Tischler, M. A., Katsuyama, T., and El-Masry, N. A. (1985). *Appl. Phys. Lett.* **47**, 51.
Cadoret, R. (1980). In "Current Topics in Materials Science" (E. Kaldis, ed.) **5**, 219, North-Holland, Amsterdam.
Cadoret, R., and Cadoret, M. (1975). *J. Cryst. Growth* **31**, 142.

Chang, L. L., Esaki, L., Howard, W. E., Ludeke, R., and Schul, G. (1973). *J. Vac. Sci. Technol.* **10**, 655.
Cho, A. Y., and Panish, M. B. (1972). *J. Appl. Phys.* **43**, 5118.
Cho, A. Y., and Reinhart, F. K. (1974). *J. Appl. Phys.* **45**, 1812.
Cox, A., and Pool, B. (1967). *J. Chem. Engn. Data* **12**, 247.
Doi, A., Aoyagi, Y., Iwai, S., and Namba, S. (1986). *In* "Extended Abstracts of the 18th (1986 International) Conf. Solid State Devices and Materials, Tokyo", p. 739.
Drathen, P., Ranke, W., and Jacobi, K. (1978). *Surf. Sci.* **77**, L162.
Flaim, T. A., and Ownby, P. D. (1971). *J. Vac. Sci. Technol.* **8**, 661.
Foxon, C. T., and Joyce, B. A. (1975). *Surf. Sci.* **55**, 434.
Foxon, C. T., and Joyce, B. A. (1977). *Surf. Sci.* **64**, 293.
Foxon, C. T., Harvey, J. A., and Joyce, B. A. (1973). *J. Phys. Chem. Solids* **34**, 1693.
Goodman, C. H. L., and Pessa, M. V. (1986). *J. Appl. Phys.* **60**, R65.
Harris, J. J., Joyce, B. A., and Dobson, P. J. (1981). *Surf. Sci.* **103**, L90.
Hellman, E. S., and Harris, Jr., J. S. (1986). *In* "Proc. 4th MBE Conf." p. 100.
Hirose, K., Ohata, K., Mizutani, T., Itoh, T., and Ogawa, M. (1986). *In* "Gallium Arsenide and Related Compounds 1985", p. 529, Institute of Physics, London.
Hiyamizu, S., Saito, J., Nanbu, K., and Ishikawa, T. (1983). *Japan. J. Appl. Phys.* **22**, L609.
Keii, T. (1970). *In* "Shokubai Han-nou Sokudoron" (T. Keii, ed.), p. 66, Chijin Shokan, Tokyo.
Kobayashi, N., Makimoto, T., and Horikoshi, Y. (1986). *Japan. J. Appl. Phys.* **24**, L962.
Korec, J. (1982). *J. Cryst. Growth* **60**, 297.
Lewis, B. F., Grunthaner, F. J., Madhukar, A., Lee, T. C., and Fernandez, R. (1985). *J. Vac. Sci. Technol.* **B3**, 1317.
Matysik, K. J. (1976). *J. Appl. Phys.* **47**, 3826.
Mizutani, T. (1988). *J. Vac. Sci. Technol.* **B6**, 1671.
Mizutani, T., and Hirose, K. (1985). *Japan. J. Appl. Phys.* **24**, L119.
Mizutani, T., and Watanabe, H. (1982). *J. Cryst. Growth* **59**, 507.
Mori, K., Ogura, A., Yoshida, M., and Terao, H. (1986). *In* "Extended Abstracts of the 18th (1986 International) Conf. Solid State Devices and Materials", p. 739.
Mori, K., Yoshida, M., and Usui, A. (1988). *In* "Gallium Arsenide and Related Compounds 1987", p. 100, Institute of Physics, London.
Neave, J. H., and Joyce, B. A. (1988). *J. Cryst. Growth* **44**, 387.
Neave, J. H., Joyce, B. A., Dobson, P. J., and Norton, N. (1983). *Appl. Phys.* **A31**, 1.
Neave, J. H., Dobson, P. J., Joyce, B. A., and Zhang, J. (1985). *Appl. Phys. Lett.* **47**, 100.
Nishizawa, J., and Kurabayashi, T. (1983). *J. Electrochem. Soc.* **130**, 413.
Nishizawa, J., and Kurabayashi, T. (1986). *J. Crystallographic Soc. Japan* **28**, 133.
Nishizawa, J., Okuno, Y., and Tadano, H. (1975). *J. Cryst. Growth* **31**, 215.
Nishizawa, J., Abe, H., and Kurabayashi, T. (1985). *J. Electrochem. Soc.* **132**, 1197.
Nishizawa, J., Abe, H., Kurabayashi, Y., and Sakurai, N. (1986). *J. Vac. Sci. Technol.* **A4**(3), 706.
Ogawa, M. (1986). *In* "Gallium Arsenide and Related Compounds 1985", p. 103, Institute of Physics, London.
Okamoto, A., Sunakawa, H., Terao, H., and Watanabe, H. (1984). *J. Cryst. Growth* **70**, 140.
Phillips, J. C. (1981). *J. Vac. Sci. Technol.* **19**, 545.
Sakaki, H. (1980). *Japan. J. Appl. Phys.* **19**, L735.
Sakamoto, T., Funabashi, H., Ohta, K., Nakagawa, T., Kawai, N. J., and Kojima, T. (1984). *Japan. J. Appl. Phys.* **23**, L657.
Shaw, D. W. (1975). *J. Cryst. Growth* **31**, 130.

SpringThorpe, A. J., and Mandeville, P. (1986). *J. Vac. Sci. Technol.* **B4**, 853.
Suntola, T., Antson, J., Pakkala, A., and Lindfors, S. (1980). *In* "SID 80 Digest", p. 108.
Theeten, J. B., and Hottier, F. (1976). *Surf. Sci.* **58**, 583.
Tischler, M. A., and Bedair, S. M. (1986). *Appl. Phys. Lett.* **47**(2), 95.
Tischler, M. A., Anderson, N. G., and Bedair, S. M. (1987). *In* "Gallium Arsenide and Related Compounds 1986", p. 135, Institute of Physics, London.
Tsang, W. T., Logan, R. A., and Ditzenberger, J. A. (1982). *Electron. Lett.* **16**, 845.
Usui, A. (1987). *In* "Extended Abstracts of the 19th Conf. Solid State Devices and Materials", p. 471, Tokyo.
Usui, A., and Sunakawa, H. (1986). *Japan. J. Appl. Phys.* **25**, L212.
Usui, A., and Sunakawa, H. (1987). *In* "Gallium Arsenide and Related Compounds 1986", p. 129, Institute of Physics, London.
Usui, A., and Watanabe, H. (1987). *In* "Proc. 10th Int. Conf. Chemical Vapor Deposition; CVD-X", p. 10, Hawaii.
Usui, A., Sunakawa, H., Mori, K., and Yoshida, M. (1987). *In* "Proc. Symp. on Compound Semiconductor Growth, Processing, and Devices for the 1990's; Japan/U.S. Perspective", p. 10.
Watanabe, H. (1975). *Japan. J. Appl. Phys.* **14**, 1451.
Watanabe, H., and Usui, A. (1987). *In* "Gallium Arsenide and Related Compounds 1986", p. 1, Institute of Physics, London.
Watanabe, M. O., Tanaka, A., Udagawa, T., Nakanishi, T., and Zohta, Y. (1983). *Japan. J. Appl. Phys.* **22**, 923.
Yoshida, M., Uesugi, F., and Watanabe, H. (1985). *J. Electrochem. Soc.* **130**, 413.

CHAPTER 2

Characteristics of Two-Dimensional Electron Gas in III-V Compound Heterostructures Grown by MBE

S. Hiyamizu*

FUJITSU LABORATORIES
ATSUGI, JAPAN

I.	INTRODUCTION.	53
II.	MODULATION DOPING IN GaAs/AlGaAs SUPERLATTICES	55
III.	SELECTIVELY DOPED GaAs/n-AlGaAs HETEROSTRUCTURES	58
	1. Introduction to 2DEG in GaAs/n-AlGaAs Heterostructures	58
	2. Electronic State in GaAs/n-AlGaAs Heterostructures	65
	3. Temperature Dependence of 2DEG Mobility	67
	4. Spacer-Layer Thickness Dependence	70
	5. Electron Concentration Dependence	72
	6. AlAs Mole Fraction Dependence	75
	7. Transport Properties of 2DEG under High Electric Fields	79
IV.	IMPROVED SELECTIVELY DOPED HETEROSTRUCTURES	82
	8. Stopper Layer Structure	83
	9. n-AlGaAs/GaAs/n-AlGaAs Quantum Well Heterostructures	85
	10. GaAs/AlAs Superlattice Structures	86
	11. InGaAs/n-InAlAs, InGaAs/n-AlGaAs Heterostructures	91
V.	SUMMARY.	100
	REFERENCES.	101

I. Introduction

Recently, III-V compound heterostructures for high-speed device applications have been required to have a fine multilayer structure with an atomically flat heterojunction interface and an abrupt doping profile of impurities, in addition to high crystal quality of constituent layers and interfaces. Molecular-beam epitaxy (MBE) is the most suitable epitaxial growth technique for such heterostructure materials because of its ultimately precise controllability of layer thickness and high lateral uniformity. The

*Present address: Faculty of Engineering Science, Osaka University, 1-1, Machikaneyama, Toyonaka, Osaka 560, Japan.

precision of layer-thickness control is about 1%, and lateral variation of layer thickness, alloy composition, and impurity doping concentration is typically ±1% over a 1-inch-diameter area. These characteristic features of MBE led to the development of almost all new high-speed heterostructure devices and the best performance of these devices. These include the high electron mobility transistor (HEMT) (Mimura et al., 1980), the hot electron transistor (HET) (Heiblum, 1981; Yokoyama et al., 1984a,b), and the resonant tunnelling hot electron transistor (RHET) (Yokoyama et al., 1985; Yokoyama, 1986). The heterojunction bipolar transistor (HBT) (Kroemer, 1959) is one of the most promising high-speed heterostructure devices, and it has a rather long history of development with the use of liquid-phase epitaxy (LPE) or vapor-phase epitaxy (VPE). The best device performance, however, has been recently achieved for an MBE-grown GaAs/AlGaAs HBT (Ishibashi et al., 1987).

In this chapter, selectivity doped heterostructures grown by MBE are described. They are excellent two-dimensional electron gas (2DEG) systems for HEMT applications. In 1979, Störmer succeeded in forming two-dimensional electron gas (2DEG) with high mobility in a selectively doped GaAs/n-AlGaAs heterostructure grown by MBE, which opened the window to basic research on 2DEG with extremely high mobility as well as to device applications of this material. The HEMT was developed using an MBE-grown GaAs/n-AlGaAs heterostructure by Mimura and Hiyamizu in 1980. Now, seven years after the birth of the HEMT, this device is being established as one of the most promising high-speed devices. Figure 1 shows a schematic

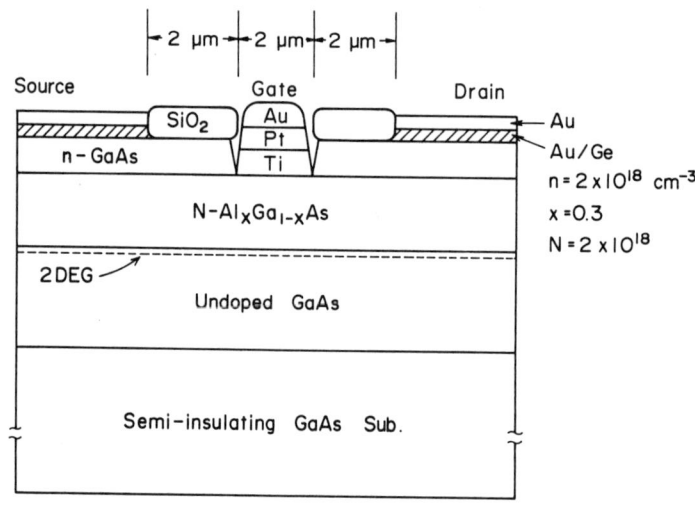

FIG. 1. Schematic cross-section of a HEMT grown by MBE.

cross-section of a HEMT grown by MBE. It consists of an undoped, high-purity GaAs layer grown on a Cr-doped, semi-insulating GaAs substrate, followed by a Si-doped $Al_xGa_{1-x}As$ ($x = 0.3$) layer, and a Si-doped GaAs cap layer. The 2DEG forms on the GaAs side of the GaAs/AlGaAs heterojunction interface and exhibits very high mobility, especially at low temperatures. Electron concentration of the high-mobility 2DEG, i.e., source–drain current, can be controlled by applying gate bias. HEMT characteristics depend on electrical properties of 2DEG (mobility and electron concentration) at low and high electric fields. These properties of 2DEG are described in this chapter.

II. Modulation Doping in GaAs/AlGaAs Superlattices

Semiconductor materials that exhibit higher electron mobility than Si have been investigated for applications to high-speed, post-silicon devices. GaAs is the most popular of these materials. The electron mobility of typical GaAs used for MESFETs (metal-semiconductor field effect transistors) is about 4800 cm^2/Vsec at room temperature with an electron concentration of 1×10^{17} cm^{-3}. This mobility is about seven times larger than that of corresponding Si. Much higher electron mobility could also be obtained in GaAs. Figure 2 shows the electron mobility as a function of temperature in an

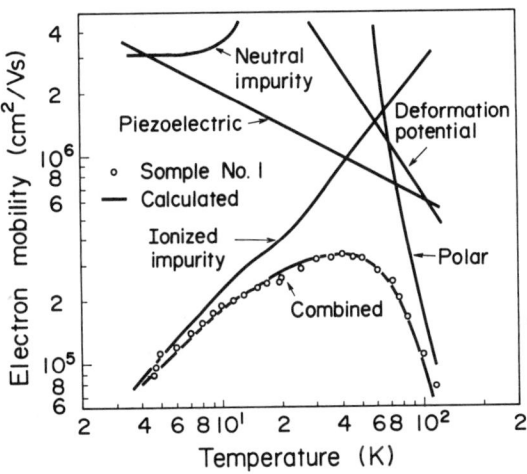

FIG. 2. Temperature dependence of electron mobility observed in high-purity GaAs with electron concentration of 2.7×10^{13} cm^{-3} (open circles). Solid lines indicate electron mobility curves calculated for each scattering process (polar optical phonon scattering, deformation potential and piezoelectric acoustic phonon scattering, ionized impurity scattering, and neutral impurity scattering) and the calculated combined mobility curve. [From Wolfe et al. (1970).]

extremely pure n-GaAs epitaxial layer with an electron concentration of 2.7×10^{13} cm^{-3} grown by VPE (Wolfe et al., 1970). Electron mobility increases with decreasing temperature from 300 K and becomes as great as 300,000 cm^2/Vsec at around 40 K because of the decrease of polar-optical and piezoelectric-acoustic phonon scattering. With a further decrease in temperature from 40 K, electron mobility decreases steeply and becomes 9000 cm^2/Vsec at about 4.5 K, which is primarily caused by ionized impurity scattering as shown in the figure. Although this peak mobility itself (300,000 cm^2/Vsec) is very attractive for device applications, the conductivity [=(mobility) × (carrier concentration)] is almost two orders of magnitude lower than that of an active layer of a GaAs MESFET, which actually make it difficult to apply this material to practical high-speed devices. Much effort has been expended searching for semiconductor materials with both high mobility and high carrier concentration, in addition to a reasonable band-gap energy (more than about 0.7 eV) for thermal stability of device performance at room temperature.

In 1969, Esaki and Tsu proposed the basic idea of modulation-doped superlattices in which impurities are doped only in wider band-gap material layers of a superlattice, to improve electron mobility by spatial separation between electrons (in narrower band-gap material layers) and their parent ionized impurities. In 1977, Sakaki reported 2DEG formed in Sn-doped GaAs/AlGaAs superlattices grown by MBE. This is the first observation of 2DEG confined in a III–V compound heterostructure, but electron mobility was not high (less than 2000 cm^2/Vsec), possibly because of uniformly distributed Sn impurities.

In 1978, based on their own idea of modulation doping, Dingle and Störmer clearly demonstrated for the first time the characteristic enhancement of electron mobility in modulation-doped GaAs/AlGaAs superlattices grown by MBE. A key technology for their success is the use of a Si doping source, which is the most suitable for modulation-doped superlattices with an extremely abrupt doping profile because doping of Si can be abruptly initiated and terminated (Cho, 1975)). Moreover, the diffusion constant of Si in AlGaAs is very small (Tatsuta et al., 1985), which results in local confinement of Si impurities in the AlGaAs region in a modulation-doped GaAs/AlGaAs superlattice even after MBE growth at 550°C for about one hour. Figure 3 shows an energy band diagram of the modulation-doped (MD) GaAs/AlGaAs superlattice, together with those of undoped and uniformly doped (UD) GaAs/AlGaAs superlattices (Dingle et al., 1978). The constituent GaAs and Al$_x$Ga$_{1-x}$As ($x = 0.3$) layers are about 200 Å thick. In the MD superlattice, Si is doped only in AlGaAs layers, and a spacer layer of undoped AlGaAs (60 Å) is introduced at every GaAs/AlGaAs interface to avoid diffusion of Si impurities into the pure GaAs region during MBE

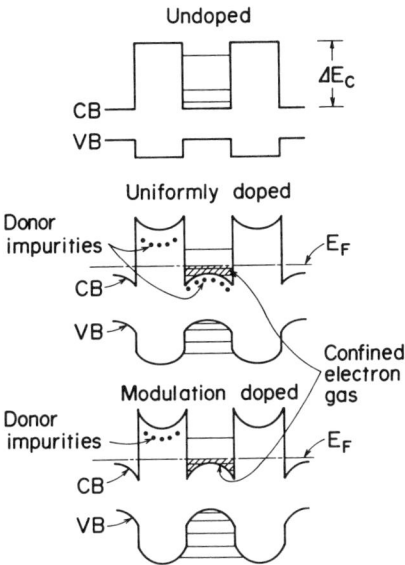

FIG. 3. Energy band diagram of (a) undoped, (b) uniformly doped and (c) modulation-doped GaAs/AlGaAs superlattices. [From Dingle et al. (1978).]

growth at 550°C. Conduction band offset at the GaAs/AlGaAs interface is about 0.3 eV (from Dingle's rule that will be discussed in Section 5). The donor level of Si impurities (about 100 meV) in $Al_xGa_{1-x}As$ ($x = 0.3$) is above the Fermi level (E_F), so electrons supplied from Si donors transfer to the GaAs quantum-well region to form 2DEG. Since these electrons are located in the undoped GaAs region where the background impurity concentration is less than 10^{15} cm^{-3}, they suffer less ionized-impurity scattering and have high mobility, especially at low temperatures.

Figure 4 shows the electron mobility in the MD GaAs/AlGaAs superlattice as a function of temperature (Dingle et al., 1978). As the temperature decreases, the electron mobility of the MD superlattice with an average electron concentration of 5×10^{16} cm^{-3} (the corresponding 2DEG concentration of 2×10^{11} cm^{-2} in each GaAs quantum well) significantly increases and reaches about 15,000 cm²/Vsec at around 50 K. In contrast, electron mobility in bulk GaAs with a slightly lower electron concentration (2×10^{16} cm^{-3}) increases very slowly from 4500 cm²/Vsec to 6100 cm²/Vsec with decreasing temperature from 300 K to 145 K, and it decreases steeply below 145 K because of ionized impurity scattering. In UD superlattices, 2DEG mobility, which is less than 3200 cm²/Vsec, is almost constant or slightly decreases with decreasing temperature. This result clearly demonstrates the characteristic enhancement of electron mobility in the MD

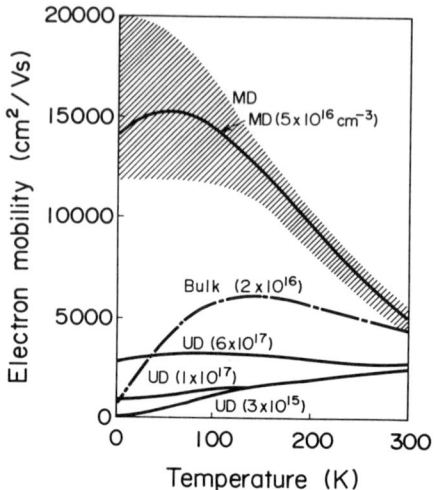

FIG. 4. Temperature dependence of electron mobility for bulk GaAs ($n = 2 \times 10^{16}$ cm^{-3}), uniformly doped (UD) and modulation-doped (MD) GaAs–AlGaAs superlattices. The crosshatched region includes most of the MD data. [From Dingle et al. (1978).]

superlattice. This sophisticated technology of modulation doping became the "breakthrough" for realizing new and attractive semiconductor materials that have both high mobility and high carrier concentration.

III. Selectively Doped GaAs/n-AlGaAs Heterostructures

Electrical properties of 2DEG systems in selectively doped GaAs/n-AlGaAs heterostructures are described.

1. INTRODUCTION TO 2DEG IN GaAs/n-AlGaAs HETEROSTRUCTURES

In 1979, 2DEG was reported to exist in a very simple, selectively doped GaAs/n-AlGaAs heterostructure grown by MBE (Störmer et al., 1979a). The heterostructure consisted of undoped GaAs and Si-doped AlGaAs. Although 2DEG mobility was low (5000 cm^2/Vsec at 4.2 K) in the first study, this GaAs/n-AlGaAs heterostructure become the original structure for subsequent studies of 2DEG in selectively doped heterostructures and of device applications. The two-dimensionality of the electric system is most conveniently demonstrated by the angular dependence of the magnetoresistance (Stern and Howard, 1967). Figure 5 shows the magnetoresistance of the GaAs/n-Al$_x$Ga$_{1-x}$As heterostructure ($x = 0.29$) at 4.2 K (Störmer et al., 1979a). Clear Shubunikov–de Haas (SdH) oscillations are obtained when

2. CHARACTERISTICS OF 2DEG COMPOUND HETEROSTRUCTURES

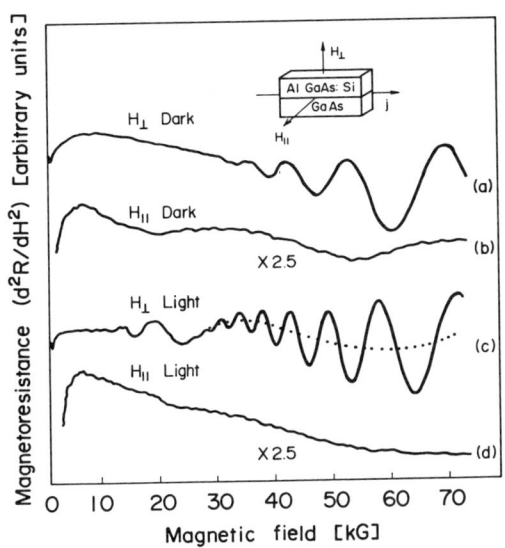

FIG. 5. Angular dependence of the magnetoresistance of the selectively doped GaAs/ n-Al$_x$Ga$_{1-x}$As ($x = 0.29$) at 4.2 K. "Light" ("dark") refers to data taken after light exposure (in the dark). [From Störmer et al. (1979a).]

the magnetic field is applied perpendicular to the interface (traces (a) and (c) in the figure), while oscillations disappear for the magnetic field applied parallel to the interface (traces (b) and (d)). This indicates the quasi-two-dimensionality of the electron system in the GaAs/n-AlGaAs single-interface heterostructure. The sheet electron concentration N_s of the 2DEG can also be obtained from the oscillation period, $\Delta(1/H)$, through the following relation:

$$N_s = 2e/[hc\Delta(1/H)], \tag{1}$$

where e is electron charge, h is Planck's constant and c is light velocity. When the SdH measurements are made in the dark (trace (a)), there is one kind of oscillation and N_s becomes 1.1×10^{12} cm^{-2}. Two kinds of oscillations appear for the SdH measurements under light illumination (trace (c)), implying two subband populations of electrons with $N_s(0) = 1.4 \times 10^{12}$ cm^{-3} (the ground subband) and $N_s(1) = 2 \times 10^{11}$ cm^{-2} (the first excited subband). This increase of 2DEG concentration is due to the persistent photoconductivity effect of the GaAs/n-AlGaAs heterostructure (which is explained in detail in Sections 4 and 5). This SdH technique has also been used to determine 2DEG concentration in selectively doped heterostructures at temperatures around 4.2 K.

Figure 6 shows a schematic diagram of a typical selectively doped GaAs/ n-AlGaAs heterostructure grown by MBE (Hiyamizu and Mimura, 1982a),

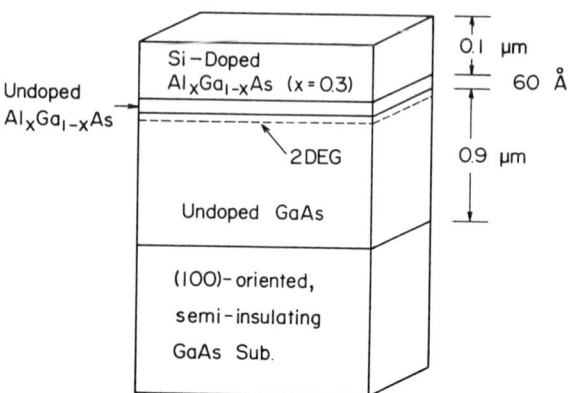

FIG. 6. Schematic diagram of a typical selectively-doped GaAs/n-Al$_x$Ga$_{1-x}$As heterostructure grown by MBE. [From Hiyamizu and Mimura (1982a).]

which consists of an undoped GaAs layer (0.9 μm thick, with a background impurity concentration of less than 10^{15} cm^{-3}) and a Si-doped Al$_x$Ga$_{1-x}$As ($x = 0.3$; 0.1 μm; Si doping concentration N_d of 1.5×10^{18} cm^{-3}) grown on a Cr-doped semi-insulating GaAs substrate. An undoped AlGaAs spacer layer (60 Å) is introduced at the GaAs–AlGaAs interface. This structure is grown at a rather high temperature (600 to 700°C) to get a high-quality AlGaAs layer.

An energy band diagram of the GaAs/n-Al$_x$Ga$_{1-x}$As ($x = 0.3$) heterostructure is shown in Fig. 7. There are two depletion layers in AlGaAs: a surface depletion layer and an interface depletion layer. Electrons supplied from Si impurities in the interface depletion region transfer to the GaAs region and accumulate in an approximately triangular potential well at the GaAs/AlGaAs interface to form 2DEG. The thicknesses of the 2DEG layer and the interface depletion layer are about 100 Å. The conduction band offset, ΔE_c, at the interface is estimated to be 0.22 eV based on Miller's rule ($\Delta E_c = 0.6 \Delta E_g$, where ΔE_g is the difference between the energy band gap of GaAs and AlGaAs, E_g(AlGaAs) $- E_g$(GaAs). The band-offset rule will be described in more detail in Section 6).

Electron mobility and sheet electron concentration N_s observed by Hall measurements in this GaAs/n-AlGaAs heterostructure are shown as a function of temperature in Fig. 8 (Hiyamizu and Mimura, 1982a). As the temperature decreases, the electron mobility increases steeply from 8030 cm^2/Vsec at 300 K to 117,000 cm^2/Vsec at 77 K ($N_s = 4.92 \times 10^{11}$ cm^{-2}), and finally to 244,000 cm^2/Vsec at 5 K with $N_s = 4.88 \times 10^{11}$ cm^{-2}, in contrast with the electron mobility of highly pure GaAs (Fig. 2). This suggests that ionized impurity scattering is significantly reduced in the selectively

2. CHARACTERISTICS OF 2DEG COMPOUND HETEROSTRUCTURES

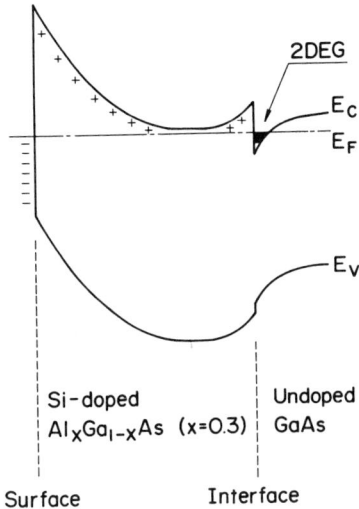

FIG. 7. Energy band diagram of a selectively doped GaAs/n-Al$_x$Ga$_{1-x}$As heterostructure. Two-dimensional electron gas (2DEG) accumulates in an approximately triangular potential well as the heterojunction interface. [From Hiyamizu and Mimura (1982a).]

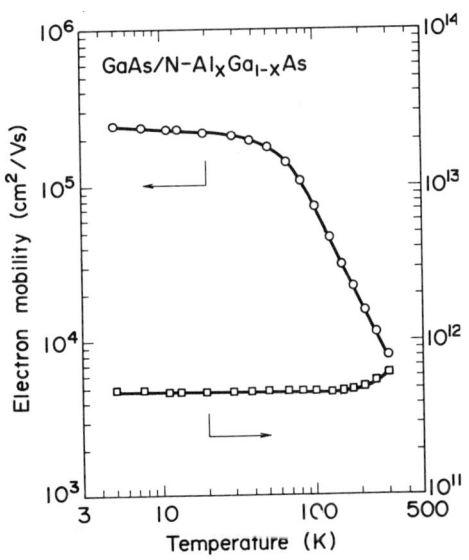

FIG. 8. Temperature dependence of electron mobility (O) and sheet electron concentration (□) observed in a selectively doped GaAs/n-Al$_x$Ga$_{1-x}$As heterostructure ($x = 0.3$) grown by MBE. [From Hiyamizu and Mimura (1982a).]

doped heterostructure. The sheet electron concentration decreases with decreasing temperature from 300 K until it becomes almost constant below about 150 K. This constant value corresponds to the 2DEG concentration and agrees well with the 2DEG concentration determined by SdH measurements at 4.2 K. Apparent excess carriers above 150 K are attributed to free electrons thermally excited from deep donors (Si) with the effective ionization energy of about 70 to 100 meV in an electrically neutral AlGaAs region between the two depletion layers. Hence, for $T > 150$ K, we have parallel conduction (2DEG at the interface and free electrons in the AlGaAs). In such a case, observed Hall mobility (μ) and electron concentration (n) are given by Petritz's parallel layer model (Petritz, 1958) as follows:

$$\mu = \frac{n_1 \mu_1^2 + n_2 \mu_2^2}{n_1 \mu_1 + n_2 \mu_2}, \quad (2)$$

$$n = \frac{(n_1 \mu_1 + n_2 \mu_2)^2}{n_1 \mu_1^2 + n_2 \mu_2^2}, \quad (3)$$

where μ_1 (μ_2) and n_1 (n_2) are the electron mobility and electron concentration of 2DEG (free electrons in n-AlGaAs). The electron mobility of $Al_xGa_{1-x}As$ ($x = 0.3$), μ_2, in this temperature range, however, is about 1000 cm^2/Vsec (almost independent of temperature) which is more than one order of magnitude lower than the 2DEG mobility, μ_1. Hence, the observed mobility μ can be considered approximately equal to 2DEG mobility μ_1 even for $T > 150$ K, if n_2 is not much higher than n_1.

The important parameters that determine the 2DEG properties (2DEG mobility and electron concentration) of GaAs/n-Al$_x$Ga$_{1-x}$As heterostructures are

(1) the spacer-layer thickness (Δt),
(2) the background impurity concentration in the GaAs region (N_i),
(3) the Si doping concentration in n-AlGaAs (N_d), and
(4) the Al mole fraction in Al$_x$Ga$_{1-x}$As (x).

These parameters have been optimized to obtain high 2DEG mobility and/or high electron concentration in GaAs/n-AlGaAs heterostructures, as will be described in the following sections (3 to 5).

Figure 9 shows the growth of 2DEG mobility in GaAs/n-AlGaAs heterostructures prepared by MBE over the past years. The 2DEG mobility (4.2 K) first obtained in 1979 was 5000 cm^2/Vsec (Störmer et al., 1979a). After the development of the HEMT in 1980, intensive efforts have been made to improve this heterostructure, which caused a rapid increase of mobility in 1981 and 1982. This improvement was partly supported by the progress in MBE technology for growing high-quality GaAs/AlGaAs heterostructures

2. CHARACTERISTICS OF 2DEG COMPOUND HETEROSTRUCTURES

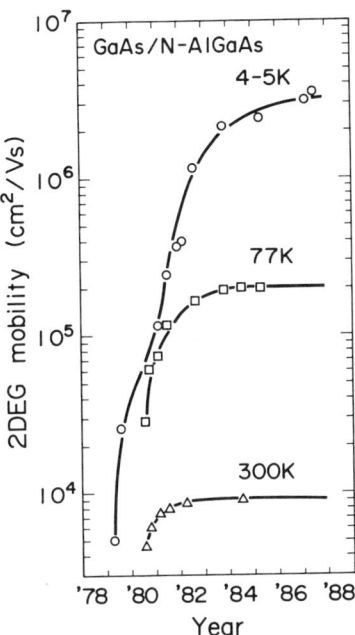

FIG. 9. Growth of 2DEG mobility (at 4–5 K, 77 K, and 300 K) in selectively doped GaAs/n-Al$_x$Ga$_{1-x}$As heterostructures prepared by MBE over the past years.

that had been achieved for developing GaAs/AlGaAs double heterostructure lasers by Tsang in 1979 and 1980. The highest 2DEG mobility (observed at temperatures of 4 to 5 K) reached 1,150,000 cm^2/Vsec ($x = 0.3$, $\Delta t = 150$ Å, $N_d = 2 \times 10^{18}$ cm^{-3}) in 1982 (Hiyamizu et al.), and 2,120,000 cm^2/Vsec ($x = 0.3$, $\Delta t = 200$ Å, $N_d = 1 \times 10^{18}$ cm^{-3}) in 1983 (Hiyamizu et al.), and finally in 1987, 3,100,000 cm^2/Vsec ($x = 0.32$, $\Delta t = 400$ Å, $N_d = 1.3 \times 10^{18}$ cm^{-3}) (Harris et al.) and 3,500,000 cm^2/Vsec ($x = 0.3$, $\Delta t = 750$ Å, $N_d = 2.1 \times 10^{17}$ cm^{-3}) (English et al.). At still lower temperatures (below 2 K), the highest 2DEG mobility ever observed (5,000,000 cm^2/Vsec) was achieved by English et al. in 1987. These electron mobilities (above 2,000,000 cm^2/Vsec) are not only higher than those of any semiconductor material, but also are two orders of magnitude higher than that of 2DEG in Si MOS inversion layers (about 10,000 cm^2/Vsec at 4.2 K). Selectively doped GaAs/n-AlGaAs heterostructures also have been grown by MOCVD techniques, but the highest 2DEG mobility of the MOCVD materials is 510,000 cm^2/Vsec at 4.2 K (Kobayashi and Fukui, 1984), which is almost $\frac{1}{6}$ of that of the MBE-grown heterostructures. These results indicate that high quality of selectively doped GaAs/n-AlGaAs heterostructures can

TABLE I

2DEG MOBILITIES AND ELECTRON CONCENTRATIONS ACHIEVED IN SELECTIVELY DOPED GaAs/n-AlGaAs HETEROSTRUCTURES

Year	300 K Mobility (cm²/Vsec)	77 K (dark) Mobility (cm²/Vsec)	77 K (dark) Concentration (cm⁻²)	4.2 K (dark) Mobility (cm²/Vsec)	4.2 K (dark) Concentration (cm⁻²)	4.2 K (light) Mobility (cm²/Vsec)	4.2 K (light) Concentration (cm⁻²)	References
1979				5,000	1.1×10^{12}	25,000	1.0×10^{12}	Störmer et al. (1979a) Störmer et al. (1979b)
1980	4,600 6,190	28,600 61,000	5.7×10^{11}					Morkoç et al. Hiyamizu et al.
1981	7,450 8,030	74,200 117,000	4.9×10^{11}	115,000 244,000	4.9×10^{11}	365,000	1.1×10^{12}	Witkowski et al. Hiyamizu et al. (1981c) Störmer et al. (1981b)
1982	8,720	122,000 165,000	5.6×10^{11} 3.4×10^{11}	256,000 218,000 540,000	5.5×10^{11} 5.5×10^{11} 3.4×10^{11}	396,000 1,150,000	1.1×10^{12} 4.7×10^{11}	Hiyamizu et al. (1982b) Hiyamizu et al. (1982b) Hiyamizu
1983		195,000	3.0×10^{11}	1,250,000	3.0×10^{11}	2,120,000	5.0×10^{11}	Hiyamizu et al.
1984	9,200	200,000	2.2×10^{11}	1,060,000	2.2×10^{11}	1,700,000	3.9×10^{11}	Heiblum et al.
1985		200,000	3.6×10^{11}	1,550,000	3.6×10^{11}	2,400,000	5.3×10^{11}	Hiyamizu et al. (1985a)
1987				1,680,000	1.3×10^{11}	3,100,000 3,500,000 5,000,000 (2 K)	3.1×10^{11} 1.6×10^{11} 1.6×10^{11}	Harris et al. English et al. English et al.

2. CHARACTERISTICS OF 2DEG COMPOUND HETEROSTRUCTURES

be realized by MBE, and we have the most nearly ideal 2DEG system in these GaAs/n-AlGaAs heterostructures. Such extremely high 2DEG mobilities in GaAs/n-AlGaAs heterostructures have been used for basic research on semiconductor physics, which has led to, for instance, the discovery of the fractional quantum Hall effect (Tsui et al., 1982; Störmer et al., 1983).

The 2DEG mobility at 77 K started from 28,600 cm^2/Vsec in 1980 (Morkoç et al.) and rapidly increased in 1980, 1981, and 1982, reaching its highest value (200,000 cm^2/Vsec) in 1984 (Heiblum et al.). Generally speaking, electron mobility observed at room temperature in selectively doped GaAs/n-AlGaAs heterostructures does not necessarily give 2DEG mobility because of the parallel conduction (2DEG and free electrons in AlGaAs). To achieve high 2DEG mobility at room temperature, it is necessary to eliminate free electrons in AlGaAs by carefully etching the AlGaAs layer, as was first discussed by Hiyamizu in 1981. The highest room-temperature mobility observed is 9200 cm^2/Vsec (Heiblum et al., 1984), which is considered to be very close to the intrinsic 2DEG mobility in a GaAs/n-AlGaAs heterostructure at room temperature.

The record of 2DEG mobility over the past years is listed in Table I.

2. Electronic State in GaAs/n-AlGaAs Heterostructures

Ando calculated electronic states in a selectively doped GaAs/n-Al$_x$Ga$_{1-x}$As ($x = 0.3$) heterostructure at absolute zero in 1982. The Schrödinger equation for the envelope wave function of electrons in an approximately triangular quantum well at the GaAs–AlGaAs heterojunction interface is

$$\left[-\frac{\hbar^2}{2m}\frac{d^2}{dz^2} + \frac{\hbar^2}{2m}k^2 + V(z) \right]\zeta_{ik}(z) = E_i(k)\zeta_{ik}(z), \tag{4}$$

where m is electron effective mass. The z-axis is chosen to be perpendicular to the GaAs–AlGaAs interface ($z = 0$). For simplicity, the same m was used for both GaAs ($z > 0$) and AlGaAs ($z < 0$). The potential $V(z)$ is expressed as

$$V(z) = V_0 \theta(-z) + v_H(z) + v_{xc}(z), \tag{5}$$

where $\theta(z)$ is the step function and $v_H(z)$ is the Hartree potential, which is given by

$$v_H(z) = \frac{4\pi e^2}{\kappa} \int^z dz' \int^{z'} dz'' n(z'') + \frac{4\pi e^2}{\kappa} N_{\text{depl}} z, \tag{6}$$

where κ is the static dielectric constant (12.9), $n(z)$ is the electron density distribution, and N_{depl} is the concentration of fixed space charges in the GaAs depletion layer. The last term in Eq. (5), $v_{xc}(z)$, is the local exchange-correlation potential. Energy level and electron wave function for the Schrödinger equation (Eq. (4)) were obtained by numerical calculations as well as by the variation method using two kinds of trial functions. A simple trial wave function is

$$\zeta(z) = \begin{cases} (b_0^3/2)^{1/2} z \exp(-b_0 z/2) & (z > 0), \\ 0 & (z < 0), \end{cases} \quad (7)$$

where b_0 is a variational parameter. This trial function was first proposed by Fang and Howard in 1966 and vanishes at the interface ($z = 0$). This is suitable for the Si inversion layer because of the very high potential barrier (about 2 eV) of the Si–SiO$_2$ interface. Another trial function, which has non-vanishing amplitude in the AlGaAs region, is somewhat more realistic for GaAs/n-AlGaAs, of which interface the potential barrier is not so high (0.2 to 0.3 eV) and is given by

$$\zeta(z) = \begin{cases} Bb^{1/2}(bz + c) \exp(-bz/2) & (z > 0), \\ B'b'^{1/2} \exp(b'z/2) & (z < 0), \end{cases} \quad (8)$$

where c, B, and B' are parameters expressed by b and b' through boundary conditions at $z = 0$ and the normalization, and b and b' are variational parameters. In the calculation using the trial functions, the local exchange correlation potential (the last term of Eq. (5)) is neglected (Hartree approximation). Parameters b and b' for Eq. (8) (or b_0 for Eq. (7)) are determined to minimize electron energy.

Figure 10 shows the energy level, wave functions and potentials calculated for $x = 0.3$ ($V_0 = 300$ meV, by Dingle's rule), $N_s = 5 \times 10^{11}$ cm^{-2} (2DEG concentration), and $N_{\text{depl}} = 5 \times 10^{10}$ cm^{-2} (Ando, 1982b). The value of $N_{\text{dep}} = 5 \times 10^{10}$ cm^{-2} corresponds to an impurity concentration of 10^{14} cm^{-3} in GaAs. All calculated wave functions have their peak amplitude at around $z = 50$ Å, indicating 2DEG is located very close to the GaAs-AlGaAs interface. The more realistic wave function of Eq. (8) (solid line) exhibits an apparent shift of peak position toward the interface (about 20 Å) compared with that of the Fang–Howard function of Eq. (7) (broken line), and agrees well with the corresponding wave function numerically calculated (dotted line), especially in the AlGaAs region. The ground subband energy (E_0) obtained from Eq. (8) agrees well with the numerical result.

Subband energy separations, $E_{i0} = E_i - E_0$, in the quantum well at the interface, which are calculated numerically using Eqs. (4) and (5), are plotted as a function of 2DEG concentration, N_s, at absolute zero in Fig. 11 (Ando, 1982a). The corresponding results for the Hartree approximation are illustrated by broken lines, indicating that the many body effect (exchange and correlation) is not crucial. It is noteworthy that $E_{i0} (= E_i - E_0)$ increases

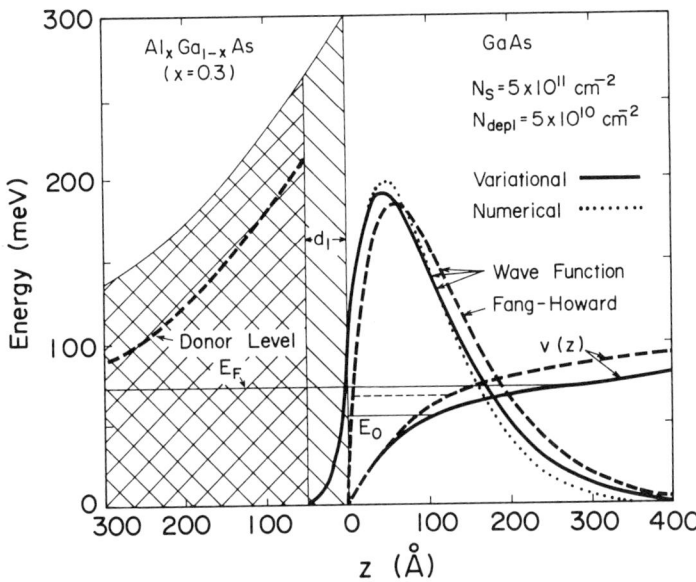

FIG. 10. Self-consistent potential, the bottom of the lowest subband and the electron wave function calculated by using variational wave functions for selectively doped GaAs/ n-Al$_x$Ga$_{1-x}$As ($x = 0.3$, spacer layer thickness $\Delta t = 50$ Å, binding energy of donor levels in Al$_x$Ga$_{1-x}$As $E_B = 50$ meV). The dotted line indicates the wave function calculated numerically. [From Ando (1982b).]

considerably with increasing N_s. The Fermi energy E_F measured from the bottom of the ground subband (E_0) increases with increasing N_s by the relation $E_F = N_s/D(E)$, where $D(E)$ ($= m/\pi\hbar^2$, including spin degeneracy of 2) is the two-dimensional density of states for a single subband (2.8×10^{10} cm^{-2}/meV for a GaAs/n-AlGaAs system). The first excited subband begins to be occupied by electrons at $N_s = 7.3 \times 10^{11}$ cm^{-2}. Above this point, energy levels of the excited subbands increase steeply with increasing N_s. Störmer et al. (1979b) observed the concentrations of 2DEG populated in the ground subband and first excited subband (N_0 and N_1) for the GaAs/n-Al$_x$Ga$_{1-x}$As heterostructure ($x = 0.29$) by SdH measurements at 4.2 K. Their observed values ($N_0 = 1.4 \times 10^{12}$ cm^{-2} and $N_1 = 2 \times 10^{11}$ cm^{-2} at $N_s = 1.6 \times 10^{12}$ cm^{-2}) agree excellently with the calculated results ($N_0 = 1.39 \times 10^{12}$ cm^{-2} and $N_1 = 2.1 \times 10^{11}$ cm^{-2}).

3. TEMPERATURE DEPENDENCE OF 2DEG MOBILITY

Temperature dependence of 2DEG mobility in GaAs/n-AlGaAs was analyzed theoretically and was found to compare excellently with experimental

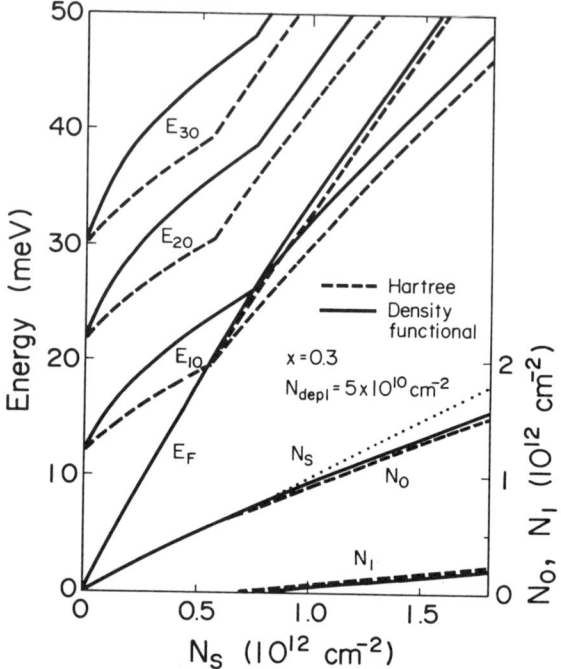

FIG. 11. Subband energy separation, $E_{i0} = E_i - E_0$, in the approximately triangular quantum well for GaAs/n-Al$_x$Ga$_{1-x}$As ($x = 0.3$, $N_{\text{depl}} = 5 \times 10^{10}$ cm^{-2}) as a function of 2DEG concentration, N_s. These are calculated numerically by using Eqs. (4) and (5) from the text. The Fermi energy E_F measured from the bottom of the lowest subband (E_0) and the 2DEG concentrations of the lowest and first excited subbands (N_0 and N_1) are also plotted. [From Ando (1982a).]

results by Walukiewicz et al. in 1984. In the calculation, they considered all major scattering processes of electrons such as deformation-potential acoustic and piezoelectric acoustic-phonon scattering, ionized-impurity scattering (by remote Si impurities in n-AlGaAs and by background impurities in GaAs), alloy-disorder scattering and polar-optical phonon scattering. In Fig. 12, the calculated 2DEG mobility (indicated by the solid line labeled "Total") is shown together with experimental data (solid circles) obtained in the dark for a high-mobility GaAs/n-Al$_x$Ga$_{1-x}$As sample ($x = 0.3$, $\Delta t = 200$ Å, $N_d = 2 \times 10^{18}$ cm^{-3}) (Hiyamizu et al., 1983). The component mobilities are also illustrated by dotted broken lines. Parameters used in the calculation are listed in Table II. The absolute limit of 2DEG mobility (dotted line) is determined by phonon scattering (polar-optical phonon, deformation-potential acoustic, and piezoelectric acoustic phonon scattering) and alloy-disorder scattering. Especially in the limit of $T \sim 0$ K, the absolute

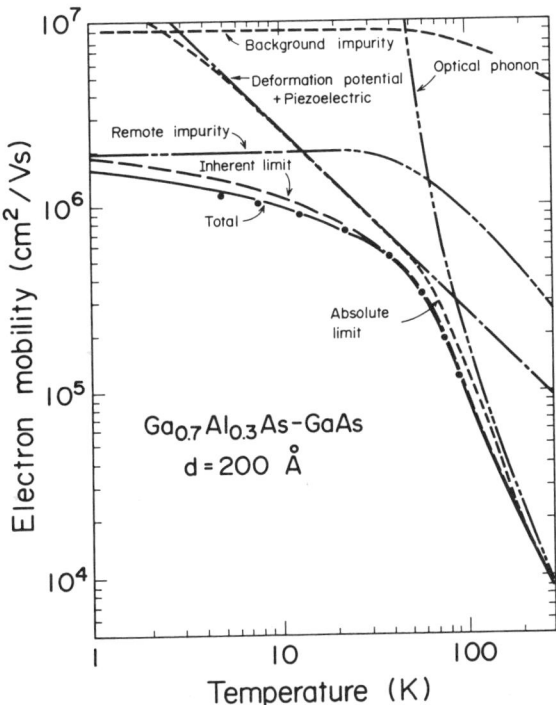

FIG. 12. Temperature dependence of calculated 2DEG mobility ("Total") for a high-quality GaAs/n-Al$_x$Ga$_{1-x}$As heterostructure with 2DEG concentration of 3×10^{11} cm^{-2} ($x = 0.3$, $\Delta t = 200$ Å, remote ionized impurity concentration 8.6×10^{16} cm^{-3}, background impurity concentration 9×10^{13} cm^{-3}). Solid circles are experimental data obtained by Hiyamizu et al. (1983). The component mobilities (deformation potential acoustic and piezoelectric acoustic phonon scattering, optical phonon scattering, remote impurity scattering, and background impurity scattering) are also represented by dotted broken lines. [From Walukiewicz et al. (1984b).]

limit is given by the alloy-disorder scattering. Remote-impurity scattering is always present in a selectively doped GaAs/n-AlGaAs system. Hence, the inherent limit of 2DEG mobility (broken line) illustrates a further reduced mobility form the absolute limit value because of the additional remote-impurity scattering. It can be seen in the figure that at high temperatures polar-optical phonon scattering is dominant, while at low temperatures acoustic-phonon scattering, remote-impurity scattering, and alloy-disorder scattering become dominant. The observed 2DEG mobilities at low temperatures are slightly lower than the inherent limit (about 1.8×10^6 cm^2/Vsec at 1 K). It is likely that other scattering mechanisms, such as interface-roughness scattering and interface-charge scattering, which are important

TABLE II

PARAMETERS EMPLOYED IN CALCULATION OF ELECTRON MOBILITY IN A
GaAs/n-AlGaAs HETEROSTRUCTURE[a]

Parameter	Value	References
Electron effective mass	$0.076 m_0$	Störmer et al. (1979a)
Deformation potential D (eV)	7	Rode (1975)
Elastic constant c_l (dyn/cm^2)	13.97×10^{11}	Rode (1975)
Piezoelectric constant P	0.064	Blakemore (1982)
Static dielectric constant	12.9	Rode (1975)
High-frequency dielectric constant	10.9	Rode (1975)
Optical-phonon energy (meV)	36	Rode (1975)
Alloy-disorder parameter V (eV)	1	Chadi (1975)
Conduction-band energy offset V_0 (eV)	0.3	Ando (1982b)

[a] From Walukiewicz et al., 1984b.

for 2DEG in a Si inversion layer (a Si–SiO$_2$ heterostructure), are not significant. The 2DEG system in the MBE-grown GaAs/n-AlGaAs heterostructure is thus almost ideal. The difference between the total mobility (solid line) and the inherent limit is attributed to the background impurity scattering in the pure GaAs region. The background impurity concentration assumed for the calculation is 9×10^{13} cm^{-3}. The remote-impurity scattering from impurities in the undoped AlGaAs spacer layer as well as from interface charges was neglected in calculating total mobility. As can be seen in Fig. 12, remote-impurity scattering is the dominant factor limiting 2DEG mobility in this heterostructure ($\Delta t = 200$ Å) at temperatures lower than about 15 K. This is why the highest 2DEG mobility has been achieved by increasing Δt as shown in Table I (and also in Fig. 13 in Section 4).

4. SPACER-LAYER THICKNESS DEPENDENCE

The spacer-layer thickness (Δt) is the most effective parameter for increasing 2DEG mobility in GaAs/n-AlGaAs heterostructures at low temperatures since the ionized impurity scattering, which is the dominant scattering factor at low temperature, is directly reduced by introducing a spacer between 2DEG and Si-doped AlGaAs.

Figure 13 shows 2DEG mobilities at temperatures in the range of 4 to 5 K as a function of Δt (Hiyamizu et al., 1983, 1985a; Heiblum et al., 1984; Harris et al., 1987; English et al., 1987). Solid circles, squares, and triangles indicate 2DEG mobility observed in the dark, and open circles, squares, rhombuses, and triangles illustrate mobilities measured under illumination (in this case, 2DEG concentration increases to about twice the original

2. CHARACTERISTICS OF 2DEG COMPOUND HETEROSTRUCTURES

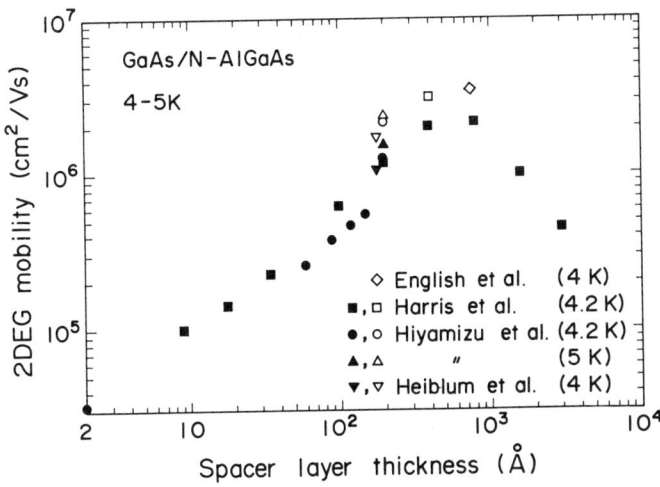

FIG. 13. 2DEG mobilities in GaAs/n-Al$_x$Ga$_{1-x}$As heterostructures in the range of 4 K to 5 K as a function of spacer layer thickness, Δt: (●, ○) Hiyamizu et al., 1983 (4.2 K); (▲, △) Hiyamizu et al., 1985 (5 K); (▼, ▽) Heiblum et al., 1984 (4 K); (■, □) Harris et al., 1987 (4.2 K); (◇) English et al., 1987 (4 K). Solid (open) marks are data obtained in the dark (after light illumination).

concentration because of the persistent photoconductivity effect). 2DEG mobility increases monotonically with increasing Δt and exhibits a broad peak (2,000,000 to 3,000,000 cm²/Vsec) in the range of 200 Å < Δt < 800 Å. As Δt increases further from 800 Å, electron mobility decreases. This result suggests that there exist two limiting factors of 2DEG mobility in GaAs/n-AlGaAs heterostructures: remote impurity scattering and 2DEG concentration. The remote impurity scattering of 2DEG by Si impurities in n-AlGaAs is reduced as Δt increases, resulting in increased electron mobility. The 2DEG concentration decreases with increasing Δt as shown in Fig. 14, resulting in enhanced ionized-impurity scattering and reduced electron mobility (see Section 5). At around Δt = 400 Å, these two mechanisms contribute equally in high-quality GaAs/n-AlGaAs heterostructures. Such a high-quality GaAs/n-AlGaAs heterostructure can be realized by MBE after careful outgassing of the MBE growth system (molecular-beam sources and the substrate holder) in an ultrahigh vacuum chamber. In the early stage of 2DEG study for GaAs/n-Al$_x$Ga$_{1-x}$As heterostructures (x = 0.25, N_d = 3 × 10^{17} cm^{-3}), 2DEG mobility began to decrease only at Δt = 75 Å (Drummond et al., 1981b), which is possibly caused by many electron-scattering centers (unintentionally doped oxygen and/or carbon impurities) existing in the AlGaAs layer. The effect of the scattering centers in an undoped AlGaAs spacer layer on 2DEG mobility is not reduced by increasing

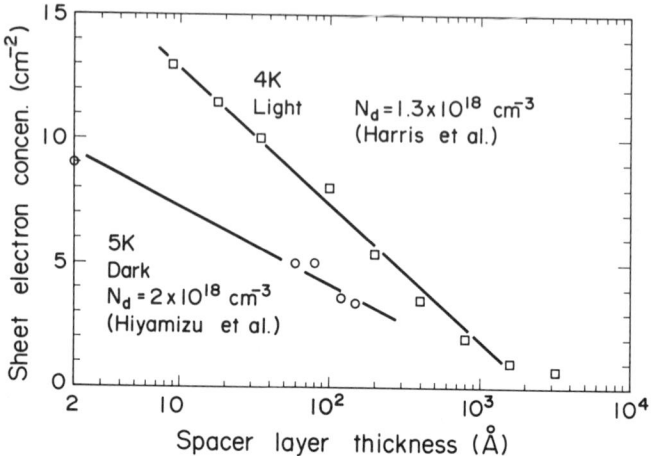

Fig. 14. 2DEG concentration observed in GaAs/n-Al$_x$Ga$_{1-x}$As heterostructures ($x = 0.3$–0.32) as a function of spacer layer thickness for GaAs/n-Al$_x$Ga$_{1-x}$As heterostructures with Si doping concentrations of $N_d = 2 \times 10^{18}$ cm^{-3} (Hiyamizu et al., 1983) and $N_d = 1.3 \times 10^{18}$ cm^{-3} (Harris et al., 1987). Open circles (squares) indicate N_s observed in the dark (light illumination) condition.

the spacer-layer thickness. Hence, the ionized impurity scattering became greatly enhanced in this heterostructure, and the two limiting factors of electron mobility balanced at a much smaller Δt value.

5. Electron Concentration Dependence

The 2DEG concentration N_s in GaAs/n-AlGaAs heterostructures can be changed (1) by changing Si doping concentration in n-AlGaAs, (2) by the persistent photoconductivity (PPC) effect of n-AlGaAs, which is caused by light illumination at low temperatures, or (3) by applying gate voltage for a HEMT structure.

Figure 15 shows 2DEG concentration observed by Hall measurements at 77 K in GaAs/n-Al$_x$Ga$_{1-x}$As heterostructures ($x = 0.3$, $\Delta t = 60$ Å) grown at 600°C as a function of Si doping concentration N_d (Hiyamizu et al., 1981a). The 2DEG concentration increases approximately in proportion to the square root of N_d and reaches 9×10^{11} cm^{-2} at $N_d = 5 \times 10^{18}$ cm^{-3}.

The 2DEG mobility observed at 77 K in the dark is shown as a function of N_s in GaAs/n-Al$_x$Ga$_{1-x}$As heterostructures ($x = 0.3$, $\Delta t = 60$ Å) with various N_d in Fig. 16 (Hiyamizu et al., 1982b). The 2DEG mobility is almost constant at around 100,000 cm^2/Vsec for $N_s < 6 \times 10^{11}$ cm^{-2} and decreases sharply above 6×10^{11} cm^{-2}. This sharp decrease in 2DEG mobility is probably due to Si impurities diffused into the GaAs channel region during

2. CHARACTERISTICS OF 2DEG COMPOUND HETEROSTRUCTURES

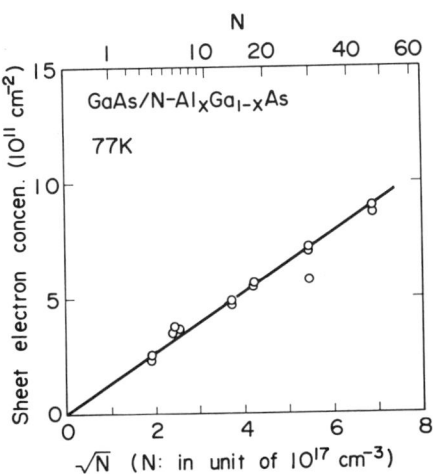

FIG. 15. 2DEG concentration observed in GaAs/n-Al$_x$Ga$_{1-x}$As heterostructures ($x = 0.3$, $\Delta t = 60$ Å) grown at 600°C as a function of Si doping concentration N_d (N in this Figure). [From Hiyamizu et al. (1981a).]

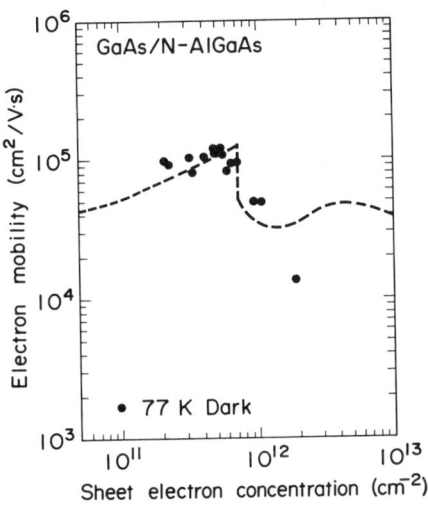

FIG. 16. 2DEG mobility observed at 77 K in GaAs/n-Al$_x$Ga$_{1-x}$As heterostructures ($x = 0.3$, $\Delta t = 60$ Å) as a function of 2DEG concentration N_s (Hiyamizu et al., 1982). Broken line represents 2DEG mobility calculated for modulation-doped GaAs–AlGaAs superlattice at 0 K [From Mori and Ando (1980).]

MBE growth at a temperature of about 600°C. The concentration of diffused Si impurities in the 2DEG region is supposed to be in the range of 10^{17} cm^{-3}. This critical N_s value corresponds to N_d of about 2×10^{18} cm^{-3} (Fig. 15). Hence, the optimum doping concentration of Si for a selectively doped GaAs/n-AlGaAs heterostructure is 1 to 2×10^{18} cm^{-3}. By the way, the maximum electron concentration observed in Si-doped Al$_x$Ga$_{1-x}$As ($x = 0.25$) grown by MBE is 4×10^{18} cm^{-3} (Drummond et al., 1982). Hence, for $N_d > 4 \times 10^{18}$ cm^{-3} all Si impurities are not electrically active in AlGaAs.

Many Si impurities form deep donors in Al$_x$Ga$_{1-x}$As($x > 0.2$) that are called "DX centers" (Lang et al., 1979). (Si donors in Al$_x$Ga$_{1-x}$As will be explained in detail in Section 6.) Si-doped AlGaAs exhibits persistent photoconductivity (PPC) at low temperature because of the DX centers: at low temperatures (lower than about 100 K), the number of free electrons excited from DX centers in AlGaAs by light illumination remains constant even after light illumination is stopped, which results in a persistent increase in conductivity of AlGaAs. A similar PPC effect was also observed for selectively doped GaAs/n-Al$_x$Ga$_{1-x}$As heterostructures ($x = 0.29$) (Störmer et al., 1979a). Utilizing the PPC effect, the 2DEG concentration, N_s, dependence of electron mobility has been investigated in GaAs/n-AlGaAs heterostructures at low temperatures ($T = 4.2$, 5, and 77 K) (Störmer et al., 1981b; Hiyamizu et al., 1983).

It is also instructive for understanding the N_s dependence of 2DEG mobility to see the behavior of the component mobilities (described in Section 3) as a function of N_s. Figure 17 shows calculated 2DEG mobility in a GaAs/n-Al$_x$Ga$_{1-x}$As heterostructure ($x = 0.3$, $\Delta t = 150$ Å, $N_d = 2 \times 10^{18}$ cm^{-3}) as a function of N_s (Walukiewicz et al., 1984b), together with observed 2DEG mobility (5 K and 77 K) (Hiyamizu et al., 1983). In the experiment, N_s in the GaAs/n-AlGaAs heterostructure was increased by controlling the intensity of illumination at low temperatures (the PPC effect). According to the calculation by Walukiewicz, the dominant factors limiting mobility are remote- and background-impurity scattering, and electron mobilities can be expressed by $\mu \propto (N_s)^\gamma$ with γ of 1.4 (remote) and 1.2 (background). Alloy-disorder scattering exhibits very strong N_s dependence, which is caused by the penetration of the 2DEG into the AlGaAs alloy region when N_s increases. This penetration of the 2DEG is significantly increased when the first excited subband is occupied by the electrons, because the electron wave function of the first excited state has non-vanishing amplitude in the AlGaAs region, in contrast with zero amplitude of the wave function for the ground state. This results in a large enhancement of alloy-disorder scattering and a decrease in "total" mobility for $N_s > 7 \times 10^{11}$ cm^{-2}. This critical value of N_s agrees well with the calculated

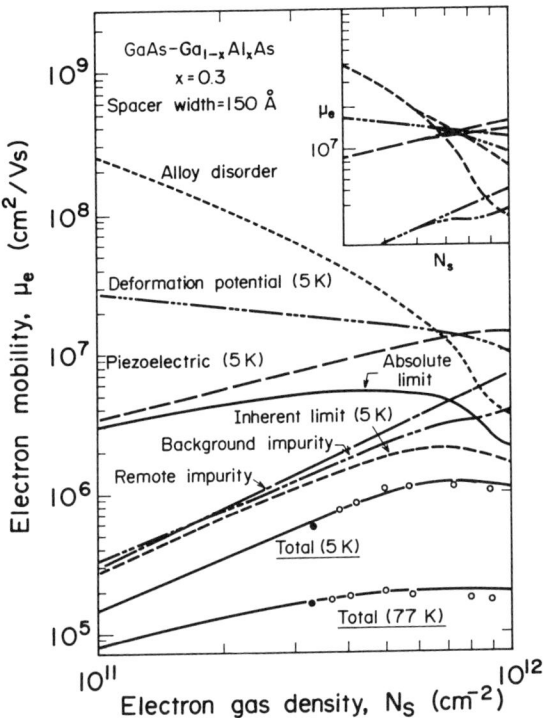

FIG. 17. Calculated 2DEG mobility and experimental data (circles, Hiyamizu et al., 1983) in a GaAs/n-Al$_x$Ga$_{1-x}$As heterostructure ($x = 0.3$, $\Delta t = 150$ Å, $N_d = 2 \times 10^{18}$ cm^{-3}) as a function of 2DEG concentration (or electron gas density) N_s. [From Walukiewicz et al. (1984b).]

N_s (7.3×10^{11} cm^{-2}) at which electrons begin to occupy the first excited subband (Ando, 1982a), which was mentioned in Section 2.

N_s dependence of 2DEG mobility has also been studied by applying gate bias for GaAs/n-AlGaAs HEMT structures (Störmer et al., 1981a, 1982; Tsui et al., 1981).

6. AlAs Mole Fraction Dependence

Two important quantities of GaAs/n-Al$_x$Ga$_{1-x}$As heterostructures for determining 2DEG properties are functions of the AlAs mole fraction, x: the conduction band offset, $\Delta E_c(x)$, at the GaAs/AlGaAs interface, and the ionization energy of Si impurities, $E_D(x)$, in Al$_x$Ga$_{1-x}$As.

The conduction band offset, $\Delta E_c(x)$, of the GaAs/Al$_x$Ga$_{1-x}$As heterojunction interface for $0 < x < 0.45$ (direct band-gap region) is a linear function of x and is given by

$$\Delta E_c(x) = \alpha \times [E_g(x) - E_g(0)] = \alpha \times 1.247x \text{ (eV)}$$

at room temperature, where $E_g(x)$ is the band gap energy of $Al_xGa_{1-x}As$ and α is a partition ratio (0.60 to 0.65). This ratio was first determined to be 0.85 from the optical absorption spectrum of the GaAs/AlGaAs quantum wells (Dingle et al., 1974). Since then, this 85% partition rule (Dingle's rule) has been widely used for GaAs/AlGaAs systems. In 1984, however, Miller et al., analyzed exciton recombination energies for GaAs-AlGaAs quantum wells and concluded that α is 0.57. After this report, many studies have been made to confirm this partition ratio by various methods, and the following values have been reported: 0.62 (Watanabe et al., C-V measurements, 1984), 0.65 (Arnold et al., I-V measurements, 1984), 0.60 (Batey et al., I-V measurements, 1985), 0.67 (Okumura et al., C-V measurements, 1985), 0.63 (Hickmott et al., I-V measurement, 1985), 0.62 (Heiblum et al., internal photoemission, 1985). Hence, recently it has been concluded that partition ratio of the energy band gap is beween 0.60 and 0.65 for a GaAs/AlGaAs system (Miller's rule).

The ionization energy of Si in $Al_xGa_{1-x}As$ changes in a complicated manner as a function of x. Figure 18 shows the apparent ionization energy, E_D, of Si impurities in $Al_xGa_{1-x}As$ as a function of x, obtained by Hall measurements (Ishibashi et al., 1982; Ishikawa et al., 1982). In these studies, E_D is defined by $n \exp(E_D/kT)$, where n is the electron concentration at temperature T, and k is the Boltzmann constant. Si doping concentrations in AlGaAs were 1×10^{18} cm^{-3} (Ishibashi) and 3×10^{17} cm^{-3} (Ishikawa). When x is less than 0.2, E_D is almost equal to that of Si donors in GaAs

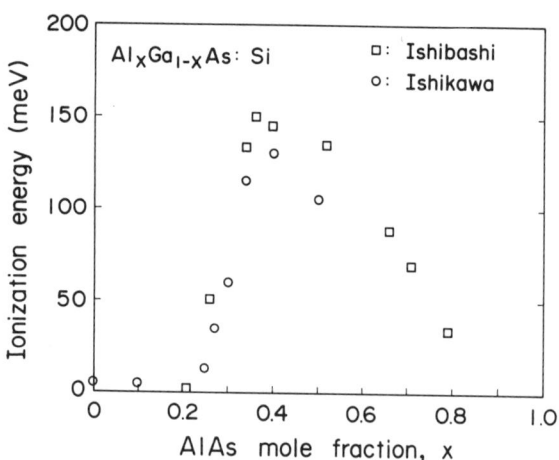

FIG. 18. Ionization energies of Si impurities in $Al_xGa_{1-x}As$ grown by MBE as a function of Al mole fraction, x, as obtained by Hall measurements: (□) from Ishibashi et al., 1982; (O) Ishikawa et al., 1982. Si doping concentrations are 1×10^{18} cm^{-3} (Ishibashi et al.) and 3×10^{17} cm^{-3} (Ishikawa et al.).

2. CHARACTERISTICS OF 2DEG COMPOUND HETEROSTRUCTURES 77

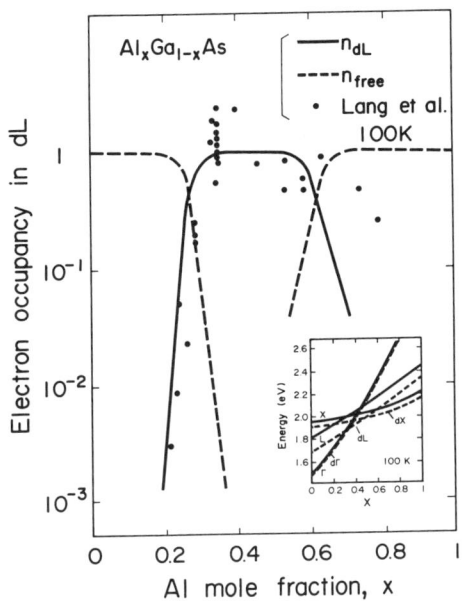

FIG. 19. The fractional electron occupancy (solid line) in dL states (deep donors associated with L-band minimum) and the free carrier concentration (broken lines) in $Al_xGa_{1-x}As$ as a function of AlAs mole fraction x. The solid circles indicate the DX center concentration in $Al_xGa_{1-x}As$ observed by Lang et al. (1979). The inset illustrates the location of the conduction band and its donor states. [From Tachikawa et al. (1985).]

(6 meV). As x increases from 0.2, however, E_D increases steeply until it becomes maximum (about 150 meV) at $x = 0.36$, then decreases slowly. This complicated behavior of a Si donor in AlGaAs can be understood by the coexistence of three kinds of Si donors (Schubert and Ploog, 1984; Tachikawa et al., 1985): one hydrogen-like shallow donor (donor level, $E_{d\Gamma} = E_\Gamma - 6$ meV) associated with the Γ-band minimum (E_Γ), and two deep donors associated with the L-band minimum ($E_{dL} = E_L - (140$ to 150 meV)) and the X-band minimum ($E_{dX} = E_X - (40$ to 50 meV)), as shown in Fig. 19 (Tachikawa et al., 1985). In the region of $0 < x < 0.4$, electrons predominantly populate two kinds of donors: the shallow donor ($E_{d\Gamma}$) and the deep donor (E_{dL}). The activation energies of these donors are almost independent of x, and only the concentration ratio of shallow donors to deep donors that electrons populate changes with x. For $x = 0.2$, the fraction of the shallow donor is almost 100%, while for $x = 0.3$, that of the deep donor becomes more than 90%. This result in the peculiar change of apparent ionization energy of Si impurities in $Al_xGa_{1-x}As$ seen in Fig. 18. The deep donor is a DX center, which is the origin of the persistent photoconductivity

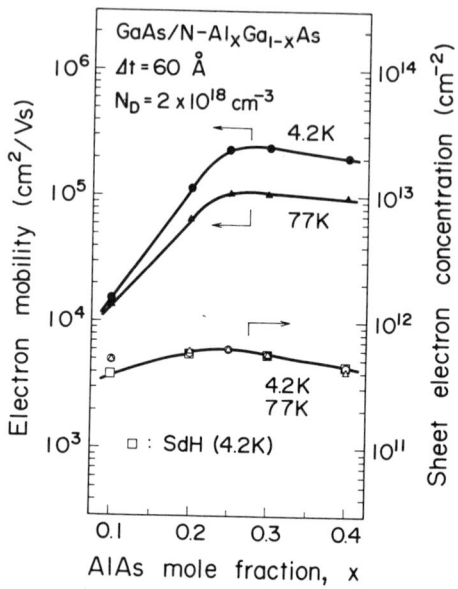

FIG. 20. 2DEG mobility and sheet electron concentration observed by Hall measurements at 4.2 K and 77 K in GaAs/n-Al$_x$Ga$_{1-x}$As heterostructures ($\Delta t = 60$ Å, Si doping concentration $N_d = 2 \times 10^{18}$ cm^{-3}) as a function of AlAs mole fraction x. 2DEG concentrations obtained by Shubnikov–de Haas (SdH) measurements at 4.2 K are also indicated by open squares, and agree well with Hall measurement data.

(PPC) effect of Al$_x$Ga$_{1-x}$As itself and also of GaAs/n-Al$_x$Ga$_{1-x}$As for $x > 0.2$.

Figure 20 shows 2DEG mobility and sheet electron concentration of GaAs/n-Al$_x$Ga$_{1-x}$As as a function of x, obtained by Hall measurements at 4.2 K and 77 K (Saito et al., 1983). The spacer-layer thickness is 60 Å, and the Si doping concentration N_d is 2×10^{18} cm^{-3}. It can be seen that maximum values of 2DEG mobility and electron concentration N_s are obtained at $x = 0.3$ and at $x = 0.25$, respectively, indicating that the optimum value of x for GaAs/n-Al$_x$Ga$_{1-x}$As heterostructures is in the range of $0.25 < x < 0.3$. Hence, GaAs/n-AlGaAs heterostructures with x in this range ($0.25 < x < 0.3$) have been most often investigated and have been widely used to fabricate HEMT devices.

The observed N_s increases slightly from 3.8×10^{11} cm^{-2} as x increases from 0.1 and reaches a maximum (6.1×10^{11} cm^{-2}) at around $x = 0.25$ (Fig. 20). This result indicates that there are two opposing factors for determining N_s as a function of x: effective ionization energy, $E_D(x)$, of Si impurities, and conduction band offset, $\Delta E_c(x)$. In the small region of x where $E_D(x)$ is almost constant (6 meV), N_s increases with increasing $\Delta E_c(x)$

since the thickness of the interface depletion layer increases as $[\Delta E_c(x)]^{1/2}$. For $x > 0.2$, however, $E_D(x)$ increases rapidly with increasing x (Fig. 18), which fully compensates for the effect of increased $\Delta E_c(x)$ at $x = 0.25$. This leads to a peak value of N_s at $x = 0.25$.

It is rather difficult to determine 2DEG concentration (N_s) for $x < 0.2$ by Hall measurements because free carriers in AlGaAs are not frozen out even at low temperatures because of shallow donors ($E_D = 6$ meV) in AlGaAs. Usually we have two conducting layers (AlGaAs neutral layer and 2DEG layer at the interface) for $x < 0.2$, so it is necessary to eliminate the conducting AlGaAs layer carefully by etching the n-AlGaAs layer to obtain reliable N_s values. In these experiments, N_s values observed by Hall measurements at 77 K and 4.2 K agree well with the 2DEG concentration obtained by Shubnikov–de Haas (SdH) measurements at 4.2 K, indicating that the free carrier concentrations in AlGaAs are sufficiently reduced.

The 2DEG mobility increases considerably as x increases from 0.1 to 0.25 and reaches a maximum value (243,000 cm^2/Vsec at 4.2 K, 107,000 cm^2/Vsec at 77 K) at $x = 0.3$. Low mobility (30,000 cm^2/Vsec at 77 K with $N_s = 4 \times 10^{11}$ cm^{-2} for $x = 0.15$) in the range of small x can be attributed primarily to the enhanced impurity scattering and alloy scattering caused by the penetration of the electron wave function into the AlGaAs region with low barrier height, $\Delta E_c(x)$. This interpretation is strongly supported by the much higher electron mobility (77,000 cm^2/Vsec at 77 K) with $N_s = 4.4 \times 10^{11}$ cm^{-2} obtained for a GaAs/n-AlGaAs heterostructure ($x = 0.15$, $N_d = 2 \times 10^{18}$ cm^{-3}) having a higher Al$_y$Ga$_{1-y}$As interface barrier ($y = 0.3$, 60 Å thick) (Ishikawa et al., 1984). The slight decrease of mobility for $x > 0.3$ is possibly due to increased scattering centers in degraded AlGaAs with larger Al content grown by MBE and/or due to enhanced impurity scattering for lower N_s.

7. Transport Properties of 2DEG under High Electric Fields

The Hall mobility of 2DEG is usually measured under low electric fields (10^{-1}–10^{-2} V/cm), whereas in a practical device such as a HEMT, electrons transit under much higher electric fields of about 10^4 V/cm. Hence, it is very important to know the transport properties of 2DEG in GaAs/n-AlGaAs heterostructures in such high electric fields for device applications.

In 1981, Inoue first observed the hot electron effect of 2DEG in a selectively doped GaAs/n-AlGaAs heterostructure by pulsed I-V measurements and pulsed-electric-field SdH measurements at 4.2 K. Above 5 V/cm, hot electron effects became evident. The electron mobility apparently decreased, and the electron temperature rose higher than the lattice

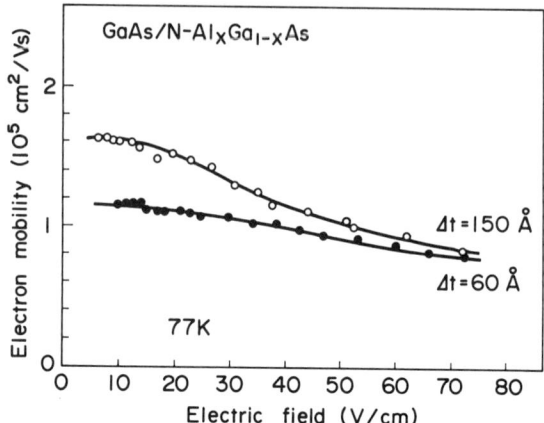

FIG. 21. Electron mobility observed at 77 K as a function of applied electric field in two GaAs/n-Al$_x$Ga$_{1-x}$As heterostructures ($x = 0.3$) with different spacer layer thicknesses ($\Delta t = 60$ Å and 150 Å) [From Inoue et al. (1983).]

temperature. This agrees qualitatively with the results for 2DEG in modulation-doped GaAs–AlGaAs multi-quantum-well structures observed by Drummond et al. (1981a).

The 2DEG mobility (77 K) for high electric fields in two GaAs/n-Al$_x$Ga$_{1-x}$As ($x = 0.3$) heterostructures with different spacer-layer thicknesses ($\Delta t = 60$ Å and 150 Å) are compared in Fig. 21 (Inoue et al., 1983a). Initial Hall mobilities were 163,000 cm^2/Vsec for $\Delta t = 150$ Å and 115,000 cm^2/Vsec for $\Delta t = 60$ Å. With increasing electric field strength, the mobility difference between the two samples was reduced and almost eliminated at $E = 70$ to 80 V/cm. This result indicates that in such moderate electric fields, the remote impurity scattering loses its effect and phonon scattering becomes dominant. In other words, for the higher-mobility sample, electrons can be heated more rapidly by electric fields, and the mobility of hot 2DEG decreases at lower electric field strength due to polar–optical phonon emission ($\hbar\omega = 36$ meV for GaAs).

The electric field dependences of drift velocity of electrons at 4.2 K in two GaAs/n-Al$_x$Ga$_{1-x}$As heterostructures ($x = 0.2, 0.3$) with $\Delta t = 60$ Å and $N_d = 2 \times 10^{18}$ cm^{-3} are shown (Inoue et al., 1983b), together with data for a bulk GaAs sample (Inoue et al., 1972), in Fig. 22. These data were deduced from the electron mobility and electron concentration obtained by the pulsed Hall measurement in high electric fields. Both heterostructure samples show a very similar field dependence of drift velocity. The drift velocity reached about 3.8×10^7 cm/sec at 2.4 kV/cm for $x = 0.2$, which is almost 1.3 times higher than that of GaAs. This indicates that 2DEG in GaAs/n-AlGaAs

2. CHARACTERISTICS OF 2DEG COMPOUND HETEROSTRUCTURES

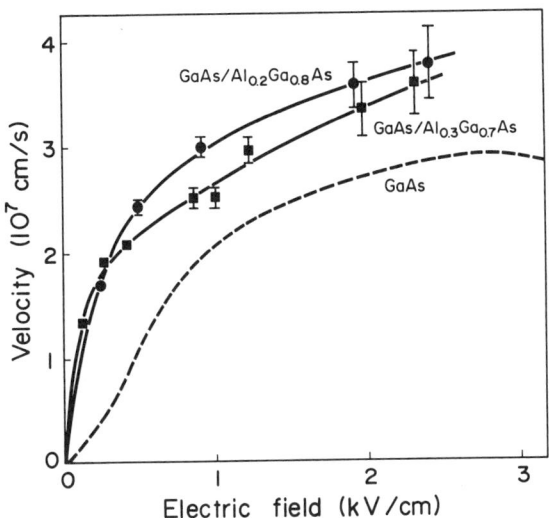

FIG. 22. Electric field dependence of electron drift velocity observed at 4.2 K in two GaAs/nAl$_x$Ga$_{1-x}$As heterostructures ($\Delta t = 60$ Å, $N_d = 2 \times 10^{18}$ cm^{-3}) with different AlAs mole fractions ($x = 0.2$ and 0.3). Observed drift velocity of electrons in bulk GaAs (Inoue et al., 1972) is also illustrated by the broken line. [From Inoue et al. (1983).]

heterostructures exhibits its superiority in velocity over GaAs even under high electric fields. This advantage was confirmed by comparing the cutoff frequencies f_T of HEMTs to those of GaAs MESFETs. The cutoff frequency is given by $f_T = v_s/(2\pi L_g)$, where v_s is the saturation velocity of electrons and L_g is the gate length. Figure 23 shows the cutoff frequency of HEMTs (Joshin et al., 1983; Berez et al., 1984; Mishra et al., 1985) and GaAs MESFETs (Chye and Huang, 1982; Feng et al., 1982; Feng et al., 1984) observed at room temperature as a function of gate length L_g in the range of 0.25 to 0.5 μm (Hiyamizu, 1986). HEMTs exhibit an f_T about 1.4 times higher than that of corresponding GaAs MESFETs, indicating that v_s of a GaAs/n-AlGaAs heterostructure is about 1.4 times higher than that of a bulk GaAs in a practical high-field condition (10^4 V/cm).

Extremely high 2DEG mobility, over 10^6 cm^2/Vsec, can be obtained in selectively doped GaAs/n-AlGaAs heterostructures with a thick spacer layer, at the cost of low electron concentration. In an actual HEMT, however, electrons transit at the saturation velocity under high electric fields. The saturation velocity of electrons is not sensitive to the low-field mobility. Therefore, high 2DEG concentration with a moderately high mobility is more desirable for device applications. Usually GaAs/n-AlGaAs heterostructures with a spacer layer of thickness less than 60 Å have been widely used for HEMT applications to obtain high 2DEG concentration. With no

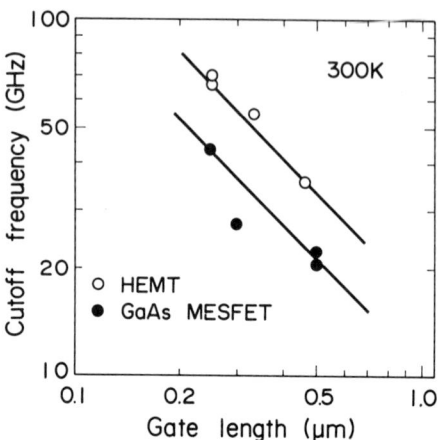

FIG. 23. Cutoff frequency of (○) HEMTs and (●) GaAs MESFETs as a function of gate length (○) after Joshin *et al.*, 1983; Berenz *et al.* 1984 and Mishra *et al.*, 1985; (●) from Chye and Huang, 1982, Feng *et al.*, 1982, and Feng *et al.*, 1984.

spacer layer, the typical 2DEG concentration of a GaAs/n-Al$_x$Ga$_{1-x}$As heterostructure ($x = 0.3$) is 1×10^{12} cm^{-2} with electron mobility of 30,000 cm^2/Vsec at 77 K. This heterostructure has been used for a discrete, low-noise HEMT as well as for a HEMT LSI (large-scale integrated circuit).

IV. Improved Selectively Doped Heterostructures

Recently, it has become clear that selectively doped GaAs/n-AlGaAs heterostructures have several limitations from the point of view of device applications. They are:

(1) the existence of DX centers in Si-doped Al$_x$Ga$_{1-x}$As ($x > 0.2$), which causes I-V collapse (a semipersistent distortion of the drain I-V characteristics) at low temperatures and thermal instability of the threshold voltage of a HEMT;

(2) a rather small conduction-band edge discontinuity, which leads to insufficient, low 2DEG concentration; and

(3) thermally unstable layer structures, especially with respect to diffusion of Si impurities from the n-AlGaAs layer into the GaAs channel region, which makes it difficult to anneal this heterostructure material.

To overcome these problems, various efforts have been made.

8. Stopper Layer Structure

In 1981, Ishikawa first studied the effect of thermal annealing on electric properties of a typical GaAs/n-Al$_x$Ga$_{1-x}$As heterostructures ($x = 0.3$, $\Delta t = 60$ Å, $N_d = 1 \times 19^{18}$ cm^{-3}) grown at 600°C by MBE. He found that 2DEG mobility at 77 K is significantly reduced from an initial value of 75,000 cm^2/Vsec to 7500 cm^2/Vsec, and 2DEG concentration is increased from 4×10^{11} cm^{-2} to 8×10^{11} cm^{-2} after furnace annealing at 760°C for 15 minutes, as shown in Fig. 24. This is mainly caused by diffusion of Si impurities from the n-AlGaAs layer to the GaAs region, since the GaAs/AlGaAs interface is considered to remain as sharp as that of the as-grown sample. (The diffusion constant of Al and Ga atoms (interdiffusion) at the GaAs/AlGaAs interface is very small, and diffusion length is only 1.3 Å after annealing at 800°C for 10 hours (Fleming et al., 1980).) Hence, it seems difficult to electrically activate impurities ion-implanted in the GaAs/n-AlGaAs heterostructure by a conventional furnace annealing without diffusion of Si impurities into the 2DEG region and the resultant degradation of 2DEG properties. This limits the application of the ion implantation to a GaAs/n-AlGaAs HEMT such as the "self-aligned" technique for GaAs MESFETs (Yokoyama et al., 1981) to improve high-speed performance by reducing parasitic resistance.

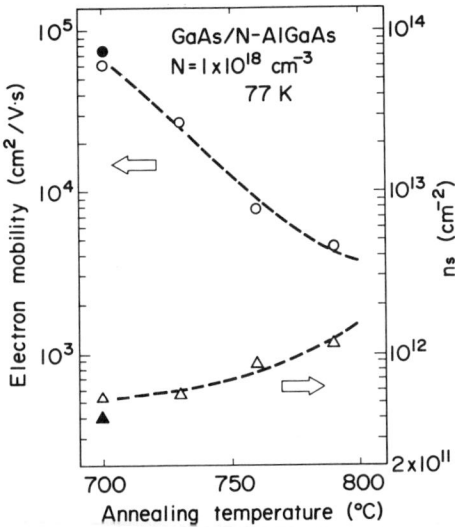

Fig. 24. Annealing temperature dependence of 2DEG mobility and sheet electron concentration observed at 77 K in GaAs/n-Al$_x$Ga$_{1-x}$As ($x = 0.3$, $\Delta t = 60$ Å, $N_d = 1 \times 10^{18}$ cm^{-3}) grown at 600°C by MBE. The solid circle and solid triangle represent "as-grown" data. [From Ishikawa et al. (1981).]

On solution is to use flash-lamp annealing (or rapid annealing), which reduces this thermal degradation of GaAs/n-AlGaAs heterostructures to some extent (Tatsuta et al., 1984; Henderson et al., 1984).

Another solution is a stopper layer structure. Lee et al. (1985) proposed a new GaAs/n-Al$_x$Ga$_{1-x}$As heterostructure ($x = 0.3$, $N_d = 2 \times 10^{18}$ cm^{-3}) with an undoped AlAs spacer layer (100 Å thick). 2DEG mobility at 77 K in the heterostructure was reduced by only 27% from the initial value after conventional furnace annealing at 800°C to 900°C for 15 minutes. Tatsuta et al. (1985) studied diffusion of Si impurities in GaAs and Al$_x$Ga$_{1-x}$As ($x = 0.3$) by C-V measurements and SIMS measurements and found that the diffusion constant is much smaller in GaAs than in AlGaAs, as shown in Fig. 25. They then demonstrated further-reduced Si diffusion in a new GaAs/n-Al$_x$Ga$_{1-x}$As heterostructure ($x = 0.3$) with an undoped GaAs stopper layer (20 Å) introduced between the undoped Al$_x$Ga$_{1-x}$As spacer layer (40 Å, $x = 0.3$) and the Si-doped Al$_x$Ga$_{1-x}$As layer ($x = 0.3$, $N_d = 1 \times 10^{18}$ cm^{-3}). In this structure, the penetration of the 2DEG wave function into the GaAs stopper layer is a low 0.4%, and no electrons accumulate in this layer because the electron energy level in the 20-Å-thick GaAs quantum well becomes higher than the Fermi level. Figure 26 shows the 2DEG mobility at 77 K in this new heterostructure, normalized by the initial mobility (100,000 cm^2/Vsec) as a function of annealing temperature for flash-lamp annealing for 10 seconds (Tatsuta et al., 1985). The reduction of 2DEG mobility in the heterostructure (solid circles) was a small 20% even at an annealing temperature of 950°C, indicating improvement over a

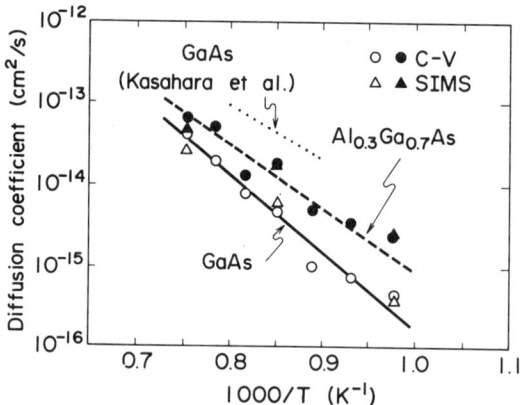

FIG. 25. Diffusion coefficient of Si impurities in GaAs (open circles and triangles) and Al$_{0.3}$Ga$_{0.7}$As (solid circles and triangles) as a function of reciprocal annealing temperature, compared with data for Si-implanted GaAs (dotted line, Kasahara et al., 1982). [From Tatsuta et al. (1985).]

2. CHARACTERISTICS OF 2DEG COMPOUND HETEROSTRUCTURES

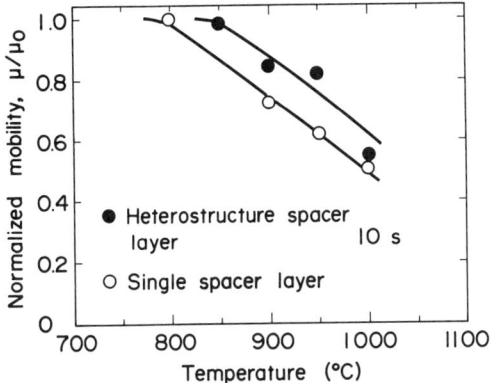

FIG. 26. Electron mobility in a new GaAs/n-Al$_x$Ga$_{1-x}$As heterostructure (solid circles) with a 20-Å-thick GaAs stopper layer introduced between an undoped Al$_x$Ga$_{1-x}$As spacer layer (40 Å, $x = 0.3$) and a Si-doped Al$_x$Ga$_{1-x}$As layer ($x = 0.3$, $N_d = 1 \times 10^{18}$ cm^{-3}) as a function of annealing temperature of flash-lamp annealing for 10 seconds, compared with data (open circles) for a conventional GaAs/n-AlGaAs heterostructure ($x = 0.3$, $\Delta t = 60$ Å, $N_d = 1 \times 10^{18}$ cm^{-3}). [From Tatsuta et al. (1985).]

conventional GaAs/n-Al$_x$Ga$_{1-x}$As heterostructure ($x = 0.3$, $\Delta t = 60$ Å) (open circles). These heterostructures proposed by Lee et al. and Tatsuta et al. are the most resistant to Si diffusion during thermal treatment, which may be partly because of the additional heterojunction interface introduced between n-AlGaAs and GaAs. These new heterostructures are promising for applications to a self-aligned HEMT.

9. n-AlGaAs/GaAs/n-AlGaAs Quantum Well Heterostructures

Inoue and Sakaki proposed in 1984 that 2DEG concentration can be increased in a selectively doped n-Al$_x$Ga$_{1-x}$As/GaAs/n-Al$_x$Ga$_{1-x}$As single-quantum-well (SQW) structure ($x = 0.3$) grown at a rather low growth temperature of 530°C (Fig. 27). Undoped AlGaAs spacer layers (100 Å thick) were introduced at both GaAs/AlGaAs interfaces. An undoped GaAs quantum well (QW) layer is 300 Å thick. The doping concentration N_d of Si in n-AlGaAs layers was 1×10^{18} cm^{-3}. Since electrons in the GaAs QW are supplied from both n-AlGaAs layers, a much-increased 2DEG concentration of 1.27×10^{12} cm^{-2} with a considerably higher electron mobility of 119,000 cm^2/Vsec was obtained at 77 K. The electronic states of 2DEG in selectively doped n-AlGaAs/GaAs/n-AlGaAs SQW structures were investigated in detail (Sasa et al., 1985b).

In 1985, Hikosaka et al. developed a microwave power HEMT with an n-AlGaAs/GaAs/n-AlGaAs SQW structure that has a 100-Å-thick GaAs

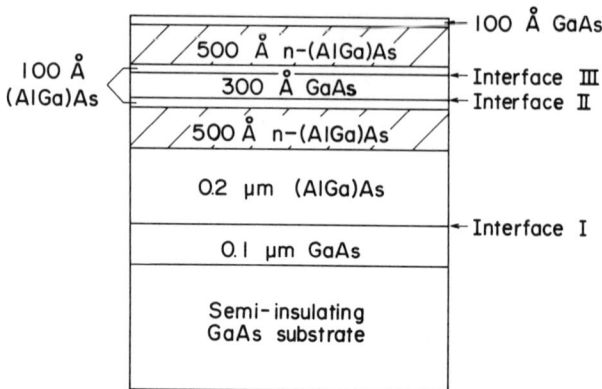

FIG. 27. Schematic cross-section of selectively doped n-Al$_x$Ga$_{1-x}$As/GaAs/n-Al$_x$Ga$_{1-x}$As single-quantum-well (or double-heterojunction) structure. [From Inoue and Sakaki et al. (1984).]

QW layer and an undoped AlGaAs spacer layer introduced only at a substrate-side GaAs/AlGaAs interface. In this device, the low doping concentration of Si in n-AlGaAs ($N_d = 5 \times 10^{17}$ cm^{-3}) was chosen to achieve a high gate breakdown voltage. The 2DEG concentration was 1.1×10^{12} cm^{-2} at 77 K (1.2×10^{12} cm^{-2} at room temperature) with an electron mobility of 48,000 cm^2/Vsec (6800 cm^2/Vsec). The measured maximum transconductance for a 1 μm gate-length HEMT was 120 mS/mm at room temperature. Power HEMTs with 2.4-mm gate periphery exhibited 1 W (0.5 W) output power with 3 dB (2.4 dB) gain and 15.5% (7%) efficiency at 20 GHz (30 GHz). Sheng et al. (1985) also fabricated HEMTs with a similar n-AlGaAs/GaAs/n-AlGaAs SQW structure having a 300-Å-thick GaAs QW layer. The Si doping concentration in n-AlGaAs was 1×10^{18} cm^{-3}, and undoped AlGaAs spacer layers (50 Å) were introduced at GaAs–AlGaAs interfaces. A maximum transconductance of 360 mS/mm at 300 K and 550 mS/mm at 77 K was obtained for a 1 μm gate-length HEMT with the SQW structure.

10. GaAs/AlAs Superlattice Structures

One of the most serious problems of selectively doped GaAs/n-AlGaAs heterostructures is a DX center in Si-doped Al$_x$Ga$_{1-x}$As ($x > 0.2$), which causes a significant shift of the threshold voltage with varying temperature (Valois et al., 1983) and I–V collapses of HEMTs (a semipersistent distortion of the drain I–V characteristics caused by large drain biases) at low temperatures (Fischer et al., 1983).

Two GaAs/AlAs superlattice structures have been proposed as an electron-supplying layer with fewer DX centers (Baba et al., 1983; Hiyamizu *et al.*, 1985b; Baba *et al.*, 1986). From the viewpoint of MBE technology, a multilayer structure, which has two kinds of $Al_xGa_{1-x}As$ layers with different values of x, is rather troublesome to grow because an additional Al source is necessary. GaAs/AlAs superlattices, however, can be grown just by opening and closing the shutters of Al and Ga sources in an MBE growth chamber with the same configuration for growth of a conventional GaAs/ *n*-AlGaAs heterostructure, so GaAs/AlAs superlattices are easily combined with rather thick GaAs and $Al_xGa_{1-x}As$ layers.

Baba *et al.* demonstrated excellent reduction of DX centers in AlAs/ *n*-GaAs superlattices. These consist of AlAs (15 Å) and GaAs (22 Å) layers and exhibit an effective band-gap energy of about 1.88 eV (almost equivalent to the E_g of $Al_xGa_{1-x}As$ ($x = 0.36$)) (Baba *et al.*, 1983). Si impurities are doped only in the GaAs regions in the superlattice. Undoped GaAs spacer layers (5.6 Å, two-monolayer thickness) are introduced at every GaAs/AlAs interface to spatially separate Si impurities from Al atoms, based on their own model of the DX center (a local configuration model of coexisting Al and Ga on the second nearest neighbor sites of a Si-occupied Ga site) (Baba *et al.*, 1986). The average doping concentration of Si over the whole superlattice is about 1×10^{18} cm^{-3} (the actual doping concentration in *n*-GaAs is 3×10^{18} cm^{-3}). Figure 28 shows electron concentration in the AlAs/*n*-GaAs

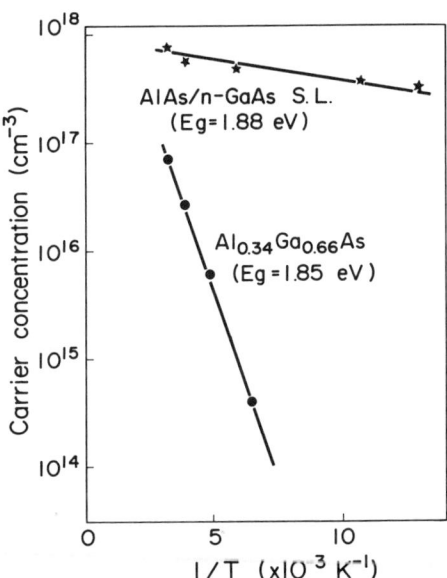

FIG. 28. Electron concentration observed in AlAs/*n*-GaAs superlattice (the average doping concentration of Si over the whole superlattice of 1×10^{18} cm^{-3}) and *n*-$Al_xGa_{1-x}As$ ($x = 0.34$, $N_d = 1 \times 10^{18}$ cm^{-3}) as a function of reciprocal temperature. [From Baba *et al.* (1986).]

superlattice as a function of reciprocal temperature, together with that of Si-doped n-Al$_x$Ga$_{1-x}$As ($x = 0.34$, $N_d = 1 \times 10^{18}$ cm^{-3}) (Baba et al., 1986). Electron concentration in the superlattice does not decrease significantly with decreasing temperature (the ionization energy of Si impurities, E_D, of less than 10 meV), in contrast with the rapid decrease of electron concentration in a conventional n-AlGaAs with E_D of about 100 meV. This indicates that most of the Si impurities in the superlattice do not form DX centers. This DX-center-free AlAs/n-GaAs superlattice was used as an electron-supplying layer for a HEMT structure (Baba et al., 1984; Baba et al., 1986). The 2DEG mobility and sheet electron concentration of the heterostructure ($\Delta t = 60$ Å, an averaged N_d of 1×10^{18} cm^{-3} in the AlAs/n-GaAs superlattice), observed by Hall measurements, are shown as a function of temperature (Baba et al., 1986), together with data for a conventional GaAs/n-Al$_x$Ga$_{1-x}$As heterostructure ($x = 0.3$, $\Delta t = 60$ Å, $N_d = 1 \times 10^{18}$ cm^{-3}) (Hiyamizu and Mimura, 1982a) in Fig. 29. Characteristic features of this heterostructure are (1) an almost constant sheet electron concentration versus temperature, and (2) an apparently high 2DEG concentration (8×10^{11} cm^{-2}) with a normal 2DEG mobility (89,000 cm^2/Vsec at 77 K). The constant sheet electron concentration in this heterostructure is possibly due to a greatly reduced number of DX centers and a very low electron mobility (150 to 300 cm^2/Vsec) in the superlattice. Even if free electrons exist in the superlattice region, the observed electron concentration, given by Eq. (3) in Section 1, is approximately equal to n_1 (2DEG concentration), since μ_2 (the mobility of electrons

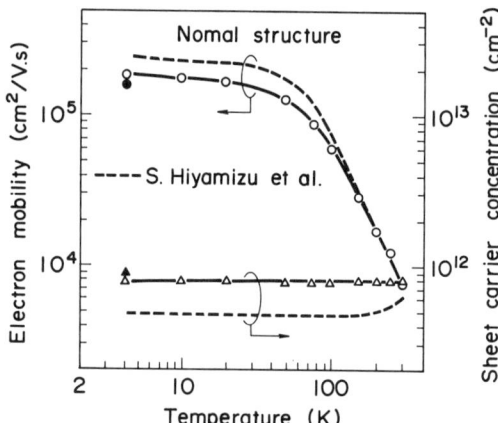

FIG. 29. 2DEG mobility and sheet electron concentration in a HEMT structure with AlAs/n-GaAs superlattice as an electron-supplying layer ($\Delta t = 60$ Å, averaged N_d of 1×10^{18} cm^{-3}) as a function of temperature, together with corresponding data (dashed line) (Hiyamizu et al., 1982) of a conventional GaAs/n-Al$_x$Ga$_{1-x}$As heterostructure ($x = 0.3$, $\Delta t = 60$ Å, $N_d = 1 \times 10^{18}$ cm^{-3}). [From Baba et al. (1986).]

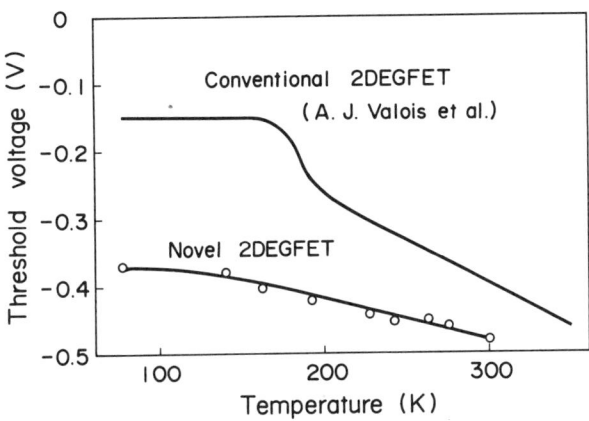

FIG. 30. Temperature dependence of threshold voltage of a HEMT (or 2DEGFET) with the AlAs/n-GaAs superlattice as an electron-supply layer, which compares favorably with data (Valois et al., 1983) for a conventional GaAs/n-Al$_x$Ga$_{1-x}$As HEMT. [From Baba et al. (1984).]

in the superlattice) is two orders of magnitude lower than μ_1 (2DEG mobility). Hence, this result indicates that the 2DEG concentration becomes independent of temperature when the Si donors are shallow (7 meV) in the electron-supplying layer of the superlattice. Another attractive feature is the high 2DEG concentration (8×10^{11} cm^{-2}), which is almost 1.5 times as high as that of a typical GaAs/n-AlGaAs heterostructure, while keeping high 2DEG mobility (89,000 cm^2/Vsec at 77 K). This is primarily caused by the shallow donor level in the electron-supplying layer. In addition, the increase in 2DEG concentration of the heterostructure by exposure to light at low temperature is less than 1×10^{11} cm^{-2} (about 10%), as illustrated by a solid triangle in Fig. 29. This is nearly one order of magnitude smaller than that for a conventional GaAs/n-AlGaAs heterostructure. This result also suggests the low concentration of DX centers in the AlAs/n-GaAs superlattice. A HEMT (or 2DEGFET) with the AlAs/n-GaAs heterostructure exhibits much-improved temperature dependence of the threshold voltage. Figure 30 shows the threshold voltage V_T of this device as a function of temperature (Baba et al., 1984), together with that for a conventional GaAs/n-AlGaAs HEMT (Valois et al., 1983). For a conventional HEMT, V_T moved from -0.39 V to -0.15 V (a shift of 0.24 V) as temperature decreased from 300 K to 80 K, while the superlattice HEMT exhibited a low V_T shift of only 0.1 V in the same range of temperature.

Silicon atomic-planar doping was applied to a GaAs/AlAs quantum well (QW) structure to improve the electron-supplying layer of a 2DEG system (Hiyamizu et al., 1985b). The sheet electron concentration originating

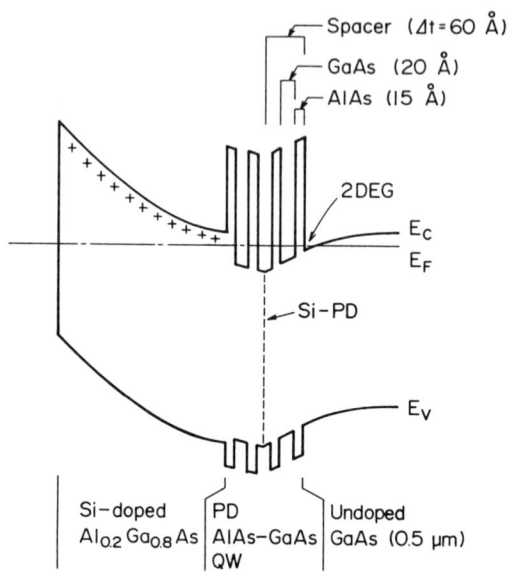

FIG. 31. Energy band diagram of an atomic-planar doped quantum well (QW) heterostructure for 2DEG system, which has a single Si-atomic-planar doped Ga plane (sheet doping concentration of Si, N_{sd}, of 2×10^{12} cm^{-2}) in an AlAs-GaAs-AlAs quantum well structure. [From Hiyamizu et al. (1985).]

from a single Si-doped atomic plane in GaAs easily reached as high as 3×10^{12} cm^{-2} (Sasa et al., 1985a), which is high enough to supply 2DEG in a selectively doped heterostructure. Figure 31 shows the energy band diagram of the atomic-planar doped QW heterostructure for 2DEG system grown at 520°C by MBE (Hiyamizu et al., 1985b). The heterostructure consists of an undoped GaAs (0.5 μm), an undoped spacer layer of AlAs (15 Å)/GaAs (20 Å)/AlAs (15 Å), Si atomic-planar doped GaAs QW (20 Å, sheet doping concentration N_{sd} of 2×10^{12} cm^{-2}), an undoped AlAs/GaAs/AlAs layer, and a contact layer of uniformly Si-doped Al$_x$Ga$_{1-x}$As (400 Å, $x = 0.2$, $N_d = 1 \times 10^{18}$ cm^{-3}). Atomic-planar doping of Si was made on the middle Ga atomic plane in the GaAs QW layer. In this case, the effective spacer layer was also 60 Å thick. The 2DEG accumulates at the interface between an undoped GaAs layer and a GaAs/AlAs multi-QW layer. The 2DEG concentration observed by Hall measurements is 1.04×10^{12} cm^{-2} with an electron mobility of 84,000 cm^2/Vsec at 77 K. When the planar doping is made in the nearest GaAs QW to the 2DEG channel layer ($\Delta t = 30$ Å), 2DEG concentration increases to 1.4×10^{12} cm^{-2} with an electron mobility of 39,000 cm^2/Vsec at 77 K. These values are the highest 2DEG concentrations obtained for selectively doped GaAs/AlGaAs heterostructures with a one-side electron supplying layer. The net increase of electron concentration due to the PPC effect at 77 K was about 1×10^{11} cm^{-2} ($\Delta N_s/N_s = 10\%$), which is almost the same as that of the AlAs/n-GaAs superlattice HEMT structure.

11. InGaAs/n-InAlAs, InGaAs/n-AlGaAs Heterostructures

An InGaAs–InAlAs heterostructure, lattice-matched to InP, has numerous properties superior to a GaAs–AlGaAs heterostructure for high-speed device applications. These properties include the following:

(1) the high electron mobility of InGaAs (13,800 cm^2/Vsec at 300 K),
(2) the light electron effective mass of InGaAs ($m^*/m_0 = 0.041$),
(3) the high saturation velocity of electrons in InGaAs (2.65×10^7 cm/sec) due to large Γ–L separation (0.55 eV) in the conduction band structure,
(4) the large conduction-band offset (0.52 eV) for the InGaAs/InAlAs heterojunction,
(5) the high carrier concentration in both n-InGaAs ($n = 5 \times 10^{19}$ cm^{-3}) and n-InAlAs ($n \geq 2 \times 10^{19}$ cm^{-3}), and
(6) the shallow donor level of Si (less than 10 meV) and lack of DX centers in Si-doped InAlAs.

There have, however, been some technical problems for MBE growth of InGaAs or InAlAs, lattice-matched to InP. This has limited the study of InGaAs–InAlAs heterostructures. Developments in the MBE technique in the 1980s now enable us to apply this attractive material widely to high-speed devices. Recent studies on selectively doped InGaAs/n-InAlAs heterostructures, lattice-matched to InP substrates, revealed that this material is the best for HEMT applications.

For GaAs/Al$_x$Ga$_{1-x}$As heterostructures grown on GaAs substrates, the lattice-matching condition (($a_{\text{AlGaAs}} - a_{\text{GaAs}})/a_{\text{GaAs}} < 0.1\%$) is automatically satisfied for almost the full range of AlAs mole fraction x, and expected electrical properties of GaAs–AlGaAs heterostructures can be easily obtained. For InGaAs/InAlAs heterostructures grown on InP, however, it is absolutely necessary to control beam intensities of In, Ga, and Al within the precision of a few percent to satisfy the lattice-matching condition, since the lattice mismatch between GaAs ($x = 0$) and InAs ($x = 2$ for In$_x$Ga$_{1-x}$As) is a large 7%. In addition, lateral uniformity of the InAs mole fraction in InGaAs (or InAlAs) over a considerable area with similar precision is necessary for device applications. In 1981, the lateral uniformity of InGaAs and InAlAs was achieved by the substrate rotation technique by Cheng. The fine beam-control technique for this heterostructure was developed by Mizutani and Matsui independently in 1985.

In 1982, Cheng *et al.* grew a selectively doped InGaAs/n-InAlAs heterostructure ($\Delta t = 80$ Å, $N_d = 1 \times 10^{17}$ cm^{-3}), lattice-matched to an InP substrate, by MBE for the first time. This heterostructure exhibited 2DEG mobilities of 8915 cm^2/Vsec with $N_s = 1.56 \times 10^{12}$ cm^{-2} at 300 K and 60,120 cm^2/Vsec with $N_s = 1.04 \times 10^{12}$ cm^{-2} at 77 K. An InGaAs/n-InAlAs HEMT was also first developed in 1982 (Chen *et al.*, 1982). Recently, InGaAs/n-InAlAs heterostructures have been much improved (Onabe *et al.*, 1986; Hirose *et al.*, 1986; Nakata *et al.*, 1987) and it became very clear that this heterostructure is the best material for HEMT applications.

Figure 32 shows 2DEG mobility and sheet electron concentration in InGaAs/n-InAlAs heterostructures ($\Delta t = 80$ Å, $N_d = 2 \times 10^{17}$ cm^{-3}) as a function of temperature (Onabe *et al.*, 1986). Si-doped n-InAlAs was 1500 Å thick and an undoped InAlAs layer was 1 μm thick. The 2DEG mobilities are 11,000 cm^2/Vsec (300 K), 82,000 cm^2/Vsec (77 K), and 110,000 cm^2/Vsec ($T < 50$ K). The saturated 2DEG mobility at low temperature, which is more than one order of magnitude lower than that of a GaAs/n-AlGaAs heterostructure, is due to the much-enhanced alloy-disorder scattering since 2DEG concentrates in a ternary InGaAs alloy region for an InGaAs/n-InAlAs heterostructure (Walukiewicz *et al.*, 1984b). The sheet electron concentration N_s decreases slightly from 9×10^{11} cm^{-2} (300 K) to 8×10^{11} cm^{-2} ($T < 40$ K) with decreasing temperature. The

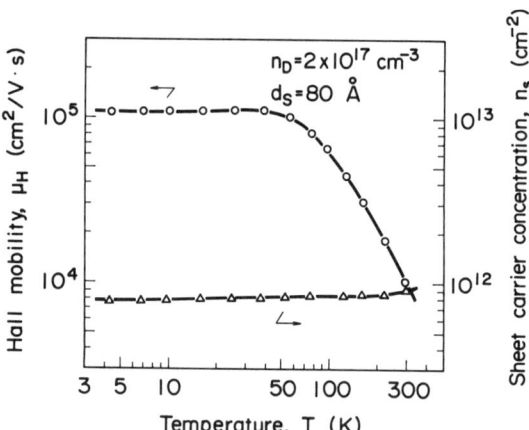

FIG. 32. (○) 2DEG mobility and (△) sheet electron concentration observed in InGaAs/n-InAlAs heterostructures, lattice-matched to InP ($\Delta t = 80$ Å, $N_d = 2 \times 10^{17}$ cm^{-3}) as a function of temperature. [From Onabe *et al.* (1986).]

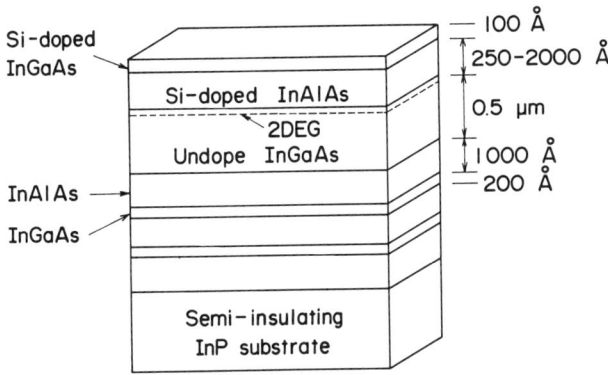

FIG. 33. Schematic cross-section of a selectively doped InGaAs/n-InAlAs heterostructure ($\Delta t = 0$ Å), lattice-matched to an InP substrate grown at 470°C by MBE. [From Nakata et al. (1987).]

change in N_s with decreasing temperature (10% of the initial N_s value) is much smaller than that for a conventional GaAs/n-AlGaAs heterostructure (27% for Fig. 8). This reduced change in N_s is almost the same as changes observed in the improved 2DEG systems with GaAs/AlAs superlattice structures mentioned in the previous section, indicating that there are much-reduced deep traps in n-InAlAs and/or at the InGaAs/InAlAs heterointerface, and almost all Si impurities form shallow donors (E_D of about 10 meV) in InAlAs.

The highest 2DEG concentration for a selectively doped, single-interface heterostructure was achieved in InGaAs/n-InAlAs heterostructures with a heavily Si-doped n-InAlAs layer (Nakata et al., 1987). Figure 33 shows a cross-section of the selectively doped InGaAs/n-InAlAs heterostructure, lattice-matched to InP, grown on an InP substrate at 470°C by MBE. It consists of a Si-doped InGaAs cap layer (100 Å, $N_d = 1 \times 10^{18}$ cm^{-3}), a Si-doped InAlAs layer (250 to 2000 Å, $N_d = 1 \times 10^{17}$ to 5×10^{18} cm^{-3}), and an undoped InGaAs channel layer (0.5 μm). No undoped InAlAs spacer layer is introduced to increase 2DEG concentration. A 3400 Å-thick InGaAs/InAlAs buffer layer was introduced between the undoped InGaAs layer and the substrate.

Figure 34 shows SdH oscillation (4.2 K) of magnetoresistance for a sample with $N_d = 5 \times 10^{18}$ cm^{-3} (Nakata et al., 1987). Three kinds of oscillation periods can be seen, indicating that three subbands [the ground ($E(0)$), the first ($E(1)$), and the second exicted subband ($E(2)$)] are occupied by electrons in the heterointerface quantum well. The observed 2DEG concentrations, which are deduced from Eq. (1) in Section 1, are $N_s(0) = 2.5 \times 10^{12}$ cm^{-2}, $N_s(1) = 7.8 \times 10^{11}$ cm^{-2}, and $N_s(2) = 1.4 \times 10^{11}$ cm^{-2}. An extremely high

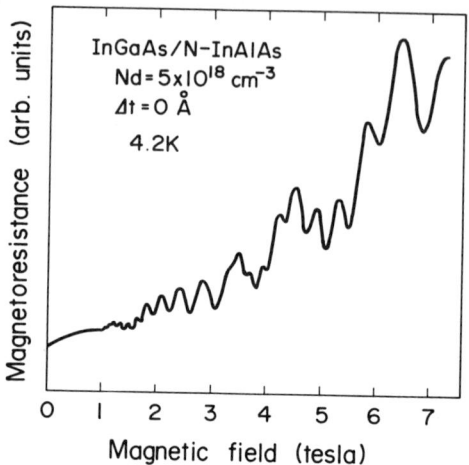

FIG. 34. Shubnikov-de Haas (SdH) oscillation of magnetoresistance observed at 4.2 K in an InGaAs/n-InAlAs heterostructure ($\Delta t = 0$ Å, $N_d = 5 \times 10^{18}$ cm^{-3}). Three kinds of oscillation periods are obtained. [From Nakata et al. (1987).]

concentration of 2DEG, $N_s = 3.5 \times 10^{12}$ cm^{-3} ($= N_s(0) + N_s(1) + N_s(2)$), can be accommodated in this single-interface heterostructure.

A calculated conduction band diagram and 2DEG distribution in the InGaAs/n-InAlAs heterostructure ($\Delta t = 0$, $N_d = 5 \times 10^{18}$ cm^{-3}) are shown in Fig. 35 (Nakata et al., 1987). A self-consistent calculation of the electronic states and 2DEG concentration in the InGaAs/n-InAlAs heterostructure was carried out by numerically solving Schrödinger's and Poisson's equations simultaneously. In the calculation, the effective mass approximation and the boundary condition of continuity of ψ_i/m_i^* at the InGaAs/n-InAlAs heterointerface (ψ_i and m_i^* are the electron wave function and effective mass in the ith layer ($i = 1, 2$), respectively) were used. This boundary condition becomes important for InGaAs/n-InAlAs heterostructures because electron effective mass m_i^* differs significantly in the two regions ($m_1^* = 0.042 m_0$ in InGaAs and $m_2^* = 0.075 m_0$ in InAlAs, where m_0 is the electron mass in vacuum). The nonparabolicity of m^* ($m^* = m_1^*(1 + 2\beta E)$) for InGaAs was also taken into account in calculating the density of states, where E is the electron energy measured from the bottom of each 2DEG subband, and $\beta = (1 - m_1^*/m_0)^2/E_g$. Fermi-Dirac statistics were used to calculate the conduction-band bending in the InAlAs electron-supplying layer in which the donor level of Si is assumed to be 10 meV. The calculated 2DEG concentration N_s reached a high 3.7×10^{12} cm^{-3} primarily because of the large conduction band offset ($\Delta E_c = 0.53$ eV) of this heterostructure. It can be seen from Fig. 35 that three subbands, $E(0)$, $E(1)$, and $E(2)$, are occupied

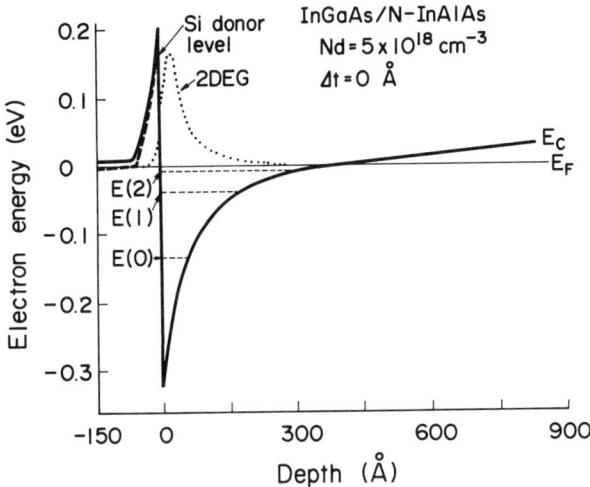

FIG. 35. Self-consistently calculated conduction band diagram and 2DEG distribution in the InGaAs/n-InAlAs heterostructure ($\Delta t = 0$ Å, $N_d = 5 \times 10^{18}$ cm^{-3}). [From Nakata et al. (1987).]

by electrons, which agrees well with results of SdH measurements. In spite of such a heavy population of electrons, the potential barrier height at the interface still remains high (more than 0.2 eV above the Fermi level E_F).

Figure 36 shows 2DEG concentrations in the InGaAs/n-InAlAs heterostructures observed by SdH measurements as a function of Si-doping concentration N_d (Nakata et al., 1987). Open circles illustrate observed 2DEG concentrations of each subband ($N_s(0)$, $N_s(1)$, and $N_s(2)$), and solid circles show the total 2DEG concentration N_s. The 2DEG concentration increases monotonically with increasing N_d and reaches 3.5×10^{12} cm^{-2} at $N_d = 5 \times 10^{18}$ cm^{-3}. Calculated N_s (solid line) and $N_s(i)$ (broken line) are also indicated as a function of N_d, showing excellent agreement with the observed results.

The 2DEG mobility values for MBE-grown InGaAs/n-InAlAs heterostructures, obtained so far by Hall measurements at 77 K and 300 K, are plotted as a function of sheet electron concentration N_s and contrasted with data for conventional GaAs/n-AlGaAs heterostructures in Fig. 37. The 2DEG mobility in InGaAs/n-InAlAs remains high (6700 cm^2/Vsec at 300 K and 22,000 cm^2/Vsec at 77 K) even for high N_s (5.0×10^{12} cm^{-3} at 300 K and 3.7×10^{12} cm^{-2} at 77 K). This N_s value is about four times higher than the highest N_s in conventional GaAs/n-AlGaAs heterostructures with similar 2DEG mobility. The advantage of an InGaAs/n-InAlAs heterostructure for HEMT applications becomes very clear, especially at room temperature.

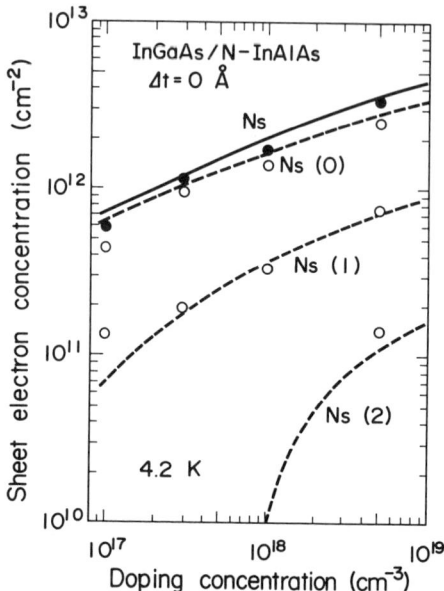

FIG. 36. 2DEG concentrations observed by SdH measurements at 4.2 K in InGaAs/n-InAlAs heterostructures as a function of Si-doping concentration, N_d: (○) 2DEG concentrations of each subband, $N_s(i)$; (●) total 2DEG concentration, N_s. Calculated N_s (solid line) and $N_s(i)$ (broken lines) exhibit excellent agreement with the observed data. [From Nakata et al. (1987).]

This characteristic feature of this material leads to excellent properties of InGaAs/n-InAlAs HEMTs (Hirose, 1986): extremely high transconductances of 440 mS/mm at room temperature and 700 mS/mm at 77 K have been achieved for a 1-μm-gate InGaAs/n-InAlAs HEMT.

Other lattice-matched (to a substrate) systems are $In_xGa_{1-x}As/n$-InP heterostructures ($x = 0.53$, InP substrate) (Guldner et al., 1982; Takikawa et al., 1983; Zhu et al., 1985) and GaAs/n-$Ga_xIn_{1-x}P$ ($x = 0.52$, GaAs substrate) heterostructures (Cox et al., 1986; Kodama et al., 1986; Tone et al., 1986; Razeghi et al., 1986). These selectively doped heterostructures mainly have been grown by VPE or by metal organic chemical vapor deposition (MOCVD) because high-quality heterostructures including P and As are difficult to grow by MBE. The conduction band offset, ΔE_c, is 0.22 eV for InGaAs/n-InP and 0.39 eV for GaAs/n-GaInP. These values are almost equal to or larger than that for a conventional GaAs/n-$Al_xGa_{1-x}As$ ($x = 0.3$) heterostructures but are smaller than ΔE_c for InGaAs/n-InAlAs. The 2DEG mobilities are 12,000 cm^2/Vsec (300 K), 83,000 cm^2/Vsec (77 K), and 92,000 cm^2/Vsec (4.2 K) with 2DEG concentrations of 3 to 3.7 × 10^{11} cm^{-2} for an InGaAs/n-InP heterostructure (Zhu et al., 1985). For a GaAs/

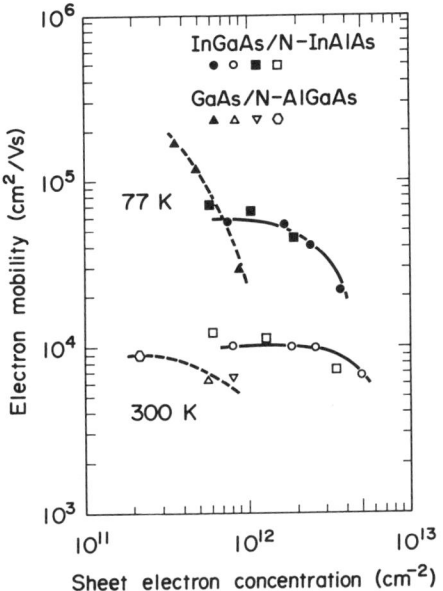

FIG. 37. 2DEG mobility observed at 77 K and 30 K in MBE-grown InGaAs/n-InAlAs heterostructures, lattice-matched to InP, as a function of sheet electron concentration, together with data for conventional GaAs/n-AlGaAs heterostructures: (●, ○) Nakata *et al.*, 1987; (■, □) Onabe *et al.*, 1986; (▲, △) Hiyamizu *et al.*, 1983; (▽) Hiyamizu *et al.*, 1985; (○) Heiblum *et al.*, 1984. [From Nakata *et al.* (1987).]

n-GaInP heterostructure, 2DEG mobilities are 6000 cm^2/Vsec (300 K), 38,000 cm^2/Vsec (77 K), and 49,000 cm^2/Vsec (4.2 K), and sheet electron concentration is 7.64×10^{11} cm^{-2} (4.2 K, dark). A slight increase in 2DEG concentration (10 to 20%) caused by the PPC effect was observed for the two systems, but the existence of DX centers in these heterostructures has not been reported yet.

For lattice-mismatched systems, there is a pseudomorphic In$_y$Ga$_{1-y}$As/ n-GaAs heterostructure grown on a GaAs substrate, which has a strained layer In$_y$Ga$_{1-y}$As ($y = 0.2$) as an active 2DEG layer and an n-GaAs electron-supplying layer. This heterostructure was first grown by MOCVD in 1983 (Zipperian). In 1985, Rosenberg reported an In$_y$Ga$_{1-y}$As/n-GaAs heterostructure ($y = 0.15$) grown by MBE. In this case, the conduction band offset ΔE_c is only about 0.15 eV. In$_y$Ga$_{1-y}$As/n-Al$_x$Ga$_{1-x}$As heterostructures ($y = x = 0.15$) were then prepared by MBE (Ketterson, 1985; Henderson, 1986; Morkoç, 1986) in which ΔE_c is increased to 0.3 eV. In the pseudomorphic heterostructures, the In$_y$Ga$_{1-y}$As layer in which 2DEG accumulates is so thin that the lattice mismatch between this layer and AlGaAs (or GaAs)

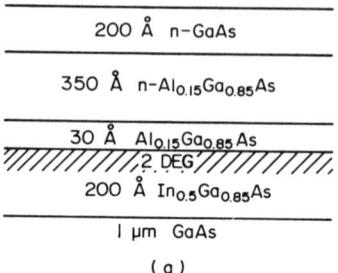

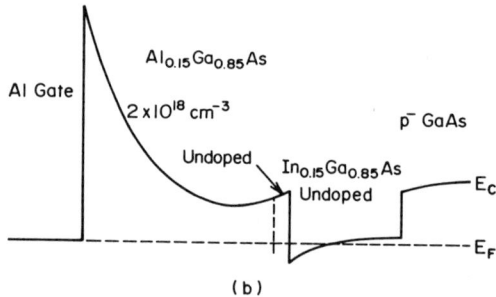

FIG. 38. Structure of a pseudomorphic InGaAs/n-AlGaAs HEMT (or MODFET) grown on a GaAs substrate (a) and its energy band diagram (b). 2DEG is formed in the strained-layer InGaAs quantum well layer. (Reprinted with permission from Ketterson, A., Moloney, M., Masselink, W. T., Klem, J., Fischer, R., Kopp, W., and Morkoc, H. (© 1985 IEEE) *IEEE Electron. Device Lett.* **EDL1-6**, 628.)

is accommodated entirely as elastic strain, as predicted by van der Merwe. The most characteristic feature of a pseudomorphic InGaAs/n-AlGaAs heterostructure is shallow donors and no DX centers in Si-doped GaAs or $Al_xGa_{1-x}As$ ($x = 0.15$).

Figure 38 shows a pseudomorphic InGaAs/n-AlGaAs heterostructure and its energy band diagram (Ketterson, 1985). It consists of n^+-GaAs (200 Å), Si-doped $Al_xGa_{1-x}As$ (350 Å, $x = 0.15$, $N_d = 2 \times 10^{18}$ cm^{-3}), an undoped $Al_xGa_{1-x}As$ spacer layer (30 Å, $x = 0.15$), pseudomorphic $In_yGa_{1-y}As$ quantum well (200 Å, $y = 0.15$), and a 1-μm-thick GaAs buffer layer. The 2DEG mobilities in the pseudomorphic InGaAs are 6000 cm^2/Vsec (300 K) and 29,000 cm^2/Vsec (77 K) with a sheet electron concentration of 1.2×10^{12} cm^{-2} (77 K). No persistent photoconductivity was observed at 77 K, indicating an absence of DX centers in this heterostructure. Electrical properties of pseudomorphic InGaAs/n-AlGaAs heterostructures are between those of a conventional GaAs/n-AlGaAs heterostructure and those of an InGaAs/n-InAlAs heterostructure lattice-matched to InP.

Electrical parameters of the improved selectively doped heterostructures are listed in Table III.

2. CHARACTERISTICS OF 2DEG COMPOUND HETEROSTRUCTURES

TABLE III

ELECTRICAL PARAMETERS OF IMPROVED SELECTIVELY DOPED HETEROSTRUCTURES

Sample	77 K (dark)		77 K (light)			References
	2DEG mobility ($cm^2/Vsec$)	2DEG concentration N_s (cm^{-2})	2DEG mobility ($cm^2/Vsec$)	2DEG concentration N_s' (cm^{-2})	Increase ratio $(N_s' - N_s)/N_s$ (%)	
n-AlGaAs/GaAs/n-AlGaAs single QW	119,000	1.27×10^{12}				Inoue & Sakaki (1984)
AlAs/n-GaAs superlattice structure	89,000	8×10^{11}				Baba et al. (1986)
	$190,000^a$	8×10^{11a}	$160,000^a$	9×10^{11a}	12.5	Baba et al. (1986)
Si-PD GaAs/AlAs QW structure	84,000	1.04×10^{11}	94,000	1.16×10^{12}	11.5	Hiyamizu et al. (1985b)
InGaAs/n-InAlAs (lattice-matched to InP)	60,120	1.04×10^{12}			$(10)^b$	Cheng et al. (1982)
	65,000	1.7×10^{12}			$(10)^b$	Onabe et al. (1986)
	55,000	1.7×10^{12}			$(10)^b$	Nakata et al. (1987)
	22,000	3.7×10^{12}			$(10)^b$	Nakata et al. (1987)
InGaAs/n-InP (lattice-matched to InP)	83,000	3.8×10^{11}				Zhu et al. (1985)
GaAs/n-GaInP (lattice-matched to GaAs)	$49,000^a$	7.64×10^{11a}		9.43×10^{11a}	23.4	Razeghi et al. (1986)
InGaAs/n-AlGaAs (pseudomorphic)	29,000	1.2×10^{12}				Ketterson et al. (1985)

[a] Values observed at 4.2 K (or 4 K).
[b] Assumed values.

V. Summary

Electrical properties of 2DEG in selectively doped GaAs/n-AlGaAs heterostructures have been well understood by current studies, and device applications of GaAs/n-AlGaAs have also been advanced. A GaAs/n-AlGaAs low-noise HEMT was confirmed to exhibit high-speed performance apparently superior to that of a conventional GaAs metal semiconductor field effect transistor (MESFET) (for instance Mishra et al., 1985), and this device has been commercially available since 1985. HEMT large-scale integrated circuits such as a 4 K static random access memory (SRAM) (Awano et al., 1987) and a 4.1 K gate alley (Kajii et al., 1987) also have been developed for ultrahigh-speed memory and logic applications. These LSIs have been made of GaAs/n-AlGaAs wafers grown by MBE because they require the excellent lateral uniformity and precisely controlled multilayer structure (layer thickness, alloy composition and doping concentration) that can be achieved only by MBE at the present stage.

These active developments of HEMTs promote the progress of MBE technology, especially the improvement of its rather small throughput. Recently, a new MBE system, which has a 190-mm diameter rotating substrate holder, was developed (Saito et al., 1987). The uniformity of layer thickness and carrier concentration of GaAs and AlGaAs was reported to be very nice ($\pm 1\%$ variation over a 180-mm diameter area). Hence, this MBE system turns out to provide 3-10 times higher throughput than a conventional MBE system. Surface defects on MBE wafers, which are one of the causes for reduced yield of HEMT fabrication, could also be reduced to 380 cm^{-2} (Saito et al., 1987). Moreover, a new MBE technology, which is called CBE (chemical-beam epitaxy) or MOMBE (metal organic MBE) or GSMBE (gas-source MBE), has been intensively studied (for instance, Panish, 1987; Tsang, 1987). In this technology, gas sources such as triethyl gallium (TEGa), triethyl aluminum (TEAl) and arsine (AsH$_3$) are used instead of conventional metal sources of Ga, Al, and As for MBE. Since this CBE/MOMBE/GSMBE has the advantages of both MBE and MOCVD at the same time, it can be expected to become an excellent crystal growth technique in the future. This technique will be used for basic research of new semiconductor materials as well as for mass production of well-established semiconductor materials such as GaAs/n-AlGaAs wafers. In 1987, an attempt was first made to grow selectively doped GaAs/n-AlGaAs by this new growth technique (Ando et al., 1987). These efforts will extend the use of sophisticated semiconductor materials with precise multilayer structures, including 2DEG materials, for the development of higher-performance semiconductor devices.

2. CHARACTERISTICS OF 2DEG COMPOUND HETEROSTRUCTURES

REFERENCES

Ando, T. (1982a). *J. Phys. Soc. Japan* **51**, 3893.
Ando, T. (1982b). *J. Phys. Soc. Japan* **51**, 3900.
Ando, H., Kondo, K., Ishikawa, H., Fujii, T., and Hiyamizu, S. (1987). *J. Electrochem. Soc.* **134**, 576C.
Arnold, D., Ketterson, A., Henderson, T., Klem, J., and Morkoç, H. (1984). *Appl. Phys. Lett.* **45**, 1237.
Awano, Y., Kosugi, M., Nagata, T., Kosemura, K., Ono, M., Kobayashi, N., Ishiwari, H., Odani, K., Mimura, T., and Abe, M. (1987). GaAs *IC Symp. Technical Digest, Portland*, p. 177.
Baba, T., Mizutani, T., and Ogawa, M. (1983). *Japan. J. Appl. Phys.* **22**, L627.
Baba, T., Mizutani, T., Ogawa, M., and Ohata, K. (1984). *Japan. J. Appl. Phys.* **23**, L654.
Baba, T., Mizutani, T., and Ogawa, M. (1986a). *J. Appl. Phys.* **59**, 526.
Baba, T., Ogawa, T., and Mizutani, T. (1986b). *Surface Sci.* **174**, 408.
Batey, J., Wright, S. L., and DiMaria, D. J. (1985). *J. Appl. Phys.* **57**, 484.
Berenz, J. J., Nakano, K., and Weller, K. P. (1984). *IEEE MTT-S Digest*, p. 83.
Blakemore, J. S. (1982). *J. Appl. Phys.* **53**, R123.
Chadi, D. J., and Cohen, M. L. (1975). *Phys. Status Solidi* **B68**, 405.
Chen, C. Y., Cho, A. Y., Cheng, K. Y., Pearsall, T. P., O'Connor, P., and Garbinski, P. A. (1982). *IEEE Electron Device Lett.* **EDL-3**, 152.
Cheng, K. Y., Cho, A. Y., and Wagner, W. R. (1981). *Appl. Phys. Lett.* **39**, 607.
Cheng, K. Y., Cho, A. Y., Drummond, T. J., and Morkoç, H. (1982). *Appl. Phys. Lett.* **40**, 147.
Cho, A. Y. (1975). *J. Appl. Phys.* **46**, 1733.
Chye, P. W., and Huang, C. (1982). *IEEE Electron Device Lett.* **EDL-3**, 401.
Cox, H. M., Hayes, J. R., Nottenberg, R. N., Hummel, S. G., and Allen, S. J. (1986). *Electron. Lett.* **22**, 73.
Dingle, R., Wiegmann, W., and Henry, C. H. (1974). *Phys. Rev. Lett.* **33**, 827.
Dingle, R., Störmer, H. L., Gossard, A. C., and Wiegmann, W. (1978). *Appl. Phys. Lett.* **33**, 665.
Drummond, T. J., Keever, M., Kopp, W., Morkoç, H., Hess, K., Streetman, B. G., and Cho, A. Y. (1981a). *Electron. Lett.* **17**, 545.
Drummond, T. J., Morkoç, H., and Cho, A. Y. (1981b). *J. Appl. Phys.* **52**, 1380.
Drummond, T. J., Lyone, W. G., Fisher, R., Throne, R. E., Morkoç, H., Hopkins, C. G., and Evans, C. A., Jr. (1982). *J. Vac. Sci. Technol.* **21**, 957.
English, J. H., Gossard, A. C., Störmer, H. L., and Bladwin, K. W. (1987). *Appl. Phys. Lett.* **50**, 1826.
Esaki, L., and Tsu, R. (1969). *IBM Research Note RC-2418*.
Fang, F. F., and Howard, W. E. (1966). *Phys. Rev. Lett.* **16**, 797.
Feng, M., Eu, V. K., Kanber, H., Watkins, E., Schellenberg, J. M., and Yamasaki, H. (1982). *Appl. Phys. Lett.* **40**, 802.
Feng, M., Kanber, H., Eu, V. K., Watkins, W., and Hackett, L. R. (1984). *Appl. Phys. Lett.* **44**, 231.
Fischer, R., Drummond, T J., Kopp, W., Morkoç, H., Lee, K., and Shur, M. S. (1983). *Electron. Lett.* **19**, 789.
Fleming, R. M., McWhan, D. B., Gossard, A. C., Wiegmann, W., and Logan, R. A. (1980). *J. Appl. Phys.* **51**, 357.
Guldner, Y., Vieren, J. P., Voision, P., and Voos, M. (1982). *Appl. Phys. Lett.* **40**, 877.
Harris, J. J., Foxon, C. T., Barnham, K. W. J., Lacklison, D. E., Hewett, J., and White, C. (1987). *J. Appl. Phys.* **61**, 1219.

Heiblum, M. (1981). *Solid State Electron.* **28**, 343.
Heiblum, M., Mendez, E. E., and Stern, F. (1984). *Appl. Phys. Lett.* **44**, 1064.
Heiblum, M., Nathan, M. I., and Eizenberg, M. (1985). *Appl. Phys. Lett.* **47**, 503.
Henderson, T., Pearah, P., Morkoç, H., and Nilsson, B. (1984). *Electron. Lett.* **20**, 371.
Henderson, T., Aksun, M. I., Peng, C. K., Morkoç, H., Chao, P. C., Smith, P. M., Duh, K.-H. G., and Lester, L. F. (1986). *IEEE Electron Device Lett.* **EDL-7**, 649.
Hickmott, T. W., Solomon, P. M., Fischer, R., and Morkoç, H. (1985). *J. Appl. Phys.* **57**, 2844.
Hikosaka, K., Hirachi, Y., Mimura, T., and Abe, M. (1985). *IEEE Electron Device Lett.* **EDL-6**, 341.
Hirose, K., Ohata, K., Mizutani, T., Itoh, T., and Ogawa, M. (1986). *Inst. Phys. Conf. Ser. No. 79; Int. Symp. GaAs and Related Compounds, Karuizawa, Japan, 1985*, p. 529.
Hiyamizu, S. (1982). *Collected Papers MBE-CST-2, Tokyo*, p. 113.
Hiyamizu, S. (1986). *Surface Sci.* **170**, 727.
Hiyamizu, S., and Mimura, T. (1982a). *J. Crystal Growth* **56**, 455.
Hiyamizu, S., Mimura, T., Fujii, T., and Nanbu, K. (1980). *Appl. Phys. Lett.* **37**, 805.
Hiyamizu, S., Mimura, T., Fujii, T., Nanbu, K., and Hashimoto, H. (1981a). *Japan. J. Appl. Phys.* **20**, L245.
Hiyamizu, S., Nanbu, K., Mimura, T., Fujii, T., and Hashimoto, H. (1981b). *Japan. J. Appl. Phys.* **20**, L378.
Hiyamizu, S., Fujii, T., Mimura, T., Nanbu, K., Saito, J., and Hashimoto, H. (1981c). *Japan. J. Appl. Phys.* **20**, L455.
Hiyamizu, S., Mimura, T., and Ishikawa, T. (1982b). *Japan. J. Appl. Phys.* **21**, Supplement *21-1*, p. 161.
Hiyamizu, S., Saito, J., Nanbu, K., and Ishikawa, T. (1983). *Japan. J. Appl. Phys.* **22**, L609.
Hiyamizu, S., Saito, J., Kondo, K., Yamamoto, T., Ishikawa, T., and Sasa, S. (1985a). *J. Vac. Sci. Technol.* **B3**, 585.
Hiyamizu, S., Sasa, S., Ishikawa, T., Kondo, K., and Ishikawa, H. (1985b). *Japan. J. Appl. Phys.* **24**, L431.
Inoue, K., and Sakaki, H. (1984). *Japan. J. Appl. Phys.* **23**, L61.
Inoue, M., Nakade, Y., Shirafuji, J., and Inuishi, Y. (1972). *J. Phys. Soc. Japan.* **32**, 1010.
Inoue, M., Hiyamizu, S., Hida, H., Hashimoto, H., and Inuishi, Y. (1981), *J. de Phys.* **C7**, 19.
Inoue, M., Hiyamizu, S., Inayama, M., and Inuishi, Y. (1983a). *Japan. J. Appl. Phys.* **22**, Suppl. *22-1*, p. 357.
Inoue, M., Inayama, M., Hiyamizu, S., and Inuishi, Y. (1983b). *Japan. J. Appl. Phys.* **22**, L213.
Ishibashi, T., Tarucha, S., and Okamoto, H. (1982). *Japan. J. Appl. Phys.* **21**, L476.
Ishibashi, T., Yamauchi, Y., Nakajima, O., Nagata, K., and Ito, H. (1987). *IEEE Electron Device Lett.* **EDL-8**, 194.
Ishikawa, T., Hiyamizu, S., Mimura, T., Saito, J., and Hashimoto, H. (1981). *Japan. J. Appl. Phys.* **20**, L814.
Ishikawa, T., Saito, J., Sasa, S., and Hiyamizu, S. (1982). *Japan. J. Appl. Phys.* **21**, L675.
Ishikawa, T., Saito, J., Sasa, S., and Hiyamizu, S. (1984). *Extended Abstracts 16th Conf. Solid State Devices and Materials, Kobe*, p. 603.
Joshin, K., Mimura, T., Niori, M., Yamashita, Y., Kosemura, K., and Saito, J. (1983). *IEEE MTT-S Digest*, p. 563.
Kajii, K., Watanabe, Y., Suzuki, M., Hanyu, I., Kosugi, M., Odani, K., Mimura, T., and Abe, M. (1987). *Prog. IEEE Custom Integrated Circuit Conf., Portland*, p. 199.
Kasahara, J., and Watanabe, N. (1982). In "Proc. Semi-insulating III-V Materials, Evian, 1982" (M. Kram-Ebeid and B. Tuck, eds.), p. 238. Shiva, Nantwich, Cheshire, UK.

Ketterson, A., Moloney, M., Masselink, W. T., Klem, J., Fischer, R., Kopp, W., and Morkoç, H. (1985). *IEEE Electron Device Lett.* **EDL-6**, 628.
Kobayashi, N., and Fukui, T. (1984). *Electron. Lett.* **20**, 887.
Kodama, K., Hoshino, M., Kitahara, K., Takikawa, M., and Ozeki, M. (1986). *Japan. J. Appl. Phys.* **25**, L127.
Kroemer, H. (1959). *Proc. IRE*, Vol. 45, p. 1535.
Lang, D. V., Logan, R. A., and Jaros, M. (1979). *Phys. Rev.* **B19**, 1015.
Lee, H., Schaff, W. J., Wicks, G. W., Eastman, L. F., and Calawa, A. R. (1985). *Int. Symp.* GaAs *and Related Compounds, Biarritz, 1984; Inst. Phys. Conf. Ser. No.* 74, p. 321.
Matsui, Y., Hayashi, H., Kikuchi, K., Iguchi, S., and Yoshida, K. (1985). *J. Vac. Sci. Technol.* **B3**, 528.
Miller, R. C., Kleinman, D. A., and Gossard, A. C. (1984). *Phys. Rev.* **B29**, 7085.
Mimura, T., Hiyamizu, S., Fujii, T., and Nanbu, K. (1980). *Japan. J. Appl. Phys.* **19**, L225.
Mishra, U. K., Palmateer, S. S., Chao, P. C., Smith, P. M., and Hwang, J. C. M. (1985). *IEEE Electron Device Lett.* **EDL-6**, 142.
Mizutani, T., and Hirose, K. (1985). *Japan. J. Appl. Phys.* **24**, L119.
Mori, S., and Ando, T. (1980). *J. Phys. Soc. Japan* **48**, 865.
Morkoç, H., Henderson, T., Kopp, W., and Peng, C. W. (1986). *Electron. Lett.* **22**, 578.
Morkoç, H., Witkowski, L. C., Drummond, T. J., Stanchak, C. M., Cho, A. Y., and Streetman, B. G. (1980). *Electron. Lett.* **16**, 753.
Nakata, Y., Sasa, S., Sugiyama, Y., Fujii, T., and Hiyamizu, S. (1987). *Japan. J. Appl. Phys.* **26**, L59.
Okumura, H., Misawa, S., Yoshida, S., and Gonda, S. (1985). *Appl. Phys. Lett.* **46**, 377.
Onabe, K., Tashiro, Y., and Ide, Y. (1986). *Surface Sci.* **174**, 401.
Panish, M. B. (1987). *J. Crystal Growth* **81**, 249.
Petritz, R. L. (1958). *Phys. Rev.* **110**, 1254.
Razeghi, M., Maurel, P., Omnes, F., Armor, S. B., Dmowski, L., and Portal, J. C. (1986). *Appl. Phys. Lett.* **48**, 1267.
Rode, D. L. (1975). *In* "Semiconductors and Semimetals" (R. K. Willardson and A. C. Beer, eds.), Vol. 10, Chap. 1. Academic, New York.
Rosenberg, J. J., Benlamri, M., Kirchner, P. D., Woodall, J. M., and Pettit, G. D. (1985). *IEEE Electron. Device Lett.* **EDL-6**, 491.
Saito, J., Nanbu, K., Ishikawa, T., and Hiyamizu, S. (1983). *Japan. J. Appl. Phys.* **22**, L79.
Saito, J., Igarashi, T., Nakamura, T., Kondo, K., and Shibatomi, A. (1987). *J. Crystal Growth* **81**, 188.
Sakaki, H., Chang, L. L., Chang, C. A., and Esaki, L. (1977). *Bull. Am. Phys. Soc.* **22**, 460.
Sasa, S., Muto, S., Kondo, K., Ishikawa, H., and Hiyamizu, S. (1985a). *Japan. J. Appl. Phys.* **24**, L602.
Sasa, S., Saito, J., Nanbu, K., Ishikawa, T., Hiyamizu, S., and Inoue, M. (1985b). *Japan. J. Appl. Phys.* **24**, L281.
Schubert, E. F., and Ploog, K. (1984). *Phys. Rev.* **B30**, 7021.
Sheng, N. H., Lee, C. P., Chen, R. T., Miller, D. L., and Lee, S. J. (1985). *IEEE Electron Device Lett.* **EDL-6**, 307.
Stern, F., and Howard, W. E. (1967). *Phys. Rev.* **163**, 816.
Störmer, H. L., Dingle, R., Gossard, A. C., Wiegmann, W., and Sturge, M. D. (1979a). *Solid State Commun.* **29**, 705.
Störmer, H. L., Dingle, R., Gossard, A. C., Wiegmann, W., and Sturge, M. D. (1979b). *J. Vac. Sci. Technol.* **16**, 1517.
Störmer, H. L., Gossard, A. C., and Wiegmann, W. (1981a). *Appl. Phys. Lett.* **39**, 493.

Störmer, H. L., Gossard, A. C., Wiegmann, W., and Baldwin, K. (1981b). *Appl. Phys. Lett.* **39**, 912.
Störmer, H. L., Gossard, A. C., and Wiegmann, W. (1982). *Solid State Commun.* **41**, 707.
Störmer, H. L., Chang, A., Tsui, D. C., Hwang, J. C. M. H., Gossard, A. C., and Wiegmann, W. (1983). *Phys. Rev. Lett.* **50**, 1953.
Tachikawa, M., Mizuta, M., Kukimoto, H., and Minomura, S. (1985). *Japan. J. Appl. Phys.* **24**, L821.
Takikawa, M., Komeno, J., and Ozeki, M. (1983). *Appl. Phys. Lett.* **43**, 280.
Tatsuta, S., Inata, T., Okamura, S., and Hiyamizu, S. (1984). *Japan. J. Appl. Phys.* **23**, L147.
Tatsuta, S., Inata, T., Okamura, S., Muto, S., Hiyamizu, S., and Umebu, I. (1985). *Mat. Res. Soc. Symp. Proc., Vol. 37*, p. 23.
Tone, K., Nakayama, T., Iechi, H., Ohtsu, K., and Kukimoto, H. (1986). *Japan. J. Appl. Phys.* **25**, L429.
Tsang, W. T. (1979). *Appl. Phys. Lett.* **34**, 473.
Tsang, W. T. (1980). *Appl. Phys. Lett.* **36**, 11.
Tsang, W. T. (1987). *J. Crystal Growth* **81**, 261.
Tsui, D. C., Gossard, A. C., Kanimsky, G., and Wiegmann, W. (1981). *Appl. Phys. Lett.* **39**, 712.
Tsui, D. C., Störmer, H. L., and Gossard, A. C. (1982). *Phys. Rev. Lett.* **48**, 1559.
Valois, A. J., Robinson, G. Y., Lee, K., and Shur, M. S. (1983). *J. Vac. Sci. Technol.* **B1**, 190.
Walukiewicz, W., Ruda, H. E., Lagowski, J., and Gatos, H. C. (1984a). *Phys. Rev.* **B29**, 4818.
Walukiewicz, W., Ruda, H. E., Lagowski, J., and Gatos, H. C. (1984b). *Phys. Rev.* **B30**, 4571.
Watanabe, M. O., Yoshida, J., Mashita, M., Nakanishi, T., and Hojo, T. (1984). *Extended Abstracts, 16th Conf. Solid State Devices and Materials, Kobe,* p. 181.
Witkowski, L. C., Drummond, T. J., Barnett, S. A., Morkoç, H., Cho, A. Y., and Greene, J. E. (1981). *Electron. Lett.* **17**, 126.
Wolfe, C. M., Stillman, G. E., and Lindley, W. T. (1970). *J. Appl. Phys.* **41**, 3088.
Yamada, S., Taguchi, A., and Sugimura, A. (1985). *Appl. Phys. Lett.* **46**, 675.
Yokoyama, N. (1986). *Extended Abstracts, 18th Conf. Solid State Devices and Materials, Tokyo,* p. 347.
Yokoyama, N., Mimura, T., Fukuda, M., and Ishikawa, H. (1981). *IEEE ISSCC, Digest of Technical Papers,* p. 218.
Yokoyama, N., Imamuara, K., Oshima, T., Nishi, H., Muto, S., Kondo, K., and Hiyamizu, S. (1984a). *Japan. J. App. Phys.* **23**, L311.
Yokoyama, N., Imamura, K., Oshima, T., Nishi, H., Muto, S., Kondo, K., and Hiyamizu, S. (1984b). *IEDM '84, San Francisco,* p. 532.
Yokoyama, N., Imamura, K., Muto, S., Hiyamizu, S., and Nishi, H. (1985). *Japan. J. Appl. Phys.* **24**, L853.
Zhu, L. D., Sulewski, P. E., Chan, K. T., Muro, K., Ballantyne, J. M., and Sievers, A. J. (1985). *J. Appl. Phys.* **58**, 3145.
Zipparian, T. E., Dawson, L. R., Osbourn, G. C., and Fritz, I. J. (1983). *IEDM Tech. Dig.* 696.

CHAPTER 3

Metalorganic Vapor Phase Epitaxy for High-Quality Active Layers

T. Nakanisi

RESEARCH AND DEVELOPMENT CENTER
TOSHIBA CORPORATION
KAWASAKI, JAPAN

I.	INTRODUCTION.	105
II.	GROWTH AND CHARACTERIZATION OF HIGH-QUALITY GaAs AND GaAlAs.	106
	1. Effects of Starting Materials.	106
	2. Surface Morphology.	113
	3. Effects of Substrate Orientation.	114
	4. Growth Parameter Dependences.	115
	5. Residual Shallow Impurities and Deep Levels.	120
	6. Impurity Doping.	129
III.	FABRICATION OF DEVICE STRUCTURES.	134
	7. GaAs MESFETs and GaAs ICs.	135
	8. HEMTs.	144
	9. MESFETs Fabricated on GaAs/Si Substrate Structure.	148
IV.	SUMMARY AND FUTURE PROBLEMS.	150
	REFERENCES.	152

I. Introduction

It is now widely accepted that metalorganic vapor-phase epitaxy (MOVPE) is one of the most promising epitaxial techniques for providing active layers for compound semiconductor devices, because of the following potential advantages: (1) precise layer composition and thickness controllability, (2) capability of large area growth, (3) low autodoping and (4) growth on insulating substrates, as first pointed out by Manasevit (1968). In spite of all these advantages, the technique has long been behind the conventional epitaxy techniques such as liquid-phase epitaxy (LPE) and halogen transport vapor-phase epitaxy (VPE) with its extensive application to existing devices. For example, although fabrication of microwave metal-semiconductor field effect transistors (MESFETs) on MOVPE GaAs wafers was attempted by several laboratories (Bass, 1978; Duchemin *et al.*, 1978; Morkoç *et al.*, 1979), the microwave performance was not satisfactory as compared with the conventional epitaxy techniques. It appears that the main reason for this was the difficulty of growing high-quality active layers. As a result, extensive efforts have been made to elucidate what determines the layer quality. Presently the purity of active layers has become comparable to that obtained by the conventional epitaxy techniques. Moreover, by utilizing the inherent advantages of MOVPE, many complicated heterostructures such as GaAlAs/GaAs for two-dimensional electron gas (2DEG) and GaAs on Si have been developed.

It is the purpose of this chapter to describe the limiting factors that determine the quality of MOVPE active layers and to determine how widely MOVPE active layers are applied to high-speed and high-frequency devices, with specific emphasis on GaAs- and GaAlAs-based materials.

II. Growth and Characterization of High-Quality GaAs and GaAlAs

State-of-the-art growth apparatus for MOVPE is substantially the same as to that used by Manasevit and Simpson (1969). The vertical or horizontal reactor system consists of a quartz bell jar containing a graphite susceptor, a heater and a gas-supplying system. The susceptor usually is heated by RF induction. The group III organometallic source, such as trimethylgallium (TMG), which is a volatile liquid, is kept in a stainless steel container. It is transported into the reactor by bubbling with high-purity hydrogen gas. The group V source is a hydride such as arsine (AsH_3) gas, diluted with hydrogen. It is supplied from a high-pressure vessel. The group III and group V sources are diluted with high-purity hydrogen carrier gas and introduced into the reactor. All the gas-flow rates are precisely controlled by mass-flow controllers. The pressure inside the reactor is kept either at one atmospheric pressure (atmospheric-pressure MOVPE: APMOVPE) or below (low-pressure MOVPE: LPMOVPE). The substrate preparation and growth methods are similar to those given in the literature (Bass, 1978; Duchemin et al., 1978).

1. Effects of Starting Materials

It is known experimentally that the electrical properties of GaAs epilayers vary when organometallic sources from different vendors are used (Wang et al., 1974; Ito et al., 1973). Nakanisi et al. (1981) reported on the effects of TMG source purity on the quality of epilayers. Table I lists the TMG source materials used in the experiment. Four TMG sources, 1 to 4, were synthesized and supplied by Sumitomo Chemical Co., Ltd. Sources 3 and 4 were synthesized in better environments than the other two sources.

TABLE I

TMG SOURCE MATERIALS USED [a]

TMG No.	Synthesis	Remarks
1	Same procedure	
2	Same procedure	
3	Same procedure	Synthesized in purer environments
4	Same procedure	Synthesized in purer environments

[a] (Reprinted with permission from Elsevier Science Publishers, Nakani, T., Udagawa, T., and Tanaka, A. (1981) *J. Crystal Growth* **55**, 255–262.)

In order to compare the purity of the two groups, trace metallic impurities were analyzed by using flameless atomic absorption spectrophotometry. In these analyses, the conditions for decomposition of the TMG were carefully optimized to prevent loss of Si. In addition to this, a specific analytical technique was developed to eliminate broad Ga background absorption around the Si 251.6 nm line.

Table II compares the trace impurity contents between TMG 1 and TMG 3. A large amount of Si was observed in both TMG sources. The Si content is more than one order of magnitude higher than the other metallic impurities in both groups. It is also seen that the Si content in TMG 1 is more than three times higher than that in TMG 3. Table II also shows that a small amount of column II impurities, such as Cd, Mg, and Zn, are also contained in both TMG sources. The concentrations of these impurities are higher in TMG 1 than in TMG 3. These results suggest that, in fact, TMG sources 3 and 4 are purer than TMG sources 1 and 2.

In Fig. 1, the room-temperature carrier concentration of undoped epilayers is plotted as a function of the AsH_3-to-TMG molar ratio, $[AsH_3]/[TMG]$, for the four different TMG sources. These curves clearly show that the electrical properties of the layers are closely related to the TMG purity. When TMG sources 1 and 2 are used, the molar ratio at which the type of the electrical conduction changes from p to n is around 7, as shown by the arrows in Fig. 1, while it is more than two times higher when TMG sources 3 and 4 are used.

TABLE II

TRACE METALLIC IMPURITY CONTENTS (IN WT PPM) IN TMG 1 AND TMG 3 SOURCE MATERIALS ANALYZED BY FLAMELESS ATOMIC ABSORPTION SPECTROPHOTOMETRY[a]

TMG No.	Al	Ca	Cd	Cr	Fe	Mg	Mn	Si	Zn
3	NM[b]	<0.05	NM	<0.1	1.7	0.06	<0.05	11	0.1
1	0.1	<0.05	0.1	<0.1	2.0	0.10	<0.05	37	0.2

[a] From Nakanisi *et al.* (1981).
[b] NM = not measured.

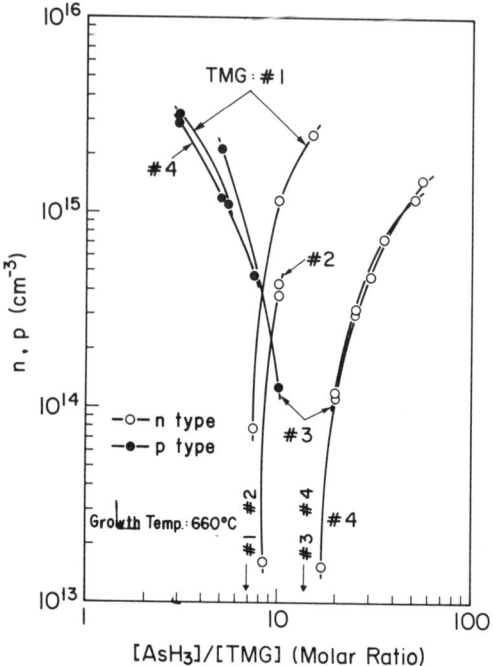

FIG. 1. Room-temperature carrier concentration of undoped epilayers as a function of AsH$_3$-to-TMG molar ratio, [AsH$_3$]/[TMG], for four different TMG sources (O, n = type; ●, p-type). [From Nakanisi et al. (1981). Copyright North-Holland Publ. Co., Amsterdam, 1981.]

In Fig. 2, the mobility at 77 K, μ_{77}, together with the room-temperature carrier concentration, is plotted as a function of [AsH$_3$]/[TMG]. It can be seen that the mobility, as well as the carrier concentration, is influenced by the TMG purity. The mobility values both for electrons and for holes decrease in the order TMG 4, 3, and 1, in consistency with the results of trace impurity analyses shown in Table II—i.e., higher mobility results from the lower impurity content. The results shown in Figs. 1 and 2 indicate that the electrical properties are extrinsic in nature and are mainly determined by the TMG purity. In Fig. 3 are plotted electron concentration n, donor and acceptor densities N_D and N_A, and total ionized impurity density $N_D + N_A$ as a function of [AsH$_3$]/[TMG]. The N_D and N_A were derived from the observed μ_{77} and n, using the Brooks–Herring formula (Blatt, 1957) and the relation $N_D - N_A = n$, after the analysis by Ikoma (1968). It can be seen that, by using TMG 4, $N_D + N_A$ is reduced by a factor of about 4, as compared with the case where TMG 1 is used. This figure shows that both N_D and N_A increase with an increase in [AsH$_3$]/[TMG]. From the molar ratio

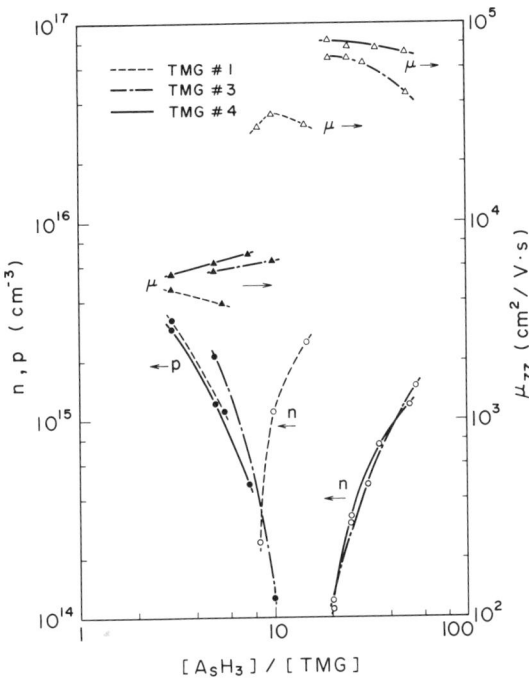

FIG. 2. Mobility at 77 K, μ_{77}, and room-temperature carrier concentration as a function of [AsH$_3$]/[TMG] for different TMG sources: (- - -) TMG 1, (— · —) TMG 3, (———) TMG 4. [From Nakanisi et al. (1981). Copyright North-Holland Publ. Co., Amsterdam, 1981.]

dependence, it may be concluded that the donors are column IV donors. As has been shown in Table II, the main impurity in the TMG sources is Si. This favors the view that the main donor impurity in the n-type region is Si. It is not clear at present in what form Si exists in the TMG sources.

Figure 3 indicates that N_A increases with an increase in [AsH$_3$]/[TMG] but does so more gradually than N_D. This trend seems to conflict with the previous finding, i.e., the main residual acceptor in the p-type region is carbon (Nakanisi and Tanaka, 1980). If carbon is still the main acceptor in the present n-type layers, N_A should decrease with an increase in [AsH$_3$]/[TMG]. To find out possible explanations for the molar ratio dependence, photoluminescence measurements of these layers were carried out at 4.2 K. It was found that, in addition to the intense donor–acceptor pair emission involving a shallow carbon acceptor that has been identified by White et al. (1973), weak emission due to Zn acceptors (Ozeki et al., 1973) was observed in some epilayers. No other pair emissions were observed. Accordingly, it could be that the gradual increase of N_A reflects the increase of the column

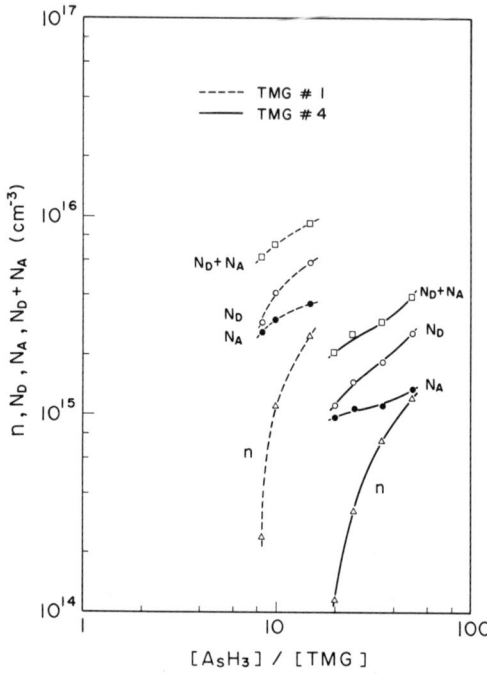

FIG. 3. Electron concentration n, donor and acceptor densities N_D and N_A and total ionized impurity density, $N_D + N_A$, vs. [AsH$_3$]/[TMG] for TMG 1 (dashed line) and 4 (solid line). [From Nakanisi et al. (1981). Copyright North-Holland Publ. Co., Amsterdam, 1981.]

II impurity densities and the decrease of carbon acceptor density. The N_A value in TMG 1 is higher than that in TMG 4, in good agreement with the results of impurity analyses in Table II.

In order to estimate effects of the AsH$_3$ gas source on the electrical properties of epilayers, different AsH$_3$ gas sources with different dates of supply were used. No meaningful change in electrical properties were found. However, Dapkus et al. (1981) studied the effects of both TMG and AsH$_3$ on the epilayer purity and found that the variation of AsH$_3$ quality from bottle to bottle also influenced the purity of the epilayers.

In the case of GaAlAs, the lot-to-lot variation of the AsH$_3$ quality has stronger effects on the electrical and optical properties of epilayers than in GaAs. Figure 4 illustrates the influence of AsH$_3$ cylinder variation on the optical quality of Ga$_{1-x}$Al$_x$As epilayers (Nakanisi, 1984). Double heterostructures (DHs) were grown under identical growth conditions, except for the use of different AsH$_3$ cylinders. Simple DH light-emitting diodes (LEDs) were fabricated and their EL intensity was measured under constant current drive. In spite of low ($x \sim 0.1$) Al composition in the active layers, the EL

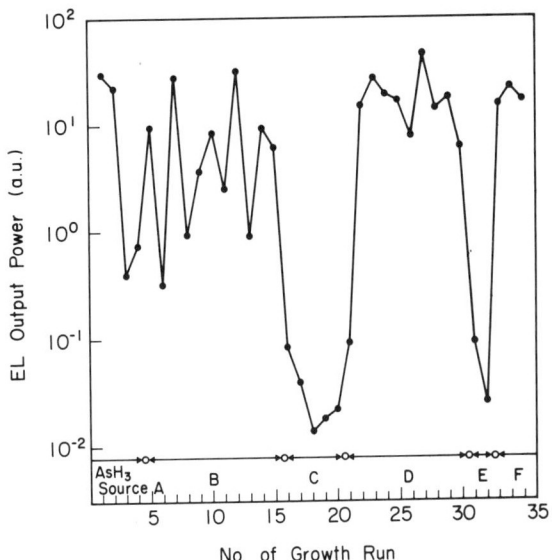

FIG. 4. EL output power variation with AsH_3 cylinders A to F. [From Nakanisi (1984). Copyright North-Holland Publ. Co., Amsterdam, 1984.]

intensity varies as much as four orders of magnitude. It must be noted that all these AsH_3 cylinders contain AsH_3 gas (10% in H_2) prepared by mixing the same 100% AsH_3 source and high-purity H_2 gas. This clearly indicates that the improper preparation of AsH_3 cylinders is the main origin of large EL intensity variation. In accordance with the decrease of EL intensity, carrier concentration of active and cladding layer ($x \sim 0.4$) decreases, indicating the occurrence of compensation. Similar phenomena are usually observed when air leaks exist in the reactor systems.

It has been shown that the degradation of electrical and optical properties is caused by the presence of $O_2 + H_2O$ in the AsH_3 source (Wagner et al., 1981: Thrush and Whiteaway, 1981). Attempts have been made to reduce such contamination by adopting heated graphite buffles (Stringfellow and Hom, 1979) and Ga-In-Al alloy mixtures (Sealy et al., 1983). Taking these findings into consideration, it is evident that one main source of the AsH_3 lot-to-lot variation is the $H_2O + O_2$ adsorbed and/or possibly chemisorbed on the inner wall of the AsH_3 cylinder, whose surface preparation varies from cylinder to cylinder. The same situation may also hold for the cylinders containing doping sources such as H_2Se. Therefore, in order to obtain high-purity GaAs and GaAlAs epilayers, care should be taken in the preparation of starting-material gas cylinders, as well as in improving the purity of 100% concentrated starting materials.

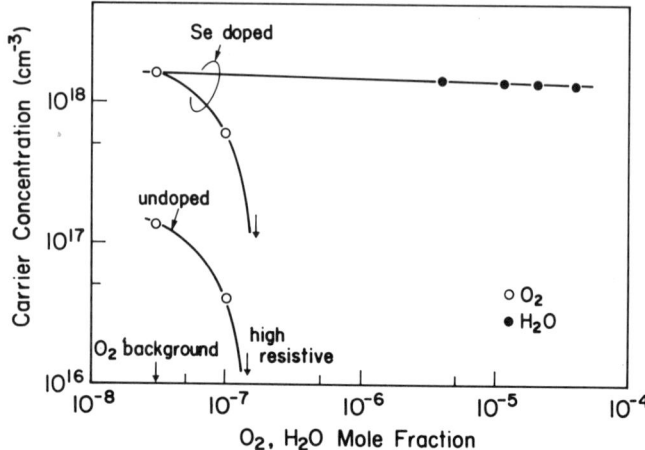

FIG. 5. Variation of carrier concentration in GaAlAs epilayers as a function of introduced O_2 (○) or H_2O (●) mole fraction in the vapor phase during growth. [From Terao and Sunakawa et al. (1984). Copyright North-Holland Publ. Co., Amsterdam, 1984.]

Recently, Terao and Sunakawa (1984) clearly demonstrated that the role of O_2 in GaAlAs MOVPE is distinctly different from that of H_2O. The rapid decrease in the carrier concentration for O_2 introduction above 0.1 ppm into the reactor is observed not only in undoped layers, but also in highly Se-doped (1.5×10^{18} cm^{-3}) ones, while H_2O addition has little effect on the Al composition (Fig. 5). Figure 6 shows the Al composition in GaAlAs as a function of introduced O_2 or H_2O mole fraction. It is evident that O_2 addition

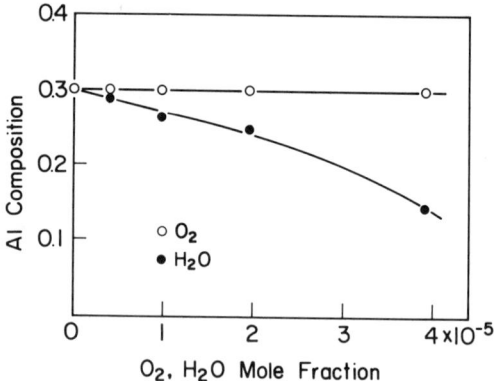

FIG. 6. Al composition in GaAlAs as a function of introduced O_2 (○) or H_2O (●) mole fraction. [From Terao and Sunakawa (1984). Copyright North-Holland Publ. Co., Amsterdam, 1984.]

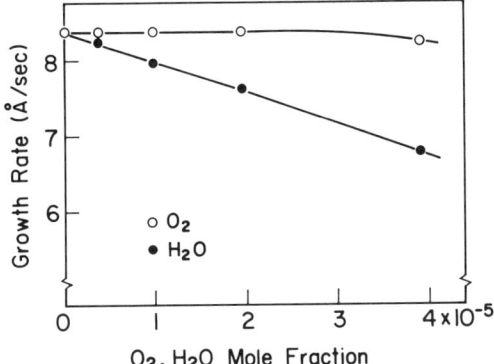

FIG. 7. Growth rate of GaAlAs as a function of introduced O_2 (○) or H_2O (●) mole fraction. [From Terao and Sunakawa (1984). Copyright North-Holland Publ. Co., Amsterdam, 1984.]

has little or no influence on Al composition. By contrast, H_2O introduction reduces the Al composition. In accordance with the decrease in Al composition of GaAlAs, the growth rate decreases with increasing H_2O mole fraction but not with O_2 (Fig. 7). From these findings, Terao and Sunakawa suggest that there exist different reactions of trimethyl aluminum (TMA) with O_2 and H_2O in the vapor phase prior to growth.

2. Surface Morphology

It is well known experimentally that GaAs epilayer surface morphology varies with substrate orientation (for example, Gottschalch et al., 1974; Nakanisi, 1984). In the case of growth on an exactly or nearly ⟨100⟩-oriented GaAs surface, a mirror-like surface as viewed by the naked eye can be obtained over a wide growth temperature range (640–780°C). However, the degree of microscopic surface smoothness depends upon many parameters, such as growth rate, total gas flow rate, the concentration of both TMG and AsH_3, etc. The surface appearance of $Ga_{1-x}Al_xAs$ on a ⟨100⟩-oriented GaAs substrate is similar to that of GaAs, but it is more difficult to attain a perfectly smooth surface. Figure 8 (Nakanisi, 1984) shows an example of the variation of $Ga_{1-x}Al_xAs$ ($x = 0.45$) surface morphology with growth temperature, together with surface roughness profiles. At 750°C, the surface appears to be specular to the naked eye; however, under high magnification with an optical microscope, a unique pattern is observed. The surface roughness profile indicates an undulation of 200 Å. At 750°C the surface pattern disappears. At 800°C the surface again tends to roughen. The results indicate that smooth and structure-free surfaces can be obtained in a relatively narrow temperature range. The reduction of the surface roughness

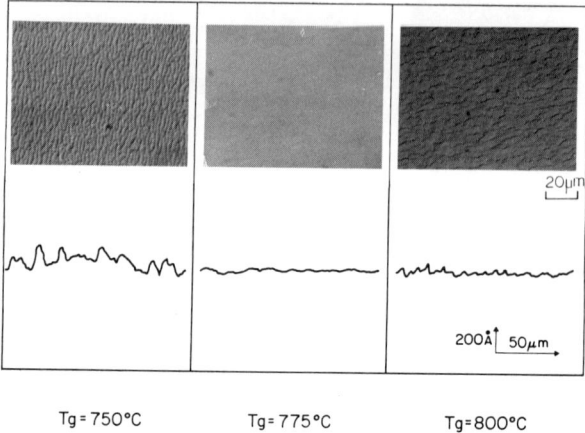

FIG. 8. Photomicrographs of $Ga_{0.55}Al_{0.45}As$ surfaces and roughness profiles grown on $\langle 100 \rangle$ GaAs at different temperatures. [From Nakanisi (1984). Copyright North-Holland Publ. Co., Amsterdam, 1984.]

is very important in fabrication of very thin heterostructures—for instance, high-electron-mobility transistors (HEMTs) and DH laser diodes (LDs)—in which active layers only several hundred Å thick are deposited successively.

3. Effects of Substrate Orientation

Unlike the large substrate-orientation dependence of background carrier concentration in the case of VPE-grown GaAs epilayers (Williams, 1964; DiLorenzo and Machala, 1971), a weaker orientation dependence has been observed in the MOVPE system. In Table III are shown the electrical properties and the growth rates of undoped GaAs MOVPE layers

TABLE III

ORIENTATION DEPENDENCE OF ELECTRICAL PROPERTIES[a]

	Orientation	Type	n or p (cm^{-3})	(77 K) (cm^2/Vsec)	$N_D + N_A$ (cm^{-3})	$N_A N_D$	Growth rate (μm/h)
[AsH$_3$]/[TMG] = 50	$\langle 100 \rangle$	n	2.0×10^{15}	29,000	1.0×10^{16}	0.68	7.8 ± 0.5
	$\langle 110 \rangle$	n	3.8×10^{14}	67,000	3.4×10^{15}	0.76	7.8 ± 0.5
	$\langle 311 \rangle_A$	n	5.3×10^{14}	50,000	4.6×10^{15}	0.80	7.8 ± 0.5
	$\langle 311 \rangle_B$	n	8.3×10^{15}	14,000	2.6×10^{16}	0.64	8.3 ± 0.5
	$\langle 111 \rangle_A$	n	1.0×10^{15}	31,000	7.5×10^{15}	0.86	7.5 ± 0.5
	$\langle 111 \rangle_B$	n?[b]	—	—	—	—	(rough surface)

[a] From Nakanisi (1984).
[b] Estimated from PL measurement at 4.2 K.

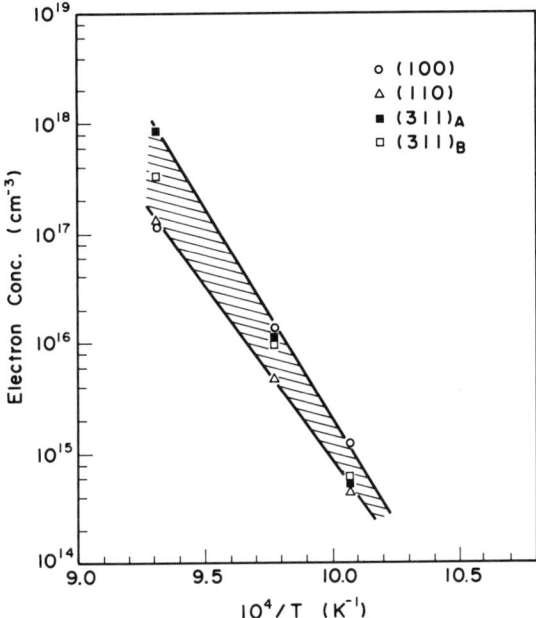

FIG. 9. Growth-temperature dependence of electron concentration for GaAs epilayers grown on differently oriented substrates: (○) ⟨100⟩, (△) ⟨110⟩, (■) ⟨311⟩$_A$, (□) ⟨311⟩$_B$.

simultaneously grown at 720°C. In contrast to the marked difference in surface appearance (Nakanisi, 1984), the carrier concentration $N_D - N_A$ and total ionized impurity density $N_D + N_A$ changed by less than an order of magnitude. The same trend still holds when the growth temperature is varied from 660 to 800°C (Fig. 9). In this temperature range, the growth rate decreases slightly with an increase of growth temperature. Figure 10 shows that slight substrate misorientation influences the electrical properties of the layers. Although the growth rates are all the same within experimental error, the electron concentration decreases with an increase in degree of misorientation. All these results indicate that the substrate orientation effects are very complicated and further study is needed to clarify them.

4. Growth Parameter Dependences

The electrical properties of GaAs epilayers vary with [AsH$_3$]/[TMG] (Ito et al., 1973; Seki et al., 1975; André et al., 1977; Dapkus et al., 1981; Nakanisi et al., 1981), as typically illustrated in Fig. 11. In Fig. 12 are shown the results for Ga$_{1-x}$Al$_x$As when carrier concentration is plotted as a function of [AsH$_3$]/([TMG] + [TMA]) (Mori et al., 1981). Here TMA means

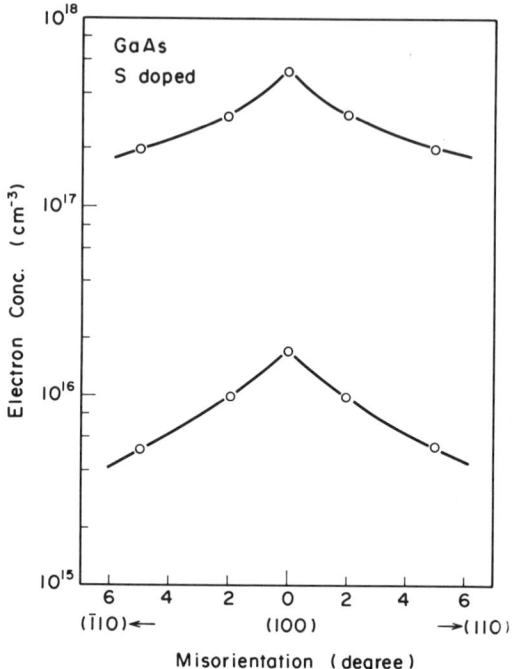

FIG. 10. Electron concentration vs. misorientation. [From Nakanisi (1984). Copyright North-Holland Publ. Co., Amsterdam, 1984.]

trimethylaluminum. Increase of acceptor concentration with an increase of x results in a shift toward a higher value of the critical molar ratio at which the p- to n-type conversion occurs. Figure 13 shows the temperature dependence of n, N_D, N_A, and $N_D + N_A$ upon growth temperature (Nakanisi et al., 1981). In the temperature range shown in this figure, n decreases and μ_{77} increases on lowering the growth temperature, indicating a decrease of both N_D and N_A. The same trend has been reported by Dapkus and coworkers (1981) in their LPMOVPE experiments. When the pressure inside the reactor is reduced to 0.1 atm., the p- to n-type conversion molar ratio becomes larger (Kobayashi et al., 1982; Takagishi, 1983).

Using high-purity TMG and AsH_3 sources and optimizing the growth parameters such as $[AsH_3]/[TMG]$ and growth temperature, attempts were made to grow high-purity GaAs epilayers in a wide electron concentration range (Nakanisi et al., 1981). Figure 14 shows the room-temperature mobility μ versus electron concentration characteristics. In the figure, open circles represent undoped layers, while solid circles represent sulfur-doped layers. Solid lines are those calculated by Rode and Knight (1971) for different

3. MOVPE FOR HIGH-QUALITY ACTIVE LAYERS

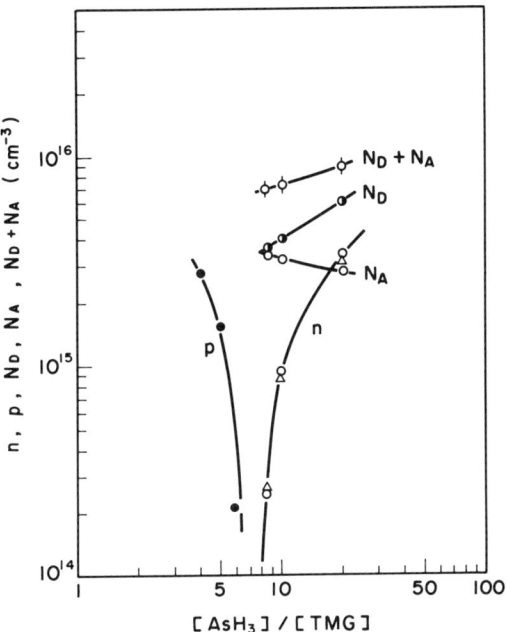

FIG. 11. Electron and hole concentrations, n and p, N_D, N_A, and $N_D + N_A$ vs. [AsH$_3$]/[TMG]. [From Nakanisi (1984). Copyright North-Holland Publ. Co., Amsterdam, 1984.]

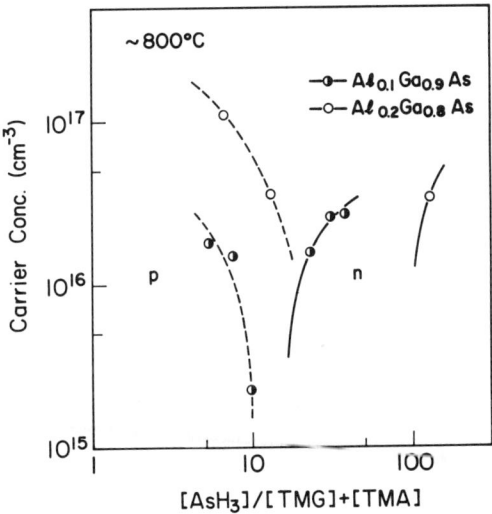

FIG. 12. Carrier concentration vs. [AsH$_3$]/([TMG] + [TMA]) relationships for GaAlAs. [From Mori et al. (1981).]

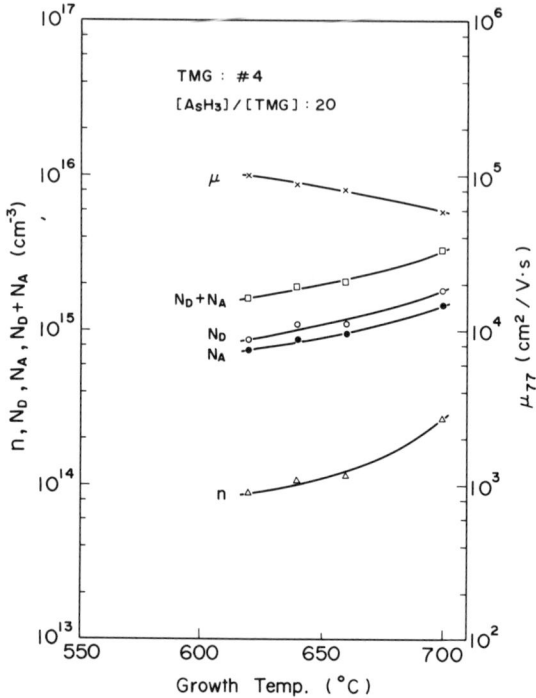

FIG. 13. Dependences of n, μ_{77}, N_D, N_A, and $N_D + N_A$ upon growth temperature. [From Nakanisi et al. (1981). Copyright North-Holland Publ. Co., Amsterdam, 1981.]

composition ratios. Typical mobility values are 8000, 7200, 6000, and 4500 cm²/Vsec at electron concentrations of 10^{14}, 10^{15}, 10^{16}, and 10^{17} cm^{-3}, respectively. Figure 15 shows the 77 K mobility, μ_{77}, versus n characteristics. At an electron concentration of around 10^{15} cm^{-3} and above, the mobility values of sulfur-doped layers are comparable to those of undoped layers. These mobilities were attained by using a high-purity H_2S doping source of four-nines purity. The use of commercial H_2S source of two-nines purity resulted in the growth of layers of lower mobility values. Even at an electron concentration of 10^{14} cm^{-3}, the mobility stays at low values because of heavy compensation. In Table IV are shown the highest mobilities obtained by using different TMG sources, together with other electrical properties. The TMG source number is the same as is given in Table I. The layer grown by using TMG 4 gives the best mobility value of 139,000 cm²/Vsec at 77 K with an electron concentration of 3.7×10^{14} cm^{-3}. The total ionized impurity density $N_D + N_A$ in the layer is more than three times lower than those of layers using TMG sources 1 and 2.

3. MOVPE FOR HIGH-QUALITY ACTIVE LAYERS

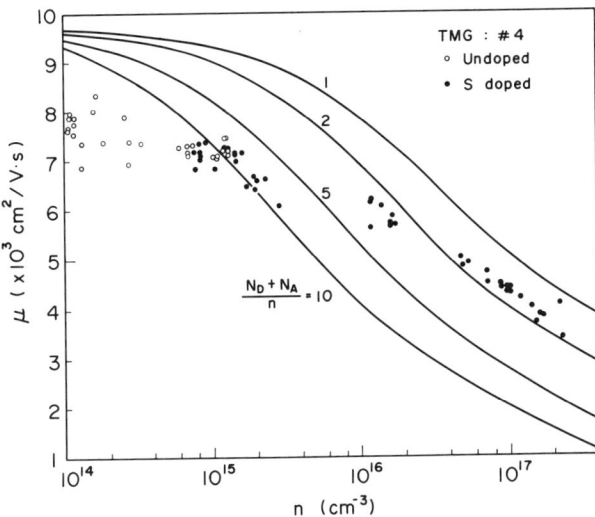

FIG. 14. Room-temperature mobility μ vs. electron concentration n for (○) undoped and (●) S-doped layers. [From Nakanisi et al. (1981). Copyright North-Holland Publ. Co., Amsterdam, 1981.]

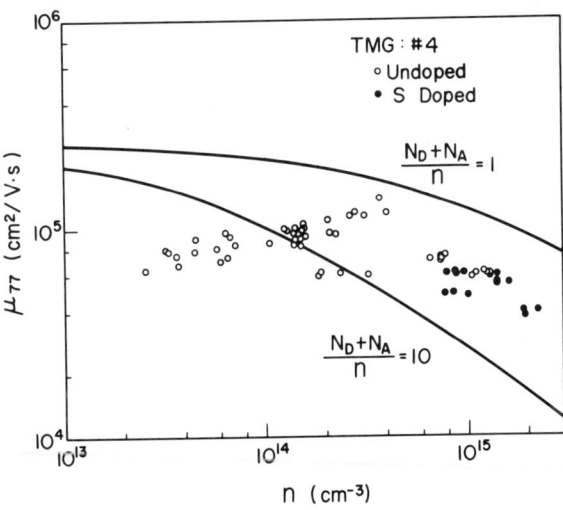

FIG. 15. Mobility at 77 K, μ_{77}, as a function of n for (○) undoped and (●) S-doped layers. [From Nakanisi et al. (1981). Copyright North-Holland Publ. Co., Amsterdam, 1981.]

TABLE IV

THE BEST MOBILITY AT 77 K AND RELATED ELECTRICAL PROPERTIES OBTAINED BY USING EACH TMG SOURCE

TMG	n (cm^{-3})	μ_{77} (cm^2/Vsec)	N_D (cm^{-3})	N_A (cm^{-3})	$N_D + N_A$ (cm^{-3})
4	3.69×10^{14}	139,000	0.81×10^{15}	0.44×10^{15}	1.26×10^{15}
3	1.92×10^{14}	99,000	1.03×10^{15}	0.75×10^{15}	1.78×10^{15}
2	4.80×10^{14}	42,000	2.86×10^{15}	2.26×10^{15}	5.08×19^{15}
1	1.64×10^{15}	44,000	3.75×10^{15}	2.19×10^{15}	5.90×10^{15}

Since the early stages of MOVPE development, it has been noticed experimentally that high-purity epilayers can be grown more easily by using triethylgallium (TEG) than by using TMG (Seki et al., 1975). Maeda (1986) has shown that when TMG and TEG are cracked in an Ar atmosphere, the residential carbon concentration in precipitated Ga is 10 to 100 times lower for TEG. This may indicate that carbon contamination is less serious in the TEG/AsH$_3$ system than in the TMG/AsH$_3$ system. Bhat et al. (1981) reported as well that there were no carbon acceptors in the epilayers grown from the TEG/AsH$_3$ system, as indicated by the measurements of photoluminescence at 4.2 K.

Table V lists the historical progress on high-mobility layers reported to date. The highest mobility has become comparable to that obtained in the conventional LPE and VPE epilayers.

5. RESIDUAL SHALLOW IMPURITIES AND DEEP LEVELS

a. Shallow Donors and Acceptors

Figure 16 illustrates how electrical properties such as carrier concentration, $N_D - N_A$, total ionized impurity density $N_D + N_A$ and compensation ratio

TABLE V

IMPROVEMENT OF THE PURITY OF GaAs EPILAYERS

Growth system	n (cm^{-3})	μ_{77} (cm^2/Vsec)	References
TMG/AsH$_3$(AP)	1.1×10^{16}	15,400	Ito et al. (1973)
TMG/AsH$_3$(AP)	4.1×10^{14}	41,520	Bass (1975)
TEG/AsH$_3$(AP)	1×10^{14}	120,000	Seki et al. (1975)
TMG/AsH$_3$(AP)	3×10^{14}	87,000	Manasevit et al. (1980)
TMG/AsH$_3$(AP)	3.7×10^{14}	139,000	Nakanisi et al. (1981)
TMG/AsH$_3$(LP)	5×10^{14}	125,000	Dapkus et al. (1981)
TMG/AsH$_3$(AP)	—	154,000	Maeda (1986)
TEG/AsH$_3$(LP)	—	185,000[a]	Norris et al. (1985)

[a] Under illumination.

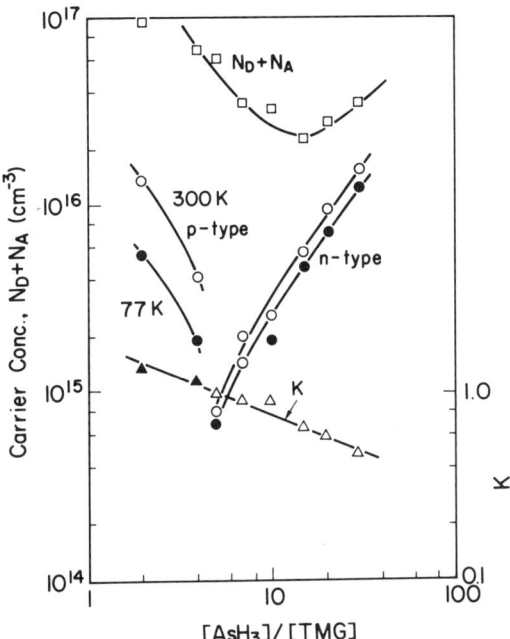

FIG. 16. Carrier concentration, $N_D + N_A$ and K ($=N_A/N_D$) as a function of [AsH$_3$]/[TMG]. [From Nakanisi (1984). Copyright North-Holland Publ. Co., Amsterdam, 1984.]

K change with the molar ratio [AsH$_3$]/[TMG]. Here [TMG] is kept constant. The small difference of carrier concentration at the two measured temperatures in both p- and n-type layers suggests that most of the donors and acceptors are shallow ones. The total ionized impurity density $N_D + N_A$ reaches a minimum at [AsH$_3$]/[TMG] ≈ 10. A log–log plot of the compensation ratio K ($=N_A/N_D$) and [AsH$_3$]/[TMG] gives a straight line. In Fig. 17 are shown the carrier concentration versus [AsH$_3$]/[TMG] relationship at two growth temperatures. The p- to n-type conversion curve is affected by the growth temperature.

The photoluminescence measurement at 4.2 K is a useful tool for obtaining further information about the donors and acceptors in high-purity GaAs metalorganic chemical-vapor deposited (MOCVD) layers. For the purpose of comparing the results with those on MOVPE epilayers, the luminescence spectra of undoped layers grown by the conventional Ga/AsCl$_3$/H$_2$ VPE system are shown in Fig. 18 (Nakanisi and Kasiwagi, 1974). The intense emission line A at around 8200 Å is due to excitons bound to neutral donors, whose donor ionization energy E_D is 5.9 meV (Rossi et al., 1970). A series of less intense lines B_1 to B_4 appear in the wavelength range from 8300 to

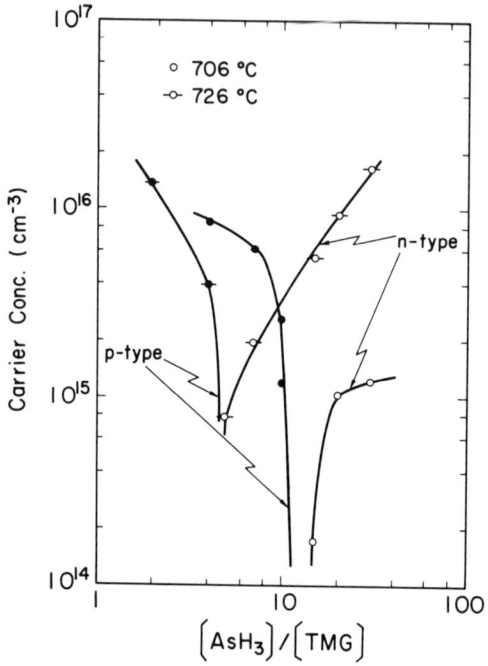

FIG. 17. Relation between carrier concentration and [AsH$_3$]/[TMG] for GaAs epilayers grown at 706°C (plain circles) and 726°C (dashed circles). [From Nakanisi (1984). Copyright North-Holland Publ. Co., Amsterdam, 1984.]

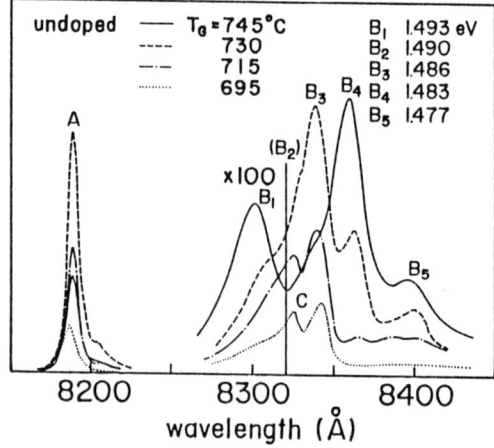

FIG. 18. 4.2 K photoluminescence (PL) spectra of GaAs epilayers grown by the conventional AsCl$_3$/Ga/H$_2$ system (———) 745°C; (– – –) 730°C; (— · —) 715°C; (······) 695°C. [From Nakanisi and Kasiwagi (1974).]

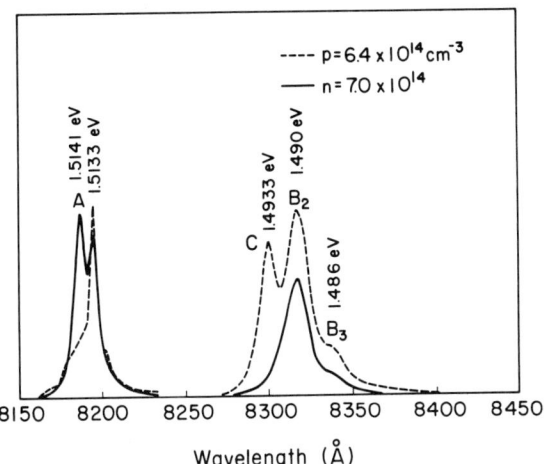

FIG. 19. PL spectra at 4.2 K of p-type (dashed line; 6.4×10^{14} cm^{-3}) and n-type (solid line; 7.0×10^{14} cm^{-3}) undoped epilayers.

8400 Å. In most samples grown by the conventional VPE method, the lines B_3 and B_4 are more intense among the four lines. Figure 19 shows the photoluminescence spectra obtained in both p- and n-type MOVPE layers. The line C centered at 1.4933 eV, which is observed only in a p-type sample, is the same as that reported to be due to the conduction band-to-acceptor transition, with acceptor ionization energy E_A being 27.1 meV (White et al., 1973). The line has been identified as due to carbon acceptors by White et al. and Ozeki et al. (1976).

In contrast with the VPE epilayers, only the line B_2 is dominant among the lines B_1–B_4. The line B_2 centered at around 1.490 eV moves towards shorter wavelengths as the donor concentration N_D increases, as shown in Fig. 20. Here, the N_D was derived from the 77 K mobility and the electron concentration by using the Brooks–Herring formula (Blatt, 1957). Also it was found that the line moved toward shorter wavelengths as optical excitation was intensified. The behavior indicates that the line B_2 arises from donor–acceptor (D–A) pair recombination. The peak energy E_p of the D–A pair emission line can be approximated by the following equation:

$$E_p = E_g - (E_D + E_A) + e^2(\pi N_D^*)^{1/3}/\varepsilon, \tag{1}$$

where N_D^* denotes the donor concentration contributing to the D–A pair emission. E_g is the band gap and E_D and E_A are the donor and acceptor ionization energies, respectively. ε is the dielectric constant. The peak energy shift $\Delta E_p = E_p - E_g + (E_D + E_A)$ is then proportional to the cubic root of N_D^*. In Fig. 21, the peak shift of the line B_2 is plotted as a function of N_D

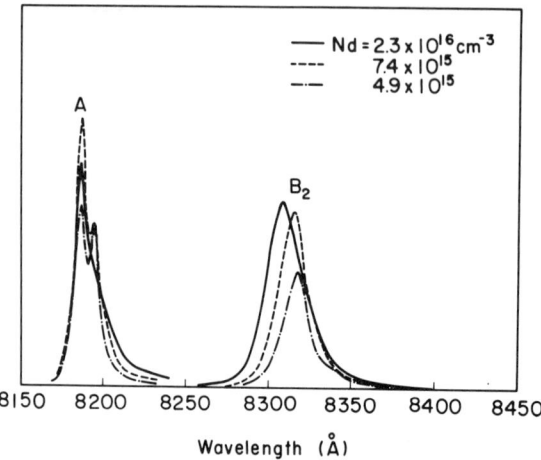

FIG. 20. PL spectra at 4.2 K of *n*-type epilayers. Note that the line B_2 moves towards shorter wavelengths with an increase of N_D derived from electrical measurements. The solid line indicates an N_D of 2.3×10^{16} cm^{-3}; the dashed line, of 7.4×10^{15} cm^{-3}; and the broken line, of 4.9×10^{15} cm^{-3}.

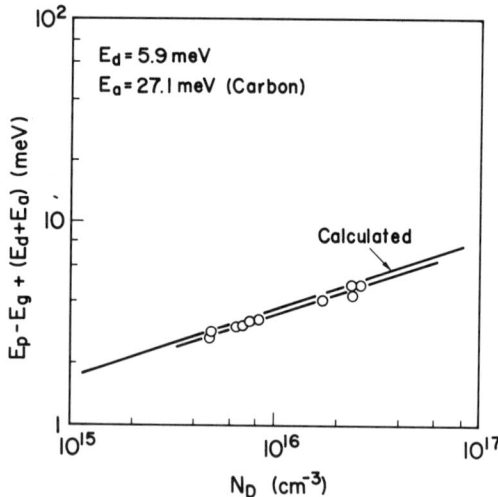

FIG. 21. Peak energy shift of line B_2, $E_p - E_g + (E_D + E_A)$ as a function of donor density derived from electrical measurements.

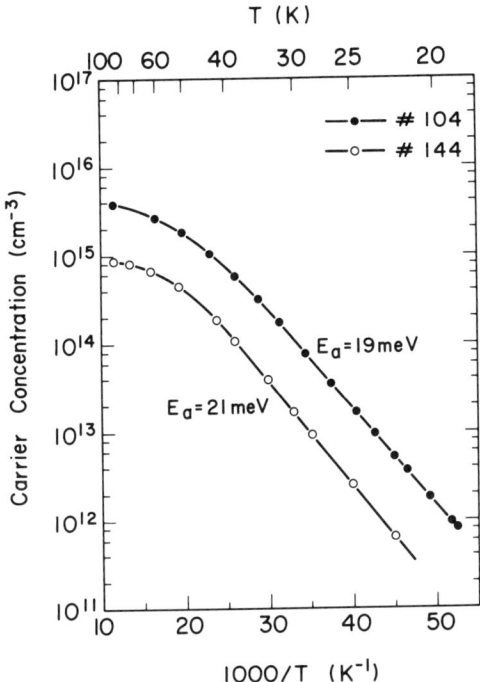

FIG. 22. Temperature dependence of hole concentration for two p-type GaAs epilayers.

derived from electrical measurements. The peak shift agrees very well with the calculation that assumes $E_D = 5.9$ meV and $E_A = 27.1$ meV. It is thus concluded that the acceptors contributing to the line B_2 are carbon (C) acceptors, with E_A being 27.1 meV. Furthermore, most of the donors are shallow ones, because the N_D derived from electrical measurements is nearly equal to N_D^*, whose donor ionization energy E_D is 5.9 meV.

Figure 22 shows the temperature dependence of hole concentration in two p-type samples. Circles indicate experimental data. Lines were obtained from curve-fitting calculations, assuming that one acceptor level was present and the acceptors were partly compensated by donors. From the curve-fitting procedure, the acceptor ionization energies were estimated to be 19 and 21 meV for each sample. These energies are close to the optical ionization energy of carbon acceptors. Thus it is concluded that most of the acceptors in p-type samples are carbon. Since the line B_2 is also intense in n-type samples, it may be reasonable to consider that carbon is also a main acceptor in n-type samples. From the results it may be postulated that the main residual donors in undoped GaAs are shallow ones with E_D being 5.7 meV, and that the dominant acceptors are carbon acceptors with E_A being 27.1 meV.

Because of the very small chemical shift, it is difficult to distinguish between different donor species from photoluminescence measurements. However, the compensation ratio K versus [AsH$_3$] relationship (Fig. 16) indicates that there is a strong correlation between the donor concentration and the carbon acceptor concentration. This may suggest that carbon acts as an amphoteric impurity in GaAs as well as in Ge and Si. On the basis of a simple equilibrium model where carbon atoms are incorporated from the vapor phase and carbon donors and acceptors occupy Ga and As sites, respectively, the following set of equations are taken into consideration:

$$e^- + h^+ = 0, \quad n_0 p_0 = n_i^2; \tag{2}$$

$$V_{Ga}^0 + V_{As}^0 = 0, \quad [V_{Ga}] = m_1 P_{As_m}^{-1/m} \quad (m = 2, 4); \tag{3}$$

$$(1/m)As_m = As_{As}^0 + V_{Ga}^0, \quad [V_{As}] = m_2 P_{As_m}^{1/m}; \tag{4}$$

$$C(g) = C_{Ga}^+ + e^- + V_{As}^0, \quad n_0[C_{Ga}^+][V_{As}^0]/[C(g)] = k_1; \tag{5}$$

$$C(g) = C_{As}^- + h^+ + V_{Ga}, \quad p_0[C_{As}^-][V_{Ga}^0]/[C(g)] = k_2; \tag{6}$$

$$n_0 + C_{As}^- = p_0 + C_{Ga}^+ \quad \text{(charge neutrality)}, \tag{7}$$

These equations can be solved analytically by approximating the charge neutrality condition. At low growth temperature and relatively high impurity concentration where Eq. (7) is approximated by $[C_{As}^-] \approx [C_{Ga}^+]$, electron and hole concentration at room temperature, n and p, and compensation ratio $K (= N_A/N_D)$ are given by:

$$n = K_1[C(g)]^{1/2} P_{As_m}^{1/2m} - K_2[C(g)]^{3/2} P_{As_m}^{-1/2m}, \tag{8}$$

$$p = K_3[C(g)]^{1/2} P_{As_m}^{-1/2m} - K_4[C(g)]^{3/2} P_{As_m}^{1/2m}, \tag{9}$$

$$K = N_A/N_D = K_5[C(g)] P_{As_m}^{-1/m}, \tag{10}$$

where K_i ($i = 1$–5) are temperature-dependent constants. When the growth temperature is high enough and impurity concentration is low, Eq. (7) can be approximated by $n_0 \approx p_0$; then n, p, and K are given by:

$$n = [C(g)]\{K_1' P_{As_m}^{1/m} - K_2' P_{As_m}^{-1/m}\}, \tag{11}$$

$$p = [C(g)]\{K_2' P_{As_m}^{-1/m} - K_1' P_{As_m}^{1/m}\}, \tag{12}$$

$$K = (m_1/m_2) P_{As_m}^{-2/m}, \tag{13}$$

where the K_i' ($i = 1, 2$) are temperature-dependent constants. In both cases, a simple equilibrium model will explain the observed [AsH$_3$]/[TMG] dependences of n, p, and K shown in Fig. 16. The model also explains that

$N_D + N_A$ reaches a minimum at around the $p-n$ conversion molar ratio. Furthermore, the model can explain how the $p-n$ conversion curve is dependent upon growth temperature (Fig. 17). From these results, it may be reasonably concluded that the dominant residual impurity in undoped GaAs is amphoteric carbon.

However, it has not been conclusively determined by any direct measurements that carbon acts as a shallow donor. Several workers attempted to identify the dominant donor species in undoped GaAs by using far-infrared photoconductivity measurements at liquid-He temperatures. However, their conclusions disagree with each other. Some workers identified the dominant photoconductivity peak as due to carbon donors (Dapkus et al., 1981; Wolfe and Stillman, 1970; Wolfe et al., 1976; Cooke et al., 1978), while other workers ascribed the same peak to Ge (Dazai et al., 1974; Asfar et al., 1980). From far-infrared laser magnetooptics with intrinsic photoexcitation measurements, it was concluded that the main donor species in undoped GaAs are Si and Ge (Ohyama et al., 1982). These discrepancies indicate that further work is necessary for the absolute identification of the dominant donor species in undoped GaAs.

Residual donors and acceptors in undoped GaAlAs have also been investigated by Mori et al. (1981). From electrical and PL measurements, they concluded that the dominant donor impurity would be Si and/or C and the dominant acceptor would be C and a column II impurity such as Mg. Therefore the conclusion is very similar to that derived for undoped GaAs.

b. Deep Levels

The presence of deep levels in GaAs and GaAlAs is known to give rise to detrimental effects on the performance of devices such as microwave MESFETs and semiconductor lasers. Therefore, the density of deep levels also determines the quality of epilayers. The dominant deep level in MOVPE GaAs is an electron trap known as EL2 with an activation energy of about 0.8 eV. Bhattacharya et al. (1980) was the first to find that the EL2 density increases with [AsH$_3$]/[TMG]. Later, Watanabe et al. (1981) reported that the EL2 density is proportional to $([AsH_3]/[TMG])^{1/4}$, while Samuelson et al. (1981) reported that the EL2 density varies with $([AsH_3]/[TMG])^{1/2}$. Watanabe et al. (1983) further investigated the discrepancies between the results and found that there are two kinds of EL2 centers, EL2a and EL2b, whose activation energies are 0.82 and 0.78 eV, respectively (Fig. 23). The EL2a density dominates at higher growth temperatures, while the EL2b density dominates at lower temperatures, which consistently explains both the results obtained by Watanabe et al. (1981) and those obtained by Samuelson et al. (1981). Calculations based upon the mass action law suggest

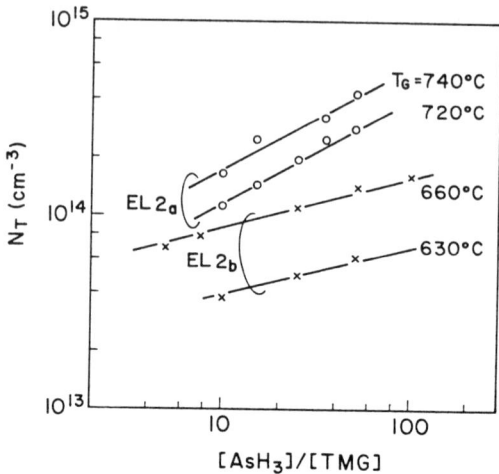

FIG. 23. Dependences of dominant deep-level densities on [AsH$_3$]/[TMG] for four growth temperatures. The straight lines have slope $\frac{1}{2}$ for EL2a and slope $\frac{1}{4}$ for EL2b. [From Watanabe et al. (1984).]

that both EL2a and EL2b are related to native defects (Watanabe et al., 1983); however, the identification of the centers with specific native defects (Fujisaki et al., 1985; Wang et al., 1986) is tentative at present and requires further investigation.

For Ga$_{1-x}$Al$_x$As with x less than 0.33, EL2 is a dominant electron trap with activation energy being invariant with x (Matsumoto et al., 1982). When the Al content exceeds 0.3, a new DLTS signal appears that is known as a DX (donor-complex) center (Lang and Logan, 1979). This center is important because the center is the origin of the persistent photoconductivity effect (Nelson, 1977) and hence gives rise to detrimental effects on the DC characteristics of MESFETs made with GaAlAs. Unfortunately, there have been almost no reports to date as to the behavior of the DX centers in MOVPE-grown GaAlAs. A preliminary result is shown in Fig. 24 (Watanabe et al., 1984). Two kinds of DLTS peaks are seen. Figure 25 (Watanabe et al., 1984) illustrates the variation of $N_D + N_{DX} - N_A$ and $N_D - N_A$ with x, where N_{DX} denotes the density of DX centers. As is seen in this figure, most of the Se impurities are incorporated as DX centers when x is larger than 0.3. Recently, Kumagai et al. (1984) measured the activation energy of DX centers doped with six different impurities (S, Se, Te, Si, Ge, and Sn). They reported that the activation energies with group IV impurities become shallower as the mass number of impurity increases, while those with group VI impurities remain constant. These chemical trends give further insight for construction of a specific model of the DX center.

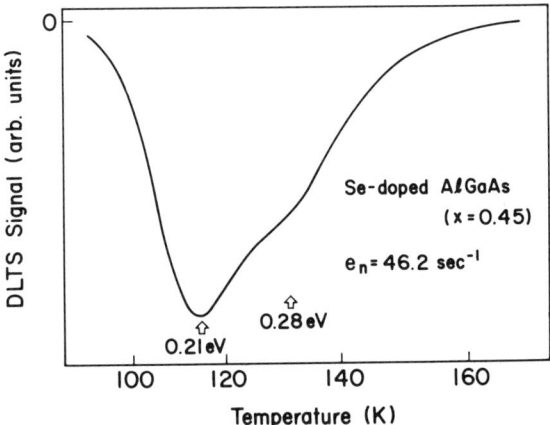

FIG. 24. DLTS spectrum of Se-doped GaAlAs ($x = 0.45$). [From Watanabe *et al.* (1984).]

6. IMPURITY DOPING

Impurity doping into GaAs MOVPE layers has been extensively studied from the practical point of view. Approximately linear relations are usually observed between carrier concentration and dopant concentration; irrespective of doping sources, however, the detailed behavior of impurity doping is rather complicated. Figure 26 compares the doping characteristics

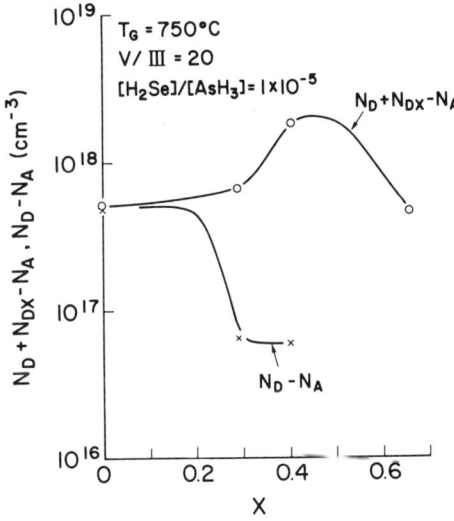

FIG. 25. $N_D + N_{DX} - N_A$ and $N_D - N_A$ vs. x relationships for Se-doped GaAlAs. [From Watanabe *et al.* (1984).]

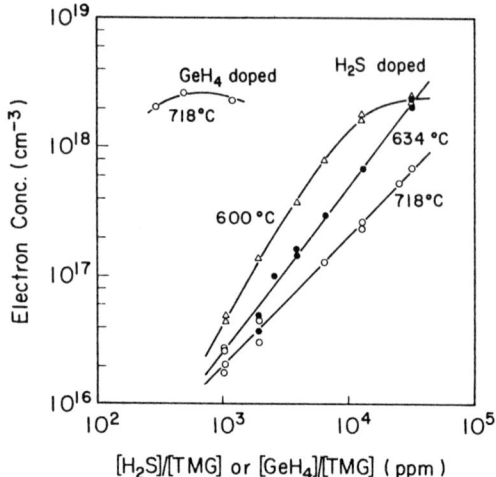

FIG. 26. Electron concentration of GaAs epilayers as a function of [H$_2$S]/[TMG] or [GeH$_4$]/[TMG].

of S and Ge in GaAs. The doping efficiency of S is about 1000 times lower than that of Ge. The doping efficiency of Se and Si has been measured to be as high as that of Ge (Tanno and Seki, 1975). It is not clear at present why S doping efficiency alone is quite different. In Fig. 27 are shown the relationships between carrier concentration and AsH$_3$ flow rate, with H$_2$S mole

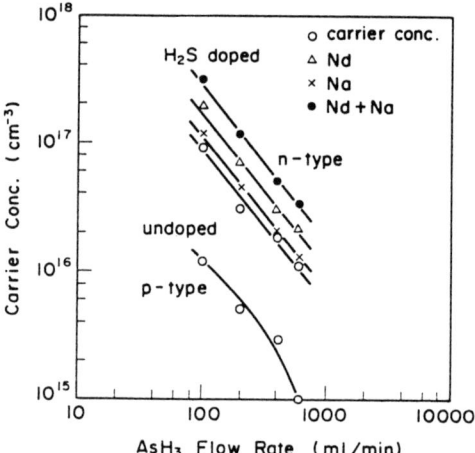

FIG. 27. AsH$_3$ flow rate dependence of carrier concentration for H$_2$S-doped and p-type undoped epilayers: (○) carrier concentration; (△) N_D; (×) N_A; (●) $N_D + N_A$. H$_2$S mole fraction is kept constant.

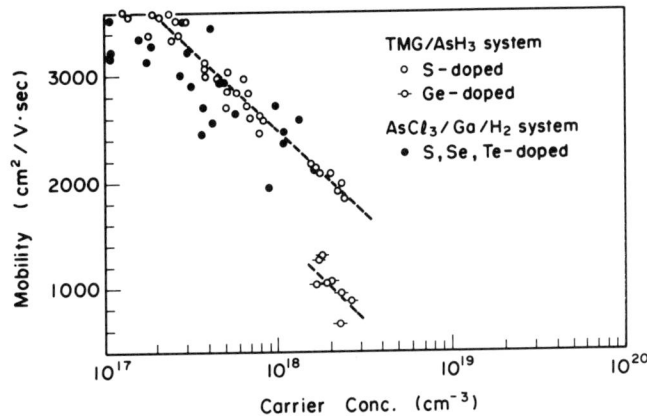

FIG. 28. Room-temperature electron mobility as a function of carrier concentration for S- (○) and Ge-doped (-○-) GaAs epilayers grown by MOVPE. Data for S-, Se-, and Te-doped epilayers by the conventional $AsCl_3/Ga/H_2$ growth system are also shown for comparison (●).

fraction being constant. The results can be qualitatively understood as indicating that the arsenic vacancy concentration, $[V_{As}]$, necessary for S incorporation decreases with $[AsH_3]$. Figure 28 shows the mobility versus carrier concentration relationships for S and Ge doping. The measured mobility values for Ge-doped layers are about half the values for S-doped layers, suggesting that Ge acts as an amphoteric impurity.

In the case of Si doping by ethyltrimethylsilane, $C_2H_5(CH_3)_3Si$, dissolved in TMG, carrier concentration decreases with $[AsH_3]/[TMG]$ (Fig. 29). A similar trend has been observed when SiH_4 is used as a doping source (Okamoto *et al.*, 1984; Kuech *et al.*, 1984; Kuech and Veuhoff, 1984; Veuhoff *et al.*, 1985). The trend cannot be explained by a simple vacancy model because of an opposite $[AsH_3]/[TMG]$ dependence. When disilane (Si_2H_6) is used instead of SiH_4, growth temperature and orientation dependence of the doping efficiency disappear, suggesting that the gas-phase mass transport is a limiting factor because of the very rapid thermal decomposition of Si_2H_6 at the growing front (Veuhoff *et al.*, 1985), as compared with SiH_4 being relatively stable up to growth temperature.

Doping with deep-level impurities is important because high-quality semi-insulating layers are required for buffer layers in GaAs MESFETs and GaAs ICs. However, there has been very little work on deep-level impurity doping in MOVPE-grown GaAs. Bass (1978) attempted to dope with Cr using hexacarbonylchromium; however, Cr doping caused epilayer surface roughening and showed a strong memory effect that made it difficult to construct the sharp carrier concentration profiles necessary for GaAs MESFET structure. To avoid these problems, Akiyama *et al.* (1984b) used

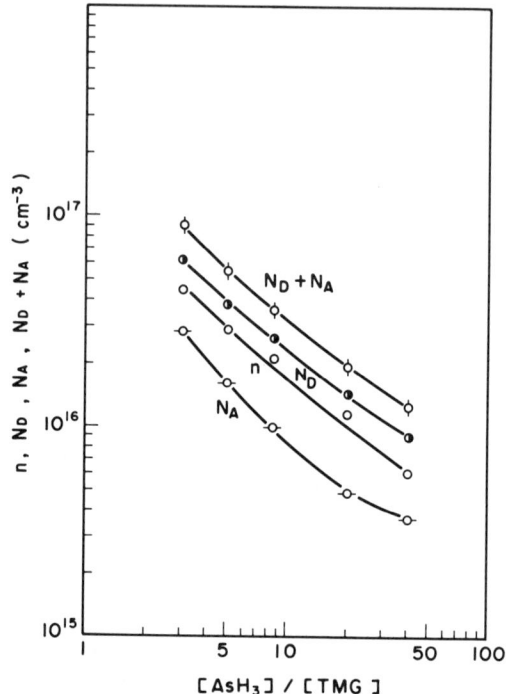

FIG. 29. n, N_D, N_A, and $N_D + N_A$ as a function of [AsH$_3$]/[TMG] for GaAs epilayers doped with C$_2$H$_5$(CH$_3$)$_3$Si.

a new vanadium-compound doping source, triethoxyvanadyl, VO(OC$_2$H$_5$)$_3$, instead of the Cr compound. Maximum deep-level density of more than 10^{18} cm^{-3} has been attained. Epilayers with resistivity higher than 10^8 Ω-cm have been obtained readily without memory effect or surface roughening.

Table VI summarizes the pressure-dependence relationships for n- and p-type dopants. There is disagreement in the results obtained by different authors.

As compared with the case of doping into GaAs, there have been few reports on doping into GaAlAs. Figure 30 shows the doping behavior of Zn in Ga$_{1-x}$Al$_x$As ($0 \leq x \leq 0.8$). The linear relation obtained is in good agreement with the report by Aebi et al. (1981). The doping efficiency decreases with an increase of acceptor ionization energy with x (Yang et al., 1982). In Fig. 31 is shown the doping behavior of Se in Ga$_{1-x}$Al$_x$As ($0 \leq x \leq 0.8$). At low Al concentrations ($x \leq 0.3$), a linear relationship between electron concentration and [H$_2$Se] is seen. The doping efficiency decreases with Al content as previously reported by other workers (Mori and Watanabe, 1981; Hallais et al., 1981). When the Al composition exceeds 0.45, the

TABLE VI

Pressure-Dependence Relationships for Dopants in GaAs

Dopant	Gas	System	Group III	Group V	Dopant	References
S	H_2S	TMG/AsH_3	$n \propto [TMG]^{0.6}$	$n \propto [AsH_3]^{-1}$	$n \propto [H_2S]^x$	Bass and Oliver (1977)
	H_2S	TMG/AsH_3	$n =$ constant at $[AsH_3]/[TMG] =$ constant		$x = 1$–1.8; temperature-dependent	Nakanisi (1984)
Se	H_2Se	TMG/AsH_3		$n \propto [AsH_3]^{-1}$	$n \propto [H_2Se]$	Mori and Watanabe (1981)
	H_2Se	TMG/AsH_3			$n \propto [H_2Se]$	Hallais et al. (1981)
	H_2Se	TMG/AsH_3	$n \propto [TMG]^{0.6}$	$n \propto [AsH_3]^{-1.1}$	$n \propto [H_2Se]$	Asai and Sugiura (1985)
Si	SiH_4	TMG/AsH_3	$n =$ constant at $[AsH_3]/[TMG] =$ constant		$n \propto [SiH_4]$	Hallais et al. (1981)
						Druminski et al. (1982)
	SiH_4	TMG/AsH_3		$n \propto [AsH_3]^{-1}$	$n \propto [SiH_4]$	Bass and Oliver (1977)
	SiH_4	TMG/AsH_3		$n \propto ([AsH_3]/[TMG])^{4/3}$ decrease with $[AsH_3]/[TMG]$	$n \propto [SiH_4]$	Okamoto et al. (1984)
	$[C_2H_5(CH_3)_3Si]$	TMG/AsH_3				Udagawa and Nakanisi (present work)
	SiH_4	TMG/AsH_3			$n \propto [SiH_4]$	Kuech and Veuhoff (1984)
	Si_2H_6	TMG/AsH_3			$n \propto [Si_2H_6]$	Kuech et al. (1984)
Sn	TESn	TMG/AsH_3	slightly decrease	constant	$n \propto $ [TESn]	Parsons and Krajenbrink (1984)
	TMSn	TMG/AsH_3			$n \propto $ [TMSn]	Roth et al. (1984)
Zn	DMZ	TMG/AsH_3	$p \propto [TMG]^{0.45}$	$p \propto [AsH_3]$	$p \propto $ [DMZ]	Bass and Oliver (1977)
	DEZ	TMG/AsH_3			nonlinear relationship	Mori and Watanabe (1981)
	DEZ	TMG/AsH_3			sublinear relationship	Aebi et al. (1981)
	DMZ	TMG/AsH_3			$p \propto [DMZ]^x, x \leq 2$	Glew (1984)
Mg	CpMg	TMG/AsH_3			$p \propto [CpMg]^x, x \leq 2$	Lewis et al. (1983)
	CpMg	TMG/AsH_3			$p \propto [CpMg]^x, x \sim 1$	Tamamura et al. (1986)
Be	DEBe	TMG/AsH_3			$p \sim [DEBe]^x, x \sim 2$	Botka et al. (1984)
Cr	HCC				Δn (reduced) $\propto $ [HCC]	Bass (1978)
V	TEV					Akiyama et al. (1984a)

TESn = tetraethyltin.
TMSn = tetramethyltin.
DMZ = dimethylzinc.
DEZ = diethylzinc.
CpMg = bis-cycropentadienylmagnesium.
DEBe = diethylberyllium.
HCC = hexacarbonylchromium.
TEV = triethoxyvanadyl.

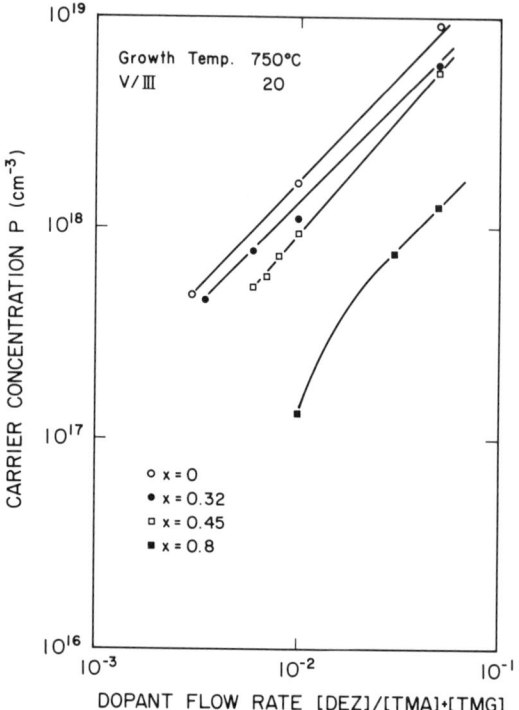

FIG. 30. Relationship of carrier concentration to [DEZ]/([TMA] + [TMG]) for GaAs (○) and GaAlAs (●, $x = 0.32$; □, $x = 0.45$; ■, $x = 0.8$). [From Nakanisi (1984). Copyright North-Holland Publ. Co., Amsterdam, 1984.]

electron concentration does not increase linearly with [H_2Se]. However, it has been reported that the electron concentration derived from conventional C-V measurements obeys a linear relationship even when the Al content is 0.45 (Nakatsuka et al., 1983). The discrepancy between the results can be explained by the fact that most of the Se impurities do not exist as shallow donors but as DX centers.

III. Fabrication of Device Structures

Figure 32 is a schematic diagram of the carrier concentration and epilayer thickness required for various electronic devices. More than four orders of magnitude of variation in carrier concentration and as great as two orders of magnitude of variation in thickness are required. The wide electron concentration range easily can be covered by sulfur doping. The electron concentration in the mid-10^{14} cm^{-3} and higher can be reproducibly controlled

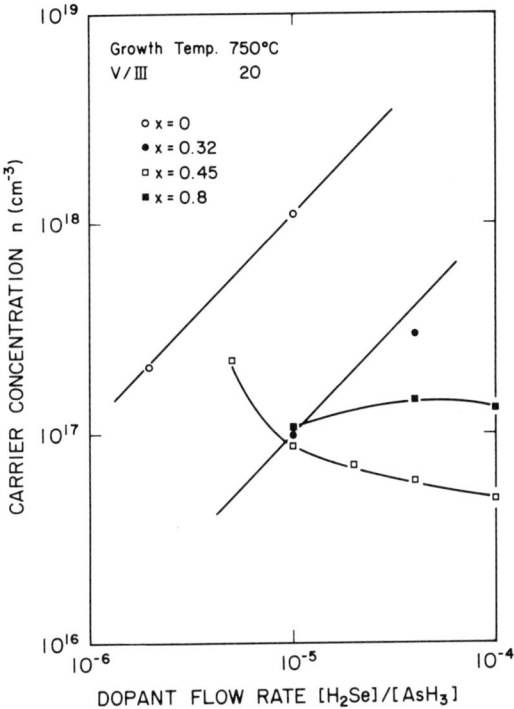

FIG. 31. Carrier concentration vs. [H$_2$Se]/[AsH$_2$] relationships for GaAs (○) and GaAlAs (●, $x = 0.32$; □, 0.45; ■, 0.8). [From Nakanisi (1984). Copyright North-Holland Publ. Co., Amsterdam, 1984.]

as shown in Fig. 33, which exhibits a run-to-run reproducibility of electron concentration and thickness. The n^- buffer layer, whose electron concentration is less than 10^{13} cm^{-3}, can be grown by adjusting the [AsH$_3$]/[TMG] molar ratio to a value at which p- to n-type conversion occurs. The [AsH$_3$]/[TMG] molar ratio at which the high-resistivity layer is obtained can be widened by a small addition of sulfur (Fig. 34).

The epilayer thickness can be adjusted reproducibly by varying the growth time (Fig. 35).

7. GaAs MESFETs and GaAs ICs

For MESFET applications, an n^-–n–n^+ epitaxial structure is required. The n^--type buffer layer grown on semi-insulating substrate is necessary to separate the n active layer from the rather impure semi-insulating substrate and from the epilayer/substrate interface, where large amounts of impurity pile-ups is usually observed. The n^+ contact layer is necessary for reducing

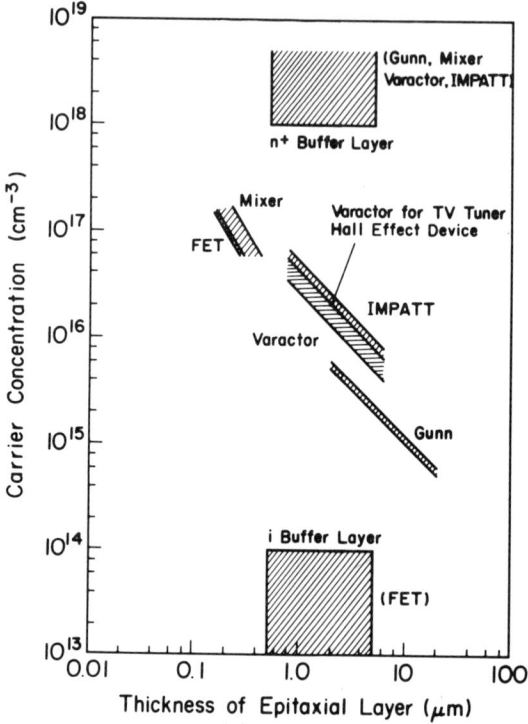

Fig. 32. Schematic representation of electron concentration vs. thickness of epitaxial layers required for electronic devices.

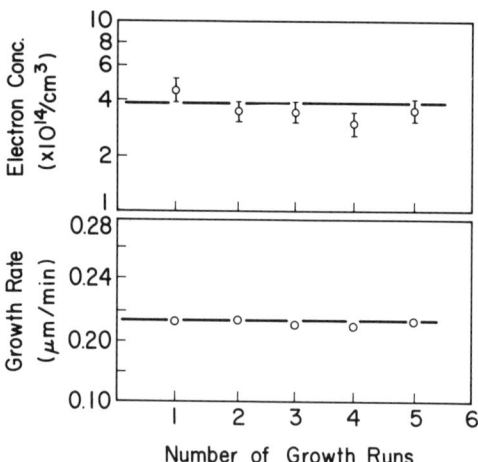

Fig. 33. Run-to-run fluctuation of epilayers for J-band Gunn diodes.

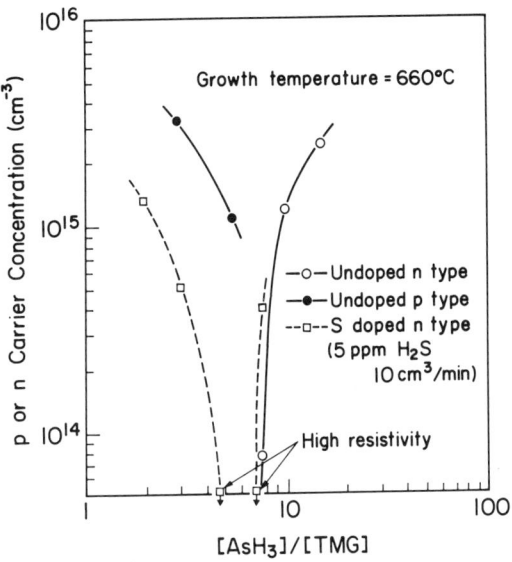

FIG. 34. Carrier concentration as a function of $[AsH_3]/[TMG]$ without doping (solid curve; open circles represent *n*-type, and solid circles, *p*-type) and with H_2S doping (dashed curve; 5 ppm H_2S, 10 cm^3/min).

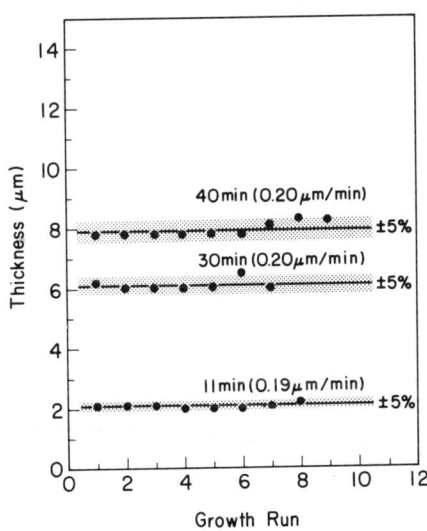

FIG. 35. Epilayer thickness vs. growth time relationships and their run-to-run fluctuation.

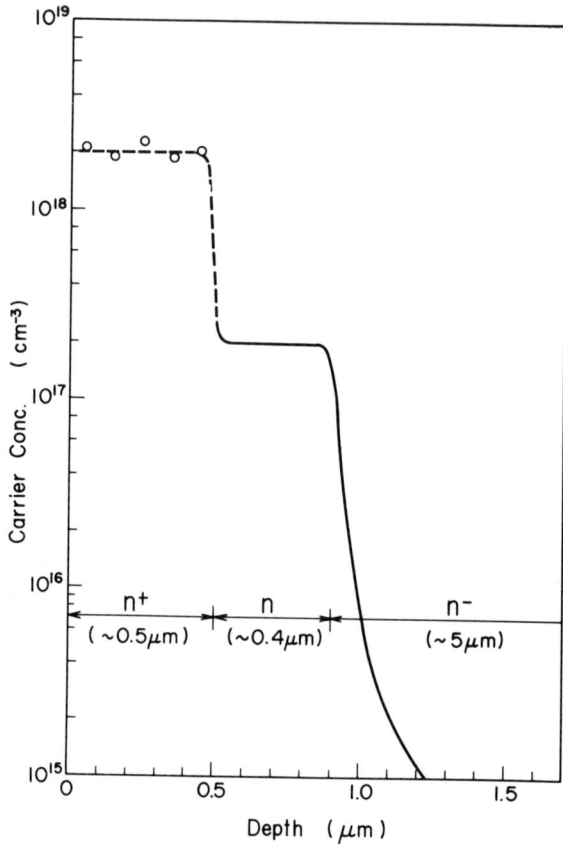

FIG. 36. Typical electron concentration profile of n^--n-n^+ epilayer for low-noise MESFET application. [From Nakanisi et al. (1981). Copyright North-Holland Publ. Co., Amsterdam, 1981.]

the series resistance due to ohmic contacts. Figure 36 (Nakanisi et al., 1981) shows a typical electron concentration profile of an n^--n-n^+ layer for low-noise MESFETs. The thickness and electron concentration of each n^-, n, and n^+ layer are, respectively: 5 μm, 10^{14} cm^{-3}; 0.4 μm, 2×10^{17} cm^{-3}; 0.5 μm, 2×10^{18} cm^{-3} in this case. From careful C-V measurements, it was found that the transition region thickness where the carrier concentration falls by one decade from the active layer to the buffer layer is only 0.07 μm. With improvements in the epilayer purity, it has become quite easy to grow high-resistivity n^- buffer layers as thick as 5 μm with good reproducibility by setting the [AsH$_3$]/[TMG] ratio to the high-resistivity "window" shown in Fig. 34.

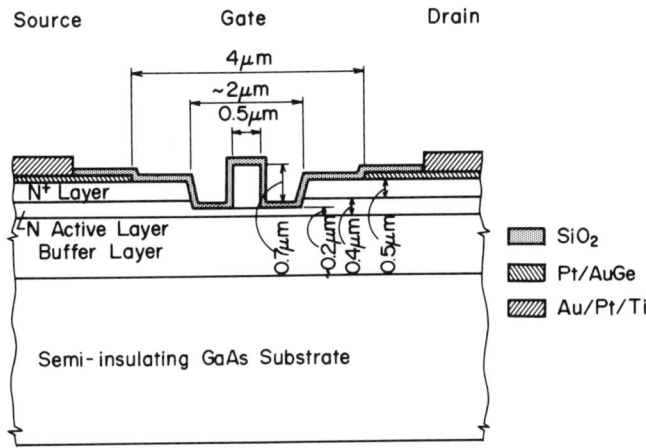

FIG. 37. Cross-sectional view of a 0.5-μm-gate MESFET. [From Nakanisi et al. (1981). Copyright North-Holland Publ. Co., Amsterdam, 1981.]

Half-micron-gate low-noise MESFETs were fabricated on n^--n-n^+ epitaxial wafers (Nakanisi et al., 1981). The MESFET structure is the same as that reported in the literature (Kamei et al., 1979). Figure 37 shows the cross-sectional view of a half-micron-gate MESFET. A Schottky gate electrode, 0.5 μm long and 300 μm wide, was formed using Al, with a thickness of 0.7 μm. Gate, source and drain electrodes were defined by a conventional photolithography and lift-off process. Source and drain ohmic contacts separated by 4 μm were formed by alloying evaporated Pt/AuGe at 460°C. Bonding pads to gate, source and drain electrodes were formed by using a Au/Pt/Ti multimetal structure to prevent intermetallic reactions. A recess with a width of 2 μm was formed to reduce parasitic resistances. Figure 38 illustrates typical drain current–voltage characteristics of the MESFET. A transconductance of 50 mS was obtained at a drain current of

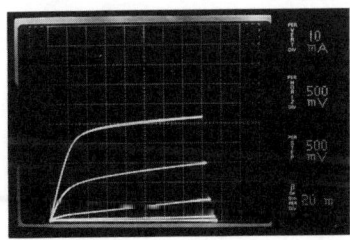

FIG. 38. Typical drain current–voltage characteristics of 0.5-μm-gate MESFET. Horizontal scale: 1 V/division; vertical scale: 10 mA/division; gate bias step; −0.5 V. [From Nakanisi et al. (1981). Copyright North-Holland Publ. Co., Amsterdam, 1981.]

30 mA. The transconductance is still higher at lower current levels, corresponding to the steep n–n^- interface profile of the epilayer. The drain current has almost none of the loops and bumps that are often observed in MESFETs without, or with thin, n^- buffer layers.

Microwave noise figure and gain measurements were performed at 12 GHz for the MESFET chips mounted on a microstrip circuit. The optimized noise figure NF and associated gain G_a as a function of drain current I_{ds} are shown in Fig. 39, where optimized gain G_{max} versus I_{ds} is also plotted. It is seen that the minimum NF is 2.0 dB with G_a of 8.2 dB. The G_{max} of more than 12 dB is obtained at an I_{ds} of 30 mA. The noise figures and gain performances are superior to those of MESFETs prepared by MOVPE by other workers (Duchemin *et al.*, 1979; Morkoç *et al.*, 1979). Quarter-micron-gate MESFETs were also fabricated (Kamei *et al.*, 1981). The structure of the quarter-micron-gate MESFET is substantially the same as that of a half-micron-gate MESFET except for the gate width (200 μm) and recess width (2 μm).

At 18 GHz, an NF of 1.75 dB with a G_a of 8.5 dB was obtained. Also, a G_{max} of 10.5 dB was measured at an I_{ds} of 30 mA. These results as well as those by Aebi *et al.* (1984) represent one of the best MESFET performances reported so far.

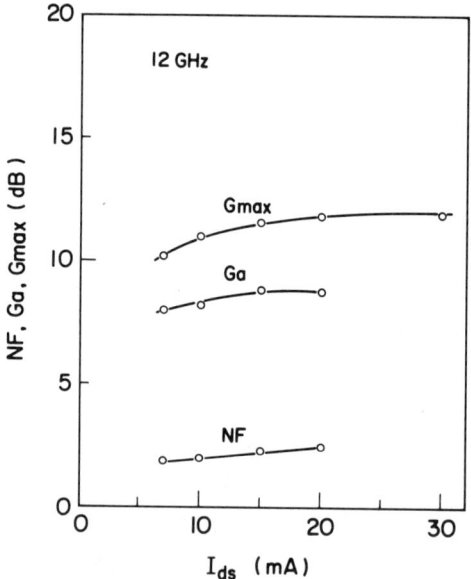

FIG. 39. Optimized microwave noise figure NF, associated gain G_a and optimized gain G_{max} as a function of drain current I_{ds}, measured at 12 GHz for MESFET chips mounted on a microstrip circuit. [From Nakanisi *et al.* (1981). Copyright North-Holland Publ. Co., Amsterdam, 1981.]

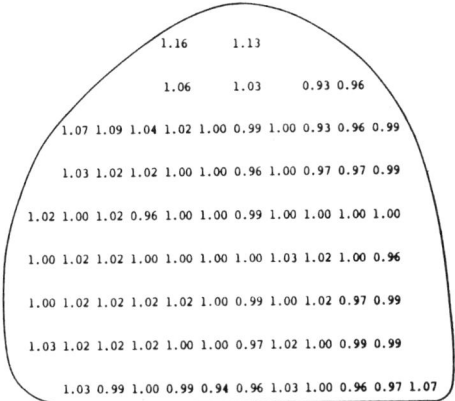

FIG. 40. Distribution of surface electron concentration n_s in units of 10^{17} cm^{-3} for a 15 cm^2 epitaxial n-n^- wafer.

Shino et al. (1981) have developed GaAs power MESFETs. Since multichip operation is indispensable for increasing output power of MESFETs, a good uniformity among chips is a prerequisite. The MOVPE technique is suitable for fabrication of power MESFETs as well, since it is superior to the conventional VPE technique in providing more uniform epiwafers, and hence less chip-to-chip variation. An example of the distribution of the surface electron concentration n_s in an n-n^- layer over a 15 cm^2 wafer is shown in Fig. 40. The fluctuation of n_s was within 5% except for the edge region of the wafer. The run-to-run fluctuation of the average n_s and thickness were kept within ±5%.

The device fabrication process for power MESFETs was almost the same as that for low-noise MESFETs reported by Kamei et al. (1981). The source and drain ohmic contacts were separated by 6 μm. The source electrodes were interconnected by Au formed by a lift-off process. A 0.5-μm-thick CVD SiO$_2$ spacer was used at the source and gate crossover. The size of an integrated MESFET chip was 1.32 × 0.555 mm, with a unit gate width of 200 μm and a total gate width of 7.2 mm. The I_{dss}, the pinchoff voltage and the transconductance were about 1.3 A, 3 V, and 400 mS, respectively. The gate-source breakdown voltage was measured to be about 20 V at a reverse current of 1 mA.

The distribution of I_{dss} within a fabricated wafer was measured and compared to that of a conventional VPE wafer in order to examine the uniformity among chips (Fig. 41 shows the histograms of I_{dss} for MOVPE wafers (a) and VPE wafers (b).) The area was chosen to be about 6 cm^2 for handling purposes. The standard deviation is 6.8% of the mean value of I_{dss} for the MOVPE wafer, whereas it is 11.7% for the VPE wafer.

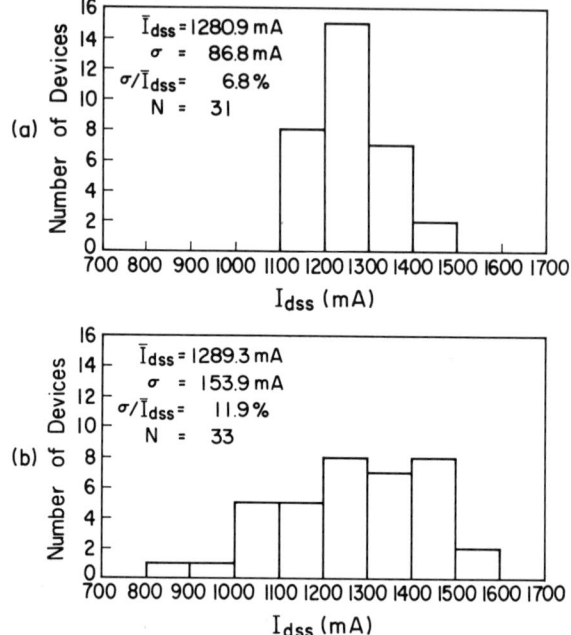

FIG. 41. I_{dss} distributions for power MESFETs prepared by (a) MOVPE and (b) VPE. [From Shino et al. (1981).]

This clearly shows that the MOVPE wafer has better uniformity than the VPE wafer, being in accordance with the uniformity of n_s shown in Fig. 40. This uniformity is believed to be more advantageous for multichip operation of GaAs power MESFETs.

In Fig. 42, the output power for one- and two-chip operation at 7.8 GHz is plotted as a function of input power at a drain voltage of 10 V. A low-pass-type internal matching circuit is used at both the input and the output port for the two-chip operation. The one-chip device delivered 4 W with 3 dB gain and 12% power-added efficiency. The results compared favorably with those reported by Duchemin et al. (1978), 1 W output power and 8 dB gain at 7 GHz, and those by Bonnet et al. (1981), 2.7 W output power and 6.5 dB gain at 8 GHz.

In order to realize GaAs digital LSIs and VLSIs in which normally-off MESFETs are utilized, high uniformity of threshold voltage, V_T is essential. One of the main origins of V_T scattering is the nonuniformity of the semi-insulating substrate where an active layer is formed by direct ion implantation onto the surface region of undoped LEC (liquid-encapsulated Czochralski) substrate. High dislocation density causes scattering of the

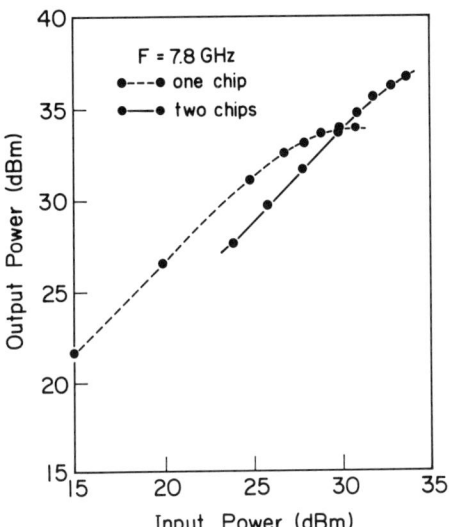

FIG. 42. Output power as a function of imput power of two-chip device (solid line) and one-chip device (dashed line). [From Shino et al. (1981).]

V_T value, whereas in a Cr-doped boat-grown substrate, nonuniform Cr distribution causes the V_T variation. In order to solve these problems, Sano et al. (1984) utilized MOVPE-grown semi-insulating buffer layer for ion-implanted MESFETs. A semi-insulating GaAs layer was grown on a Cr-doped horizontal Bridgeman (HB) semi-insulating substrate, using a low-pressure (100 torr) MOVPE growth system at growth temperatures between 640 and 710°C. By adjusting the [AsH$_3$]/[TMG] molar ratio of which p- to n-type conversion occurs, a high-resistance layer with a thickness of 3 μm was reproducibly obtained. Although the layer is not truly semi-insulating, the layer was used as a substrate for ion implantation because the surface leakage current is very low. The etch pit density (EPD) in the epilayer was almost the same as that in the substrate. The outdiffusion of Cr impurity from the substrate into the epilayer was found to be negligible up to 1.5 μm from the surface from SIMS depth profile measurements, even after annealing at 800°C for 20 min.

MESFETs were fabricated by the W–Al gate self-alignment process. Gate length was 1.5 μm and the MESFETs were arrayed by 4.8 mm × 5.6 mm pitch. The ^{29}Si ions were implanted to form an active layer at an energy of 60 keV with a dose of 1.3×10^{12} cm^{-2}, and the self-aligned n^+ source and drain regions were formed by ^{29}Si implantation at 100 keV with a dose of 1.5×10^{13} cm^{-2}. Two kinds of substrates, on MOVPE-grown one (2-inch diameter with EPD of about 3,000 cm^2) and a commercially available undoped LEC substrate (2-inch diameter and EPD of about 50,000 cm²) were used for comparison. These two wafers were processed under the same conditions. Annealing of the implanted ions was performed at 800°C for 20 min with a SiO$_2$ cap. The V_T of MESFETs was determined from the relation between V_g and $(I_{ds})^{1/2}$. In the MOVPE-grown substrate, the standard deviation of the V_T, σ_{V_T}, was only 18.5 mV in a 2-inch full wafer at a mean value of +220 mV. In the undoped LEC substrate, the σ_{V_T} was 33.5 mV at a mean value of +160 mV. The σ_{V_T} of 18.5 mV is the best value ever reported when the V_T values are measured over a 2-inch full wafer, indicating that the activation efficiency of implanted Si ions and carrier profiles are very uniform in the MOVPE-grown substrate.

Imamura et al. (1984) used selectively grown n^+ MOVPE layers to reduce gate resistance R_g and source resistance R_s for WSi/TiN/Au gate self-aligned GaAs enhancement-mode MESFETs (EFETs) and depletion mode MESFETs (DFETs). The technique was found to be effective for reducing the R_g to one-tenth and the R_s to one-fifth of the values for the conventional ion-implanted WSi gate MESFETs. It is shown that WSi/TiN/Au gate self-aligned GaAs MESFETs with an MOVPE n^+ layer are useful in fabrication of high-performance microwave MESFETs and monolithic microwave ICs. A similar attempt has been carried out by Uetake et al. (1985). They have developed high G_m GaAs MESFETs with a reduced short-channel effect for GaAs LSI applications by using MOVPE selective n^+ growth. GaAs MESFETs exhibited a high G_m (350 mS/mm) without a serious short-channel effect.

8. HEMTs

Recently a great deal of attention has been paid to selectively doped (SD) GaAs/n-GaAlAs heterostructures from the viewpoint of their applications to high-frequency and high-speed devices. This is because the two-dimensional electron gas (2DEG) accumulating at the heterointerface has extremely high mobility (Dingle et al., 1978). The SD GaAs/GaAlAs heterostructures have conventionally been grown by molecular-beam epitaxy (MBE), and a 2DEG mobility as high as 2,120,000 cm²/Vsec at 5 K has been reported (Hiyamizu et al., 1983). Attempts have been made to grow SD GaAs/n-GaAlAs heterostructures by MOVPE (Coleman et al., 1981; Hersee et al., 1982; Maluenda and Frijlink, 1983). Kobayashi et al. (1984) recently have succeeded in realizing SD GaAs/n-GaAlAs heterostructures having high electron mobility. In the fabrication of high-quality GaAlAs layers, O$_2$ contamination from an air leak in the MOVPE gas manifold or reactor system, or from starting material such as AsH$_3$, results in O$_2$ incorporation in GaAlAs. André et al. (1983) reported that 1 ppm of O$_2$ in the vapor phase

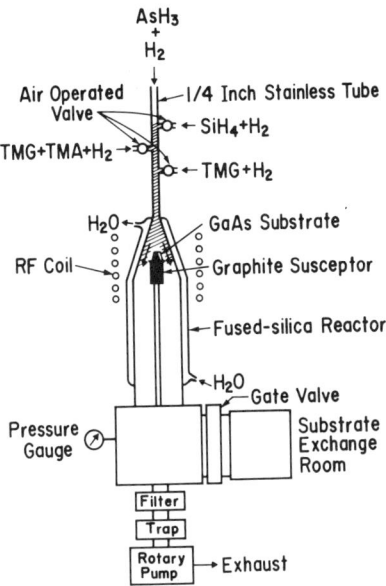

FIG. 43. Schematic view of reactor system in MOVPE apparatus. The internal volume V_0 is shown as the cross-hatched region. [From Kobayashi et al. (1984).]

leads to an oxygen concentration of 3×10^{19} cm^{-3} in the $Ga_{0.5}Al_{0.5}As$ solid phase. Oxygen-doped GaAlAs becomes semi-insulating because of the formation of deep levels (Stringfellow and Hom, 1979; Wallis et al., 1981). Therefore, a leak-proof gas manifold and reactor system are minimum requirements in order to obtain stable n-type doping in GaAlAs. Moreover, an abrupt heterointerface and doping profiles are necessary in obtaining SD GaAs/n-GaAlAs heterostructures with high 2DEG mobility. Kobayashi et al. (1984) used an MOVPE reactor system as shown schematically in Fig. 43. The MOVPE growth apparatus was helium-leak-tested to a rate of less than 10^{-8} atm cm^3/sec, and the oxygen concentration in the hydrogen carrier gas flowing through the gas manifold and the reactor from the palladium-diffused hydrogen purifier was less than 0.01 ppm. Two TMG bubblers were used, one for GaAs growth and the other for GaAlAs growth. This made it possible in the MOVPE growth of GaAs/GaAlAs heterostructures to vary [AsH$_3$]/[III] in GaAs and GaAlAs growth independently. For a rapid change in gas composition over the substrate in obtaining abrupt heterointerfaces and Si doping profiles, it was necessary to minimize the inside volume V_0 (the cross-hatched region shown in Fig. 43) and increase the total gas flow velocity. Kobayashi et al. (1984) attained a V_0 of 150 cm^3 and gas flow velocity of 60 cm/sec for a total flow of 10 standard liters per minute under a reactor pressure of 80 torr. The reactor construction enabled the gas composition to be changed within 0.1 sec.

Cr-doped semi-insulating substrates whose orientation was $\langle 100 \rangle$ tilted 2° towards $\langle 110 \rangle$ were used. The growth temperature was 650°C and the growth rate was 4.0–12.4 Å/sec. The abruptness of GaAs/n-Ga$_{0.7}$Al$_{0.3}$As heterointerface was estimated to be less than 6 Å by using Auger electron spectroscopy (AES) with Ar$^+$ ion sputtering.

The carrier concentration and the type of conduction for GaAs and GaAlAs epilayers depend on [AsH$_3$]/[III]. Carbon impurity is introduced into the epilayers for lower molar ratios, while the concentration of donor impurities such as Si increases. In order to obtain SD GaAs/GaAs heterostructures with high 2DEG mobility by MOVPE, it is necessary to minimize the total impurity concentration in both the undoped GaAs and the Ga$_{0.7}$Al$_{0.3}$As spacer layers.

The SD heterostructures consist of Si-doped n-Ga$_{0.7}$Al$_{0.3}$As (500 Å)/undoped Ga$_{0.7}$Al$_{0.3}$As (100 Å)/undoped GaAs (0.7–3 μm)/semi-insulating GaAs substrates. By adjusting the [AsH$_3$]/[TMG] and [AsH$_3$]/([TMG] + [TMA]) to be close to a value where p- to n-type conversion occurs, the highest electron mobility at 77 K of 50,000–55,000 cm^2/Vsec was obtained. Moreover, by optimizing the [AsH$_3$]/([TMG] + [TMA]) for the growth of Ga$_{0.7}$Al$_{0.3}$As undoped spacer layer, a remarkable increase in the 2DEG mobility was observed. A 2DEG mobility of 150,000 cm^2/Vsec at 4.2 K was obtained. Kobayashi and Fukui (1984) further succeeded in raising the 2DEG mobility by using triethyl organometallic sources, triethylgallium (TEG) and triethylaluminum (TEA), instead of TMG and TMA. They found that for GaAlAs grown by TMA and TMG, remarkable carbon incorporation of up to 10^{18} cm^{-3} is observed. On the other hand, the use of TEA and TEG results in less carbon incorporation (10^{16} cm^{-3}). In SD GaAs/n-Ga$_{0.7}$Al$_{0.3}$As heterostructures, 2DEG mobility of 445,000 cm^2/Vsec was obtained for the SD heterostructure with a sheet electron concentration of 5.1×10^{11} cm^{-2} at 2 K.

High electron mobility transistors (HEMTs) with GaAs/n-GaAlAs SD heterostructures have demonstrated their superior microwave performance as compared to GaAs MESFETs (Kamei *et al.*, 1984). Most work on HEMTs has been conducted using epiwafers grown by the MBE technique. However, the MBE technique contains many unsolved problems such as low throughput, generation of oval defects, etc. On the other hand, the MOVPE technique is believed to be potentially superior to MBE in providing a number of epiwafers in the same growth run (Hayafuji *et al.*, 1986; Gersten *et al.*, 1984). In spite of recent extensive work on heterostructure growth by the MOVPE technique, relatively few reports have been published for microwave HEMTs. Kawasaki *et al.* (1986) grew epiwafers for HEMTs using TMG

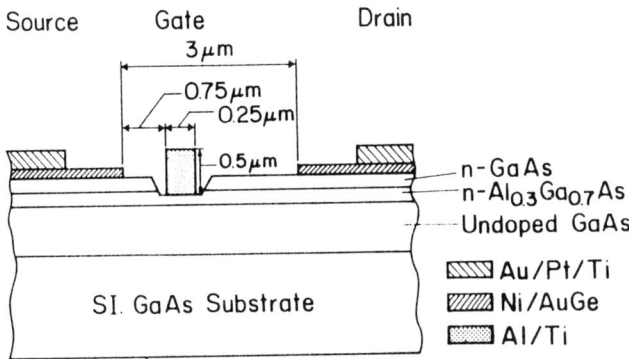

FIG. 44. Schematic cross-section of 0.25-μm-gate HEMT. (Reprinted with permission from IEICE, Kawasaki, H., Imai, I., Tanaka, A., Tokuda, H., Higashiura, M., Hori, S., Kamei, K. (1986) *IEICE Japan* **E69**, 294–295.)

and TMA as organometallic sources under atmospheric pressure. The epilayers consisted of a 0.5 μm undoped GaAs buffer layer, a 30 nm n-Ga$_{0.7}$Al$_{0.3}$As layer and a 60-nm nGaAs layer grown on a Cr-doped HB-grown GaAs substrate. The n-type layers were doped to 2×10^{18} cm^{-3} by SiH$_4$ gas. A slow growth rate of 3 Å/sec was adopted in order to obtain a steep transition between the Ga$_{0.7}$Al$_{0.3}$As layer and the GaAs buffer layer. [AsH$_3$]/[TMG] was carefully selected to reduce the impurity level in the GaAs buffer layer. The measured Hall electron mobility and sheet carrier concentration in the epiwafers for HEMT fabrication were 3,500 cm^2/Vsec and 1.5×10^{12} cm^{-2} at room temperature, respectively.

After 0.3 μm high-mesa formation, source and drain ohmic contacts, separated by 3 μm, were formed. The gate pattern was delineated closer to the source contact edge between the source and drain electrodes by electron beam lithography in order to reduce the source resistance. After recess formation to a depth of 70 nm, the gate electrodes with a width of 200 μm and length of 0.25 μm were formed in the center of the recess by Al/Ti with a thickness of 0.5 μm (Fig. 44).

Microwave noise figures and associated gain were measured at 12 GHz on the HEMT chips mounted in a microwave test fixture. In Fig. 45, the measured 12-GHz optimized noise figures and associated gain of both the 0.25-μm-gate HEMT (solid line) and the 0.5-μm-gate HEMT (dashed line) are plotted as a function of drain current, where the drain voltage is 3.5 V. The minimum noise figures of 0.75 dB and 0.97 dB with associated gains of 11.1 dB and 10.8 dB were obtained at 12 GHz for the 0.25-μm-gate and 0.5-μm-gate HEMTs, respectively. These are among the best HEMT performances every reported, developed using epiwafers grown by MOVPE technique (Takakuwa *et al.*, 1985; Tanaka *et al.*, 1986a, b; Takakuwa *et al.*, 1986).

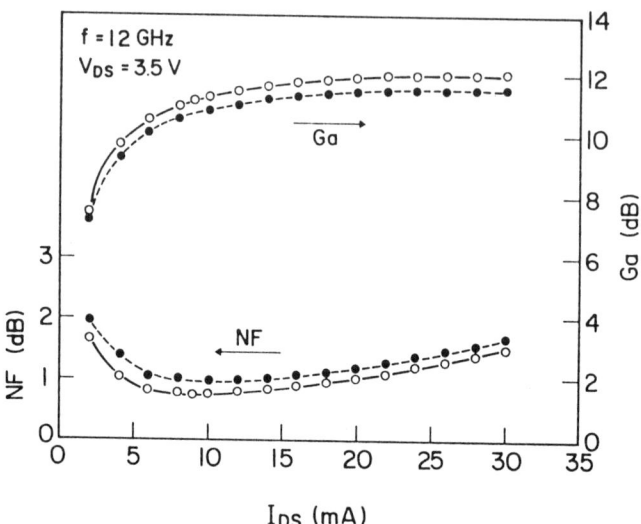

FIG. 45. Optimized noise figures (NF) and associated gain (G_a vs. drain current of 0.25-μm-gate HEMT (solid line) and 0.5-μm-gate HEMT (dashed line) at 12 GHz. (Reprinted with permission from IEICE, Kawasaki, H., Imai, I., Tanaka, A., Tokuda, H., Higashiura, M., Hori, S., Kamei, K. (1986) *IEICE Japan* **E69**, 294–295.)

9. MESFETs Fabricated on GaAs/Si Substrate Structure

In recent years, there has been much interest in growing GaAs epilayers on Si substrates because the Si wafers are less fragile than GaAs. Large-diameter Si wafers are readily available commercially, whereas even 3-inch-diameter GaAs wafers have been produced only very recently. In addition to this, the thermal conductivity of Si is about twice of that of GaAs, which is convenient as a heat sink. For this reason, attempts have been made to grow GaAs on Si substrates since 1980. However, since the crystal structure of GaAs (zincblende) is not the same as that of Si (diamond) and since there is a lattice mismatch as large as 4% between GaAs and Si, it had not even been thought to be possible to grow GaAs epilayers on Si without buffer layers to relax the lattice strain. Recently, however, Wang (1984) succeeded in growing single-domain GaAs layers directly on Si substrates by MBE.

Akiyama *et al.* (1984a) were the first to grow GaAs layers directly on Si substrate by MOVPE. They used a low-pressure (90 torr) MOVPE system with a horizontal reactor. TMG and AsH$_3$ were bubbled through H$_2$ carrier gas. The sequence of growth to obtain a single domain GaAs layer was as follows. First, the substrate was heated to above 900°C and maintained for 10 min in the flow of H$_2$ and AsH$_3$. Then, the substrate was cooled to 450°C or below, and a thin GaAs layer of less than 200 Å was deposited as the first layer. Subsequently, the substrate was reheated to the conventional growth temperature of 700–750°C, and the second GaAs layers were grown. Single-domain structures could be grown on the whole area of the 2-inch wafer. The unintentionally doped layers showed n-type conductivity with a carrier concentration of about 1×10^{16} cm^{-3}, presumably due to Si autodoping from the substrates. These layers exhibited nearly the same mobilities as those of bulk GaAs. A value as high as 5,200 cm^2/Vsec was reported at room temperature. The importance of the conditions of the heat treatment before growth and the thickness of the first layer was emphasized.

GaAs MESFETs and ring oscillators on GaAs layers on Si were fabricated also by Nonaka *et al.* (1984) by using the technique developed above. The structure of the GaAs MESFETs is shown in Fig. 46. The MESFETs were fabricated on the top undoped GaAs layer by the direct ion-implantation techniques using the W/Al gate self-alignment process. ^{29}Si ions were implanted at 60 keV with a dose of 1.1–1.3×10^{12} cm^{-2} and 2.0–2.2×10^{12} cm^{-2} for EFETs and DFETs, respectively; the dose was 1.5×10^{13} cm^{-2} at 100 keV for n^+ regions. Annealing was carried out by the capless annealing method in AsH$_3$/H$_2$/Ar atmosphere at 800°C for 20 min after W/Al gate metallization. MESFETs were frabricated with gate dimensions of $1.0\,\mu$m $\times$ $10\,\mu$m and 17-stage E/D gate ring oscillators, both with E- and DFETs of $1.0\,\mu$m $\times$ $20\,\mu$m gate. The built-in voltage was measured to be about 0.74 V, and the ideality factor was less than 1.1. The reverse breakdown voltage was more than 10 V at 10 μA. The results indicate that the V-doped epilayer has enough effect for isolation. The maximum transconductance was 240 mS/mm. The σ_{V_T} of 400 MESFETs fabricated on a 2-inch diameter was reported to be less than 100 mV.

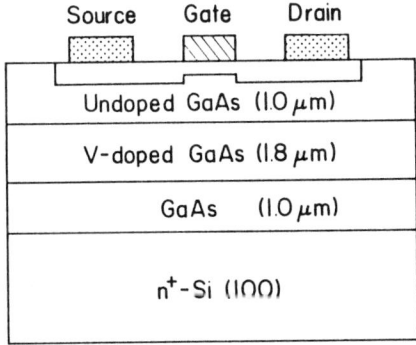

FIG. 46. Design of GaAs MESFET fabricated on GaAs/Si substrate. [From Nonaka *et al.* (1984).]

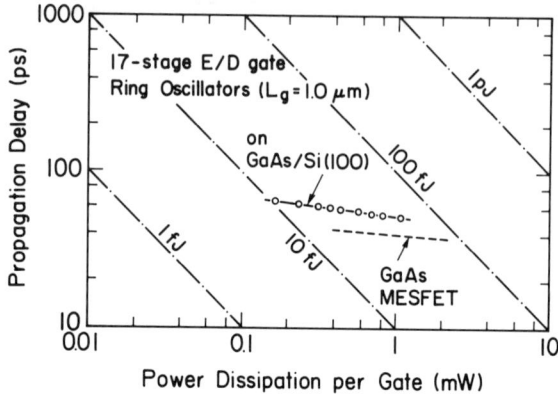

FIG. 47. Speed–power performance of E/D gate ring oscillators on GaAs/Si substrate and on conventional GaAs substrate, both having E- and DFETs 1.0 μm in gate length. [From Nonaka et al. (1984).]

Figure 47 shows the speed–power performance of the ring oscillator, indicating that the ring oscillator fabricated on Si $\langle 100 \rangle$ substrate has the same quality as that fabricated on bulk substrate.

IV. Summary and Future Problems

The history of progress in the area of the growth of high-quality MOVPE GaAs and GaAlAs epilayers and their applications to high-frequency and high-speed devices has been reviewed. Epilayers with smooth surfaces are obtained readily, although problems still remain unsolved since the detailed mechanisms of hillock formation and surface roughening are unknown at present. The important effects of the starting materials, substrate orientation, growth conditions and impurity doping on the epilayer quality have been stressed. The dominant residual impurities in MOVPE layers are estimated to be C and Si with smaller amounts of column II elements such as Zn, Mg, etc. Refinement of the starting materials together with the optimization of the growth conditions have greatly improved the epilayer purity, so that it is comparable to that grown by the conventional VPE and LPE techniques. Most of the conventional devices have been fabricated on MOVPE wafers, giving state-of-the-art performances. Accordingly, the MOVPE technique has passed through the "catch-up" phase. By utilizing a potential advantage, i.e., precise composition and thickness controllability of very thin layers, more sophisticated devices such as HEMTs have been realized. Moreover, a superlattice structure such as $(InAs)_1(GaAs)_1$ has been grown by MOVPE (Fukui and Saito, 1984), indicating that the MOVPE-technique has nearly the same status as MBE in controlling the monoatomic planer growth. This opens up the possibility of using MOVPE as a tool for developing new quantum effect devices, such as hot electron transistors, HETs (Heiblum, 1981) and resonant-tunnelling hot electron transistors, RHETs (Yokoyama et al., 1985).

It has been shown that GaAs epilayers can be grown on highly mismatched substrate such as Si. MESFETs and E/D ring oscillators fabricated by this technique showed performances comparable to those of homoepitaxial devices. This technique may remove the restrictions on using more fragile and smaller-diameter GaAs substrates. Furthermore, this technique leads to the integration of GaAs and Si devices on one chip, which would have a huge impact on future device development.

Another of the technical merits of MOVPE is that most of the compound semiconductors, and hence their alloys, can be grown, as shown experimentally by Manasevit (1981). One of the important materials systems is InGaAsP/InP, since this system has many advantages for long-wavelength electro-optical devices. The quality of InP, InGaAs, and InGaAsP epilayers has been improved year by year as reviewed by Stringfellow (1985) and Razeghi (1985). The progress strongly suggests that the MOVPE technique will be able to yield many kinds of new materials for future devices.

Although the MOVPE technique has been shown to be as effective as the MBE technique in designing and developing future devices, there still remain some problems, as stated below:

(1) Starting materials purity: Even for GaAs and GaAlAs, which have been extensively studied, there still exist impurity fluctuations caused by the inadequate impurity of starting-materials. Further purification of these materials is necessary. This is more serious for III–V materials other than GaAs and GaAlAs.

(2) Growth mechanism: In spite of the amount of work on MOVPE, relatively few reports have been presented on the growth mechanism, even on that of GaAs. The lack of detailed knowledge about the basic rate process of the growth will give rise to a more serious problem when growth of newer materials is attempted. Only the trial-and-error approach may be left to optimize the growth conditions.

(3) In-process monitoring: It appears that one of the weakest points of the MOVPE technique as opposed to MBE is the lack of *in situ* monitoring systems. The MBE system allows the installation of mass-spectrometric, RHEED and AES equipment, which enables us to observe the growth procedure directly. Furthermore, the information from *in situ* observation can be used as feedback data for real-time on-line growth control. Therefore, the development of unique in-process monitoring systems is expected for MOVPE as well, to compete with MBE.

(4) Optimization of growth apparatus: In most cases, present MOVPE reactors are nothing but laboratory-scale apparatuses. The reactors are constructed experimentally depending upon the know-how of individual researchers. This may indicate that the reactors are not completely optimized, which in turn makes it difficult to design and construct large-scale reactors for practical use.

Even with the solution of all these problems, the MOVPE technique has a definite place in compound semiconductor technology today and in the future.

REFERENCES

Aebi, V., Cooper III, C. B., Moon, R. L., and Saxana, R. R. (1981). *J. Crystal Growth* **55**, 517.
Aebi, V., Bandy, S., Nishimoto, C., and Zdasiuk, G. (1984). *Appl. Phys. Lett.* **44**, 1056.
Akiyama, M., Kawarada, Y., and Kaminishi, K. (1984a). *J. Crystal Growth* **68**, 39.
Akiyama, M., Kawarada, Y., and Kaminishi, K. (1984b). *Japan. J. Appl. Phys.* **23**, L843.
André, J. P., Gallais, A., and Hallais, J. (1977). *Inst. Phys. Conf. Ser.* **33a**, 1.
André, J. P., Schiller, C., Mittonneau, A., and Zdasiuk, G. (1983). *Inst. Phys. Conf. Ser.* **65**, 117.
Asai, H., and Sugiura, H. (1985). *Japan. J. Appl. Phys.* **24**, L815.
Asfar, M. N., Button, K. J., and McKoy, G. L. (1980). *Inst. Phys. Conf. Ser.* **56**, 701.
Bass, S. J. (1975). *J. Crystal Growth* **31**, 172.
Bass, S. J. (1978). *J. Crystal Growth* **44**, 29.
Bass, S. J., and Oliver, P. (1977). *Inst. Phys. Conf. Ser.* **33b**, 1.
Bhat, R., Connor, P. O., Temkin, H., Dingle, R., and Keramidas, V. G. (1981). *Inst. Phys. Conf. Ser.* **63**, 95.
Bhattacharya, P. K., Ku, J. W., Owen, S. J. T., Aebi, V., Cooper III, C. B., and Moon, R. L. (1980). *Appl. Phys. Lett.* **36**, 304.
Blatt, F. J. (1957). *In* "Solid State Physics" (F. Seitz and D. Turnbull, eds.), Vol. 4, p. 344. Academic Press, New York.
Bonnet, M., Visentin, N., Bessonneau, G., and Duchemin, J. P. (1981). *J. Crystal Growth* **55**, 246.
Botka, N., Sillmon, R. S., and Tseng, W. F. (1984). *J. Crystal Growth* **68**, 54.
Coleman, J. J., Dapkus, P. D., and Yang, J. J. J. (1981). *Electron. Lett.* **17**, 606.
Cooke, R. A., Hoult, R. A., Kirkman, R. F., and Stradling, R. A. (1978). *J. Phys.* **D11**, 945.
Dapkus, P. D., Manasevit, H. M., Hess, K. L., Low, T. S., and Stillman, G. E. (1981). *J. Crystal Growth* **55**, 10.
Dazai, K., Ihara, N., and Ozeki, M. (1974). *Fujitsu Sci. Technol. J.* **10**, 125.
DiLorenzo, J. V., and Machala, A. E. (1971). *J. Electrochem. Soc.* **118**, 1516.
Dingle, R., Störmer, H. L., Gossard, A. C., and Wiegmann, W. (1978). *Appl. Phys. Lett.* **33**, 665.
Druminski, M., Wolf, H. D., Zchauer, K. H., and Whittmark, K. (1982). *J. Crystal Growth* **57**, 318.
Duchemin, J. P., Bonnet, M., Koelsch, K., and Huyghe, D. (1978). *J. Crystal Growth* **45**, 81.
Duchemin, J. P., Bonnet, M., Koelsch, K., and Huyghe, D. (1979). *J. Electrochem. Soc.* **126**, 1134.
Fujisaki, Y., Takano, Y., Ishida, T., Sakaguchi, H., and Ono, Y. (1985). *Japan. J. Appl. Phys.* **24**, L899.

Fukui, T., and Saito, H. (1984). *Japan. J. Appl. Phys.* **23**, L521.
Gersten, S. W., Vendura, Jr., G. J., and Yeh, Y. C. M. (1986). *J. Crystal Growth* **77**, 286.
Glew, R. W. (1984). *J. Crystal Growth* **68**, 44.
Gottschalch, V., Petzke, W., and Butter, E. (1974). *Kristall Tech.* **9**, 209.
Hallais, J., Andre, J. P., Mircea-Roussel, A., Mahieu, M., Varon, J., and Boissy, M. C. (1981). *J. Electron. Mater.* **10**, 665.
Hayafuji, N., Mizuguchi, K., Ochi, S., and Murotani, T. (1986). *J. Crystal Growth* **77**, 281.
Herblum, M. (1981). *Solid State Electron.* **24**, 343.
Hersee, S. D., Hirtz, J. P., Baldy, M., and Duchemin, J. P. (1982). *Electron. Lett.* **18**, 1076.
Hiyamizu, S., Saito, S., Nanbu, K., and Ishikawa, T. (1983). *Japan. J. Appl. Phys.* **22**, L609.
Ikoma, H. (1968). *J. Phys. Soc. Japan* **25**, 1069.
Imamura, K., Yokoyama, N., Ohnishi, T., Suzuki, S., Nakai, K., Nishi, H., and Shibatomi, A. (1984). *Japan. J. Appl. Phys.* **23**, L342.
Ito, S., Shinohara, T., and Seki, Y. (1973). *J. Electrochem. Soc.* **120**, 1419.
Kamei, K., Tatematsu, M., Nakanisi, T., Tanaka, A., and Okano, S. (1979). *Proc. Intern. Electron Devices Meeting, Washington, D.C., 1979*, p. 380.
Kamei, K., Kawasaki, H., Chigira, T., Nakanisi, T., and Yoshimi, M. (1981). *Electron. Lett.* **17**, 450.
Kamei, K., Kawasaki, H., Higashiura, M., and Hori, S. (1984). *Inst. Phys. Conf. Ser.* **74**, 545.
Kawasaki, H., Imai, I., Tanaka, A., Tokuda, H., Higashiura, M., and Hori, S. (1986). *Trans. IECE Japan* **69**, 294.
Kobayashi, N., and Fukui, T. (1984). *Electron. Lett.* **20**, 888.
Kobayashi, N., Fukui, T., and Horikoshi, Y. (1982). *Japan. J. Appl. Phys.* **21**, L705.
Kobayashi, N., Fukui, T., and Tsubaki, K. (1984). *Japan. J. Appl. Phys.* **23**, 1176.
Kuech, T. F., and Veuhoff, E. (1984). *J. Crystal Growth* **68**, 148.
Kuech, T. F., Meyerson, B. S., and Veuhoff, E. (1984). *Appl. Phys. Lett.* **44**, 986.
Kumagai, U., Kawai, H., Mori, Y., and Kaneko, (1984). *Appl. Phys. Lett.* **45**, 1322.
Lang, D. V., and Logan, R. A. (1979). *Phys. Rev.* **B19**, 1015.
Lewis, C. R., Dietze, W. T., and Ludowise, M. J. (1983). *J. Electron. Mater.* **12**, 507.
Maeda, H. (1986). *Hyomen* **24**, 89 (in Japanese).
Maluenda, J., and Frijlink, P. M. (1983). *Japan. J. Appl. Phys.* **22**, L127.
Manasevit, H. M. (1968). *Appl. Phys. Lett.* **12**, 156.
Manasevit, H. M. (1981). *J. Crystal Growth* **55**, 1.
Manasevit, H. M., and Simpson, W. I. (1969). *J. Electrochem. Soc.* **116**, 1725.
Manasevit, H. M., Dapkus, P. D., Hess, K. L., and Yang, J. J. (1980). 22nd Annual Electronic Materials Conf., Ithaca, NY, 1980. *Pap. D-1.*
Matsumoto, T., Bhattacharya, P. K., and Ludowise, M. J. (1982). *Appl. Phys. Lett.* **41**, 662.
Mori, Y., and Watanabe, N. (1981). *J. Appl. Phys.* **52**, 2792.
Mori, Y., Ikeda, M., Sato, H., Kaneko, K., and Watanabe, N. (1981). *Inst. Phys. Conf. Ser.* **63**, 95.
Morkoç, H., Andrews, J., and Aebi, V. (1979). *Electron. Lett.* **15**, 105.
Nakanisi, T. (1984). *J. Crystal Growth* **68**, 282.
Nakanisi, T., and Kasiwagi, K. (1974). *Japan. J. Appl. Phys.* **13**, 484.
Nakanisi, T., and Tanaka, A. (1980). 22nd Annual Electronic Materials Conf., Ithaca, NY, 1980. *Pap. D-7.*
Nakanisi, T., Udagawa, T., and Kamei, K. (1981). *J. Crystal Growth* **55**, 255.
Nakatsuka, S., Ono, Y., Kajimura, T., and Nakamura, M. (1983). *Extended Abstracts 15th Conf. on Solid State Devices and Materials, Tokyo, 1983*, p. 297.
Nelson, R. J. (1977). *Appl. Phys. Lett.* **31**, 351.

Nonaka, T., Akiyama, M., Kawarada, Y., and Kaminishi, K. (1984). *Japan. J. Appl. Phys.* **23**, L919.
Norris, P., Zemon, S., Koteles, E., and Lambert, G. (1985). 1985 Annual Electronic Materials Conf., Boulder, Colorado, CO, 1985.
Ohyama, T., Otsuka, E., Matsuda, O., Mori, Y., and Kaneko, K. (1982). *Japan. J. Appl. Phys.* **21**, L583.
Okamoto, K., Onozawa, O., and Imai, T. (1984). *J. Appl. Phys.* **56**, 2993.
Ozeki, M., Nakai, K., Dazai, K., and Ryuzan, O. (1973). *Japan. J. Appl. Phys.* **12**, 478.
Ozeki, M., Nakai, K., Dazai, K., and Ryuzan, O. (1973). *Japan. J. Appl. Phys.* **13**, 120.
Parsons, J. D., and Krajenbrick, F. G. (1984). *J. Crystal Growth* **68**, 60.
Razeghi, M. (1985). In "Semiconductors and Semimetals" (W. T. Tsang, ed.), Vol. 22, Part A, p. 299. Academic Press, Orlando, Florida.
Rode, D. L., and Knight, S. (1971). *Phys. Rev.* **B3**, 2534.
Rossi, J. A., Wolfe, C. M., and Dimmock, J. O. (1970). *Phys. Rev. Lett.* **25**, 1614.
Roth, A. P., Yakimova, R., and Sundaram, V. S. (1984). *J. Crystal Growth* **68**, 65.
Samuelson, L., Omling, P., Titze, H., and Grimmeis, H. G. (1981). *J. Crystal Growth* **55**, 164.
Sano, Y., Nakamura, H., Akiyama, M., Egawa, T., Ishida, T., and Kaminishi, K. (1984). *Japan. J. Appl. Phys.* **23**, L290.
Sealy, J. R., Kreismanis, V. G., Wagner, D. K., and Woodall, J. M. (1983). *Appl. Phys. Lett.* **42**, 83.
Seki, Y., Tanno, K., Iida, K., and Ishii, E. (1975). *J. Electrochem. Soc.* **122**, 1108.
Shino, T., Yanagawa, S., Yamada, Y., Arai, K., Kamei, K., Chigira, T., and Nakanisi, T. (1981). *Electron. Lett.* **17**, 738.
Stringfellow, G. B. (1985). In "Semiconductors and Semimetals" (W. T. Tsang, ed.), Vol. 22, Part A, p. 209. Academic Press, Orlando, Florida.
Stringfellow, G. B., and Hom, G. (1979). *Appl. Phys. Lett.* **34**, 794.
Takagishi, S. (1983). *Japan. J. Appl. Phys.* **22**, L795.
Takakuwa, H., Kato, Y., Watanabe, S., and Mori, Y. (1985). *Electron. Lett.* **21**, 125.
Takakuwa, H., Tanaka, K., Mori, Y., Arai, M., Kato, Y., and Watanabe, S. (1986). *IEEE Trans. Electron Devices* **ED-33**, 595.
Tamamura, K., Ohhata, T., Kawai, H., and Kojima, C. (1986). *J. Appl. Phys.* **59**, 3549.
Tanaka, K., Takakuwa, H., Nakamura, F., Mori, Y., and Kato, Y. (1986a). *Electron. Lett.* **22**, 488.
Tanaka, K., Ogawa, M., Togashi, K., Takakuwa, H., Ohke, H., Kanazawa, M., Kato, Y., and Watanabe, S. (1986b). *IEEE Trans. Electron Devices* **ED-33**, 2053.
Tanno, K., and Seki, Y. (1975). Unpublished materials.
Thrush, E. J., and Whiteaway, J. E. A. (1981). *Inst. Phys. Conf. Ser.* **56**, 337.
Terao, H., and Sunakawa, H. (1984). *J. Crystal Growth* **68**, 157.
Uetake, K., Katano, F., Kamiya, M., Misaki, T., and Higashisaka, A. (1985). *Inst. Phys. Conf. Ser.* **79**, 505.
Veuhoff, E., Kuech, T. F., and Meyerson, B. S. (1985). *J. Electrochem. Soc.* **132**, 1958.
Wagner, E. E., Hom, G., and Stringfellow, G. B. (1981). *J. Electron. Mater.* **10**, 239.
Wallis, R. H., Poisson, M. A. F., Bonnet, M., Beuchet, G., and Duchemin, J. P. (1981). *Inst. Phys. Conf. Ser.* **56**, 73.
Wang, W. I. (1984). *Appl. Phys. Lett.* **44**, 1149.
Wang, C. C., Dougherty, F. C., Zanzucci, R. J., and McFarlane, III, S. H. (1974). *J. Electrochem. Soc.* **121**, 572.
Wang, W. L., Li, S. S., and Lee, D. H. (1986). *J. Electrochem. Soc.* **133**, 196.
Watanabe, M. O., Tanaka, A., Nakanisi, T., and Zohta, Y. (1981). *Japan, J. Appl. Phys.* **20**, L429.

Watanabe, M. O., Tanaka, A., Udagawa, T., and Nakanisi, T. (1983). *Japan. J. Appl. Phys.* **22,** 923.

Watanabe, M. O., Okajima, M., Motegi, N., Nakanisi, T., and Zohta, Y. (1984). Unpublished materials. Referenced by Nakanisi, T. (1984). *J. Crystal Growth* **68,** 282.

White, A. M., Dean, P. J., Ashen, D. J., Mullin, J. B., Webb, M., Day, B., and Greene, P. D. (1973). *J. Phys.* **C6,** L243.

Williams, F. V. (1964). *J. Electrochem. Soc.* **111,** 887.

Wolfe, C. M., and Stillman, G. E. (1970). *Inst. Phys. Conf. Ser.* **9,** 3.

Wolfe, C. M., Stillman,, G. E., and Korn, D. M. (1976). *Inst. Phys. Conf. Ser.* **33b,** 120.

Yang, J. J., Simpson, W. I., and Moudy, L. A. (1982). *J. Appl. Phys.* **53,** 771.

Yokoyama, N., Imamura, K., Muto, S., Hiyamizu, S., and Nishi, H. (1985). *Inst. Phys. Conf. Ser.* **79,** 739.

CHAPTER 4

High Electron Mobility Transistor and LSI Applications

T. Mimura

COMPOUND SEMICONDUCTOR DEVICES LABORATORY
FUJITSU LABORATORIES, LTD.
ATSUGI, JAPAN

I.	INTRODUCTION	157
II.	HEMT STRUCTURES AND FABRICATION PROCESS	158
III.	ELECTRICAL CHARACTERISTICS	160
	1. Current–Voltage Characteristics	160
	2. Capacitance–Voltage Characteristics	166
	3. High-Frequency Characteristics	169
IV.	LSI TECHNOLOGY	173
	4. Gate Performance	173
	5. Fabrication Technology	174
	6. Integrated Circuit Implementations	179
V.	SUMMARY	183
	APPENDIX A	184
	APPENDIX B	186
	APPENDIX C	187
	APPENDIX D	192
	REFERENCES	192

I. Introduction

Electronic devices with high-speed capabilities are needed for the application of future electronic systems such as supercomputers, high-speed signal processors, and satellite communications links.

The superior electron dynamics that GaAs offers over Si have long pointed to the speed advantage of field effect transistors of GaAs over those of Si. Recent advances in GaAs Schottky gate metal-semiconductor field effect transistor (MESFET) technology have enabled the demonstration of GaAs integrated circuits with high speeds and low power-consumption capability. Electron mobility in the MESFET channel with typical donor concentrations around 10^{17} cm^{-3} ranges from 4000 cm^2/Vsec to 5000 cm^2/Vsec at room temperature. The mobility in the channel at 77 K is not much higher than at room temperature. This is because of ionized impurity scattering. In undoped GaAs, electron mobility of 2 to 3×10^5 cm^2/Vsec has been obtained at 77 K. The mobility of GaAs with feasible high electron concentrations for facilitating the fabrication of devices was found to increase through the modulation-doping technique demonstrated in GaAs/AlGaAs superlattices (Dingle et al., 1978).

High electron mobility transistors (HEMTs) based on selectively doped GaAs/n-AlGaAs heterojunction structures can greatly improve the 77 K channel mobility as well as the 300 K channel mobility, offering new possibilities for high-speed, low-power, very large-scale integration (Mimura et al., 1980).

This chapter will present the basic characteristics of the HEMT and its applications to high-speed, low-power LSI circuits.

II. HEMT Structures and Fabrication Process

The cross-sectional structure of a HEMT is shown in Fig. 1. The epilayers consisting of undoped GaAs, Si-doped n-type $Al_xGa_{1-x}As$ and n-type GaAs are grown on a $\langle 100 \rangle$ semi-insulating GaAs substrate by molecular-beam epitaxy (MBE). Some epilayers were grown with a 6-nm-thick undoped $Al_xGa_{1-x}As$ spacer layer inserted between the undoped GaAs and n-$Al_xGa_{1-x}As$ layers to enhance electron mobility. The AlAs mole fraction, denoted as x, was tentatively selected as about 0.3. It can be expected that higher AlAs mole fractions would increase the maximum carrier density achievable in the heterojunctions. This reduces the sheet resistivity of the two-dimensional electron gas (2DEG), resulting in a reduction of the source–gate parasitic resistance, which, in turn, enhances the high-frequency performance of the device. Generally, however, $Al_xGa_{1-x}As$ with a high AlAs mole fraction exhibits crystal deterioration such as a slight hazy surface and an increase in deep traps. For fabrication of the prototype HEMT, the

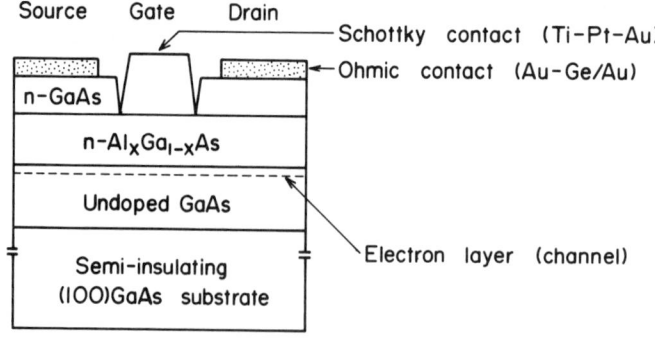

FIG. 1. Typical cross section of a HEMT.

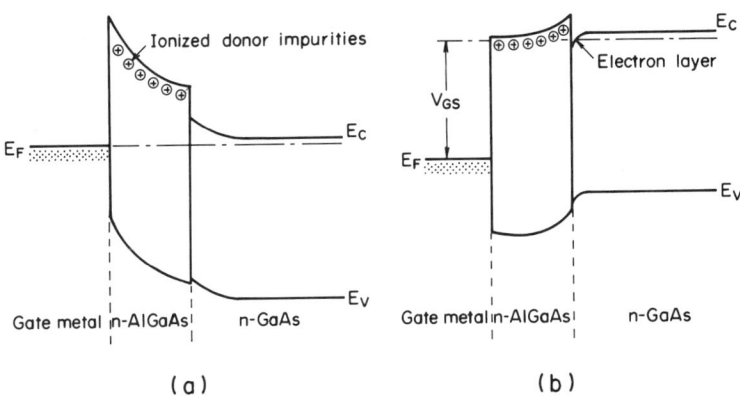

FIG. 2. Energy band diagrams (a) at thermal equilibrium and (b) with positive metal gate voltage higher than the threshold of an enhancement-mode HEMT.

doping concentration was 2×10^{18} cm^{-3}, the AlGaAs thickness was 0.06 μm and the *n*-GaAs thickness was 0.05 μm (Mimura *et al.*, 1981a).

Let us consider what happens when the AlGaAs layer becomes thinner. As the AlGaAs layer thickness is reduced, the space charges caused by donor ions in the AlGaAs layer will be insufficient to support the Schottky barrier potential.

Therefore, the Schottky barrier depletion layer will be expected to extend into the undoped GaAs layer, causing upward band bending and diminishing the 2DEG at the interface. The resultant energy band diagrams are shown in Fig. 2. In Fig. 2 we assume the undoped GaAs layer to be *n*-type (see Appendix A). The positive metal voltage, which is higher than the threshold voltage V_T, attracts electrons to the surface of the undoped GaAs layer, and the bottom of the conduction band falls below the Fermi level, as shown in Fig. 2b. This electron accumulation layer acts as the current channel, and in this case, a HEMT can operate in the enhancement mode. Thus, in the enhancement-mode (E) HEMT, donors in AlGaAs serve not to form the two-dimensional electron channel but to control the surface potential of the undoped GaAs layer.

Device fabrication starts with the etching of mesa islands down to the undoped GaAs layer adjacent to the semi-insulating substrate to isolate the active region. The source and drain regions are metallized with a Au–Ge eutectic alloy with a Au overlay alloying to form ohmic contacts to the 2DEG. Selective dry etching with CCl_2F_2 and He (Hikosaka *et al.*, 1981) was carried out to remove the top epilayer of *n*-GaAs for fabrication of a recessed gate structure. A Schottky contact for a gate was provided by deposition of Ti–Pt–Au on the surface of *n*-AlGaAs.

III. Electrical Characteristics

1. Current–Voltage Characteristics

The drain current I_D of a HEMT can be written as

$$I_D = -\mu_n W_G Q_n(x) \frac{dV(x)}{dx}, \qquad (1)$$

where μ_n is the electron mobility, W_G is the gate width, $Q_n(x)$ is the electron charge per unit area, and $V(x)$ is the potential of the surface of the channel with respect to the source end of the gate ($x = 0$). At the drain end of the gate, x becomes L_G, the gate length.

In Eq. (1), we assume that electron mobility of the 2DEG is independent of electron concentration, which changes with the gate voltage. It was experimentally observed, however, that the Hall mobility at 77 K increases with electron concentration (see Appendix B). This dependence is weak, especially for a high-quality material with high electron mobility. This means the constant-mobility assumption is acceptable.

Change of electron charge in the channel ΔQ_n can be readily found from the charge neutrality condition expressed by

$$\Delta Q_g + \Delta Q_n = 0, \qquad (2)$$

where ΔQ_g and ΔQ_n represent induced charges on the metal gate and in the channel, respectively. Equation (2) assumes that the AlGaAs layer is completely depleted (Appendix C) and no variation of charges in AlGaAs takes place for any nonequilibrium conditions under which a HEMT can operate normally.

The energy band diagrams used for device modeling are shown in Fig. 3. An undoped GaAs layer grown by MBE is experimentally found to be lightly

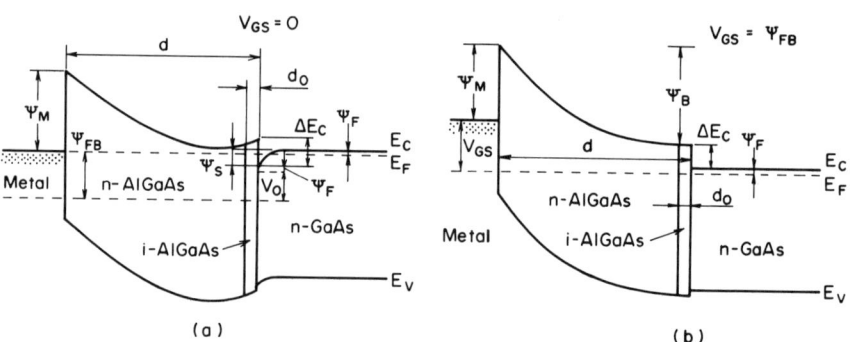

FIG. 3. Energy band diagrams used for device modeling: (a) thermal equilibrium; (b) flatband condition.

n-type and the donor concentration is about 1×10^{15} cm^{-3}, as shown in Appendix A. It is assumed that two depletion layers, one from the surface and the other from the interface, extend into the whole n-AlGaAs layer. The surface depletion region results from trapping of free electrons at surface states, and the interface depletion region results from transfer of electrons into the adjacent undoped GaAs layer. Surface states cause Fermi level pinning at the AlGaAs surface, and the Fermi level of the gate metal is matched to the pinning point. Here, the pinning point is assumed to be 1.2 eV below the conduction band in accordance with the empirical law (Mead and Spitzer, 1964).

The application of a voltage ($V_{GS} - \Psi_{FB}$) across a gate changes the potential across the AlGaAs layer by V_0 and the surface potential by Ψ_s, as shown in Fig. 3a.

Hence,

$$V_{GS} - \Psi_{FB} = V_0 + \Psi_s, \tag{3}$$

where Ψ_{FB} is flatband potential.

When $V_{GS} = \Psi_{FB}$, $Q_n(\Psi_{FB}) = 0$. The flatband potential Ψ_{FB} can be found by noting that the quasi-Fermi level is constant throughout Fig. 3b. Summing potentials around the circuit, we have

$$\Psi_{FB} = \Psi_M - (\Delta E_C + \Psi_F + \Psi_B), \tag{4}$$

where Ψ_M is the metal-semiconductor barrier potential, ΔE_C is the difference in energy between the AlGaAs and GaAs conduction band edges, Ψ_F is the Fermi potential deep in the undoped GaAs bulk and Ψ_F is the built-in potential of AlGaAs at the flatband condition.

Ψ_B is given by

$$\Psi_B = \frac{q n_D (d - d_0)^2}{2 \varepsilon_{s1}} \tag{5}$$

where n_D is a donor concentration in the AlGaAs layer, ε_{s1} is the permittivity of AlGaAs, d is the layer thickness and d_0 is the thickness of undoped AlGaAs.

The induced incremental gate charge density per unit area ΔQ_g can be expressed by

$$\Delta Q_g = C_0 V_0 = -\Delta Q_n, \tag{6}$$

where C_0 ($= \varepsilon_{s1}/d$) is the AlGaAs capacitance per unit area. According to the definition, the induced electron charge density ΔQ_n is given by

$$\Delta Q_n = Q_n(V_{GS}) - Q_n(V_{GS} = \Psi_{FB}) = Q_n(V_{GS}) \tag{7}$$

Then, from Eqs. (6) and (7) we have

$$Q_n = -C_0 V_0. \tag{8}$$

Substituting Eq. (3) into Eq. (8) yields

$$Q_n(x) = -C_0(V_{GS} - \Psi_{FB} - \Psi_s), \tag{9}$$

Therefore, we can obtain $Q_n(x)$ as

$$Q_n(x) = -C_0\{V_{GS} - V(x) - \Psi_{ss} - \Psi_{FB}\}, \tag{10}$$

where Ψ_{ss} is the surface potential at $x = 0$, the source end of the gate, or $\Psi_{ss} + V(x)$ is the surface potential at any point x along the channel.

The surface potential Ψ_{ss} can be expressed by

$$\Psi_{ss} = \phi_s + \Psi_F, \tag{11}$$

where ϕ_s denotes the surface Fermi potential.

Substituting Eqs. (4) and (11) into Eq. (10), we have

$$Q_n(x) = -C_0\{V_{GS} - V(x) - \phi_s - \Psi_M + \Delta E_c + \Psi_B\}. \tag{12}$$

Since ϕ_s, which is a function of V_{GS}, is small as compared to the other terms, Eq. (9) can be approximated as

$$Q_n(x) = -C_0\{V_{GS} - V(x) - V_T\}, \tag{13}$$

where

$$V_T = \Psi_M - \Psi_B - \Delta E_c. \tag{14}$$

The absolute value of ϕ_s falls below one-tenth the value of the quantity $(V_{GS} - \Psi_{FB})$ when $(V_{GS} - \Psi_{FB})$ becomes greater than 0.07 V at 77 K for a 0.77-μm-thick AlGaAs layer (see Appendix D).

In more sophisticated analysis, Lee et al. (1983b) have obtained an approximate solution for the variation of the Fermi potential during operation. Their approximation requires corrections to the Fermi potential (ΔE_{F0}) and to the channel depth (Δd)

$$Q_n(x) = -\frac{\varepsilon_{s1}}{(d + \Delta d)}(V_{Gs} - V(x) - V_T - \Delta E_{F0}) \tag{13'}$$

where $\Delta E_{F0} = 0$ at $T = 300$ K, $\Delta E_{F0} = 25$ mV at $T = 77$ K and below, and $\Delta d = 8$ nm for GaAs/AlGaAs systems. Substituting Eq. (13) into Eq. (1), and integrating I_D from $x = 0$ to $x = L_G$ with the boundary conditions $V(0) = 0$ and $V(L_G) = V_{DS}$, we obtain

$$\int_0^{L_G} I_D \, dx = \mu_n W_G \int_0^{V_{DS}} C_0\{V_{GS} - V(x) - V_T\} \, dV(x). \tag{15}$$

Carrying out the integration, we obtain for the drain current

$$I_D = \frac{\mu_n W_G C_0}{L_G}\{(V_{GS} - V_T)V_{DS} - \tfrac{1}{2}V_{DS}^2\}. \tag{16}$$

The last term in brackets of Eq. (16) can be negligible as compared to the others, especially for small V_{DS}. Under these conditions, Eq. (16) reduces to

$$I_D \cong \frac{\mu_n W_G C_0}{L_G}(V_{GS} - V_T)V_{DS} \qquad (V_{DS} \to 0). \tag{17}$$

It is noted in Eq. (17) that I_D becomes zero for a small but finite V_{DS} when V_{GS} is equal to V_T, which is the threshold voltage.

According to gradual channel approximation (Shockley, 1952), I_D saturates at the pinchoff voltage V'_{DS}, which can be given by

$$\left(\frac{dI_D}{dV_{DS}}\right)_{V_{DS} = V'_{DS}} = 0. \tag{18}$$

We can find

$$V_{DS} = V_{GS} - V_T, \tag{19}$$

and then, under saturation conditions,

$$I_{DS} = \frac{\mu_n W_G C_0}{2L_G}(V_{GS} - V_T)^2. \tag{20}$$

Thus, a HEMT exhibits the so-called square-law characteristics.

It is recognized that any useful analysis for short-channel GaAs FETs must properly take into account the drift velocity characteristics as a function of electric field, since the electric field in a short-channel device can reach E_m (= 3.2 kV/cm), the value at which the electron drift velocity becomes maximum, at relatively low drain voltage.

In the analysis by Turner and Wilson (1963), the velocity saturation is assumed to occur when the drain field reaches E_m, and the saturation current is calculated by imposing the boundary condition that the field at the drain be equal to E_m.

When we consider the common gate configuration to simplify the analysis, V_{DS} can be written as

$$V_{DS} = V_{GS} - V_{GD}. \tag{21}$$

Then, Eq. (16) can be written as

$$I_D = I_0(\eta^2 - \zeta^2), \tag{22}$$

where

$$I_0 = \frac{\mu_n W_G C_0 V_T^2}{2L_G}, \tag{23}$$

$$\eta^2 = \left(\frac{V_S^I - V_G^I + V_T}{V_T}\right)^2, \tag{24}$$

and

$$\zeta^2 = \left(\frac{V_D^I - V_G^I + V_T}{V_T}\right)^2. \tag{25}$$

The quantities η and ζ denote the normalized potential differences at the source and drain, respectively.

The current saturation condition is given by the boundary condition, $dU/dx = E_m$ at $x = L_G$, where U is the electrostatic potential difference between the gate and the channel. The drain saturation current I_{DS} is then given by

$$I_{DS} = -\frac{2I_0 E_m L_G \zeta_{sat}}{V_T}, \tag{26}$$

where ζ_{sat} is the value of ζ at the saturation of I_D and must also satisfy Eq. (22). That is,

$$I_{DS} = I_0(\eta^2 - \zeta_{sat}^2). \tag{27}$$

Then, we can calculate I_{DS} from Eqs. (26) and (27) linked by ζ_{sat}.

If $V_T = 0$, Eq. (16) can be written as

$$I_D = \frac{\mu_n W_G C_0}{2L_G}(V_{GS}^2 - V_{GD}^2), \tag{28}$$

and then I_{DS} can be expressed as

$$I_{DS} = -E_m W_G \mu_n C_0 V_{GDsat}, \tag{29}$$

where V_{GDsat} is the value of V_{GD} at the saturation of I_D and must also satisfy Eq. (28), yielding

$$I_{DS} = \frac{\mu_n W_G C_0}{2L_G}(V_{GS}^2 - V_{GDsat}^2). \tag{30}$$

The expression for I_{DS} can be obtained from Eqs. (29) and (30) as

$$I_{DS} = \frac{\mu_n W_G C_0}{2L_G}\left\{V_{GS}^2 - \left(\frac{I_{DS}}{E_m W_G \mu_n C_0}\right)^2\right\}. \tag{31}$$

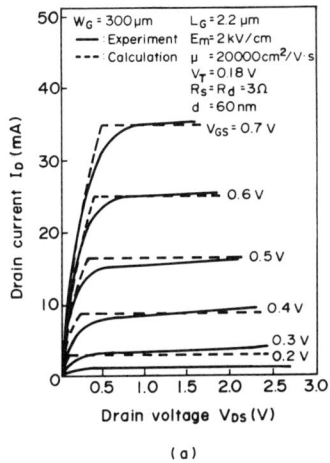
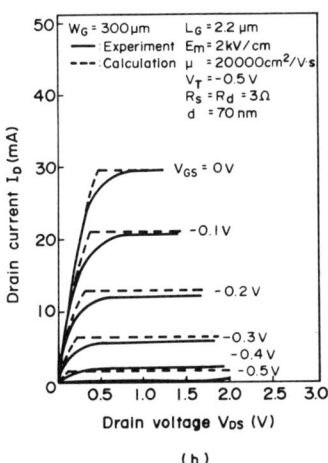

FIG. 4. Drain current–voltage characteristics at 77 K of (a) enhancement-mode HEMT, and (b) depletion-mode HEMT. Solid and dashed lines show the experimental and calculated I–V characteristics, respectively.

Considering the source and drain parasitic resistance R_S and R_d, the gate and drain voltage V_G and V_D are written as

$$V_G^I = V_G - I_D R_S$$
$$V_D^I = V_D - I_D(R_S + R_d). \qquad (32)$$

From Eqs. (22) to (32) we calculate the drain current–voltage characteristics of the HEMT.

The measured drain current–voltage characteristics obtained for the short-channel ($L_G = 2.2\,\mu$m) depletion-mode and enhancement-mode HEMTs are shown in Fig. 4. The material and device parameters used in the calculations are inserted in Fig. 4. Reasonable agreement beween experimental and calculated results is found.

It is to be noted in Fig. 4 that the short-channel D- and EHEMTs operating in a high-field regime exhibit the square-law $I_{DS} = K(V_{GS} - V_T)^2$ characteristics.

Dependence of the transconductance g_m ($g_m = \partial I_{DS}/\partial V_{GS}$) and K-value ($K = I_{DS}/(V_{GS} - V_T)^2$) of EHEMTs on gate length was measured at both 77 K and 300 K and is plotted in Fig. 5. Dashed lines indicate L_G^{-1}-dependence of K and g_m expected from the gradual channel approximation. Below 1 μm gate length at 300 K, K and g_m drop off from the L_G^{-1}-dependence. Velocity saturation effect and parasitic source resistances probably play a significant role in these results. The 0.5-μm-gate EHEMT at 77 K exhibits g_m of 0.5 S/mm, which is the highest value ever reported for any FET-type device.

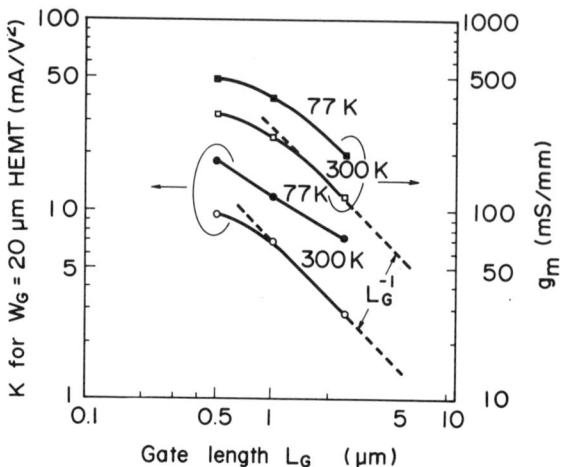

FIG. 5. Dependence of the K-value and transconductance of enhancement-mode HEMTs on the gate length.

No significant variation in threshold voltages with gate length was observed in the range from $L_G = 10\,\mu\text{m}$ to $L_G = 0.5\,\mu\text{m}$. This indicates that reducing the size of HEMTs is easily acceptable for improving the performance without causing problems with short-channel effects.

2. CAPACITANCE–VOLTAGE CHARACTERISTICS

According to the definition, the gate–source capacitance C_{gs} per unit area of the HEMT is given by

$$C_{gs} = \frac{dQ_g}{dV_{GS}}. \tag{33}$$

It can be expressed in the normalized form

$$\frac{C_{gs}}{C_0} = \left(1 + \frac{C_0}{C_s}\right)^{-1}, \tag{34}$$

where $C_s = -dQ_s/d\Psi_s$ is the carrier capacitance and C_0 is the AlGaAs capacitance.

To obtain C_{gs} as a function of $V_{GS} - \Psi_{FB}$, Q_s is first computed as a function of surface potential Ψ_s by using Eqs. (C-7) to (C-11) in Appendix C. C_s is graphically obtained for each Ψ_s. Then $V_{GS} - \Psi_{FB}$ can be found for Ψ_s, for a given AlGaAs thickness, with noting that

$$V_{GS} - \Psi_{FB} = \Psi_s + \frac{Q_s}{C_0}. \tag{35}$$

The ratio (C_{gs}/C_0) can be obtained from Eq. (34).

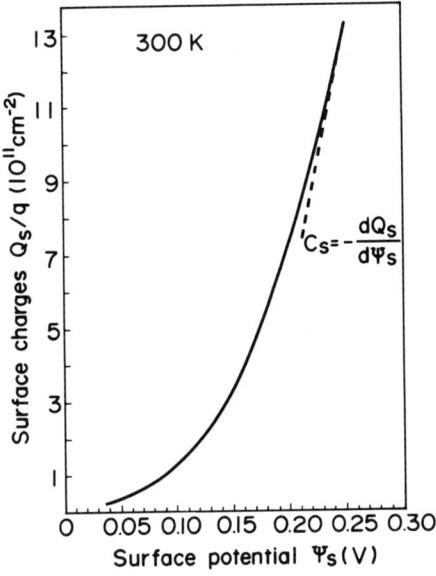

FIG. 6. Surface charge Q_s vs. surface potential Ψ_s.

The relationship is shown in Fig. 6. With the aid of Fig. 6, the carrier capacitance C_s can be obtained graphically as a function of Ψ_s. Using Eqs. (35) and (C-7) to (C-11), the relationship between V_{GS} and Ψ_s is obtained, as shown in Fig. 7.

The flatband potential Ψ_{FB} is calculated to be -0.65 V, assuming

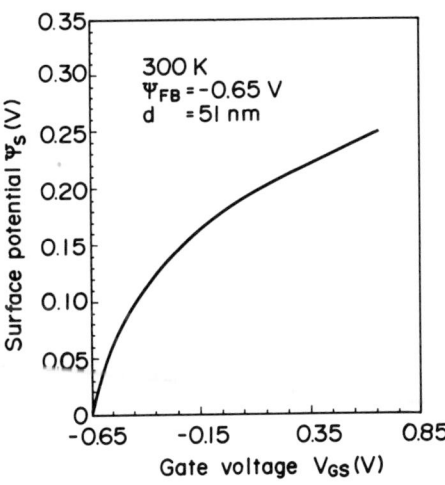

FIG. 7. Surface potential Ψ_s as a function of gate voltage V_{GS}.

that $\Psi_M = 1.2$ V, $\Delta E_c = 0.32$ eV, $N_D = 10^{15}$ cm^{-3}, d (thickness of n-AlGaAs) $= 51$ nm, d_0 (thickness of undoped AlGaAs) $= 0$, n_D (donor concentration in n-AlGaAs) $= 1.8 \times 10^{18}$ cm^{-3}, and ε_{s1} (dielectric constant of Al$_{0.3}$Ga$_{0.7}$As) $= 12\varepsilon_0$.

Figure 8 shows the normalized gate–source capacitance as a function of the gate voltage. As the gate voltage increases, the electron concentration at the undoped GaAs surface increases, causing the gate–source capacitance to approach the AlGaAs capacitance C_0. It is also found that C_{gs}/C_0 rapidly decreases as V_{GS} approaches the threshold.

When the drain current flows, the electron charge varies along the channel. This results in carrier capacitance variation along the channel. For the saturation region, the carrier capacitance at the drain end of the gate is expected to become significantly smaller than that at the source end of the gate. This causes a drop in the total gate–source capacitance C_{gs} to a value somewhat less than the prediction shown in Fig. 8.

We calculate the gate capacitance in the Shockley regime. The total channel charge Q_n is given by

$$Q_n = W_G \int_0^{L_G} Q_n(x)\, dx = W_G \int_0^{V_{DS}} Q(x) \frac{dx}{dV}\, dV. \quad (36)$$

Using Eq. (1), Q_n can be expressed by

$$Q_n = -\frac{\mu_n W_G^2}{I_D} \int_0^{V_{DS}} Q_n^2\, dV$$

$$= -\frac{C_0^2 \mu_n W_G^2}{I_D} \int_0^{V_{DS}} (V_{GS} - V_T - V(x))^2\, dV. \quad (37)$$

Integrating Eq. (37), and also using Eq. 16, we obtain

$$Q_n = \frac{C_0^2 \mu_n W_G^2}{3 I_D} [(V_{GS} - V_T - V_{DS})^3 - (V_{GS} - V_T)^3]$$

$$= \frac{C_0 W_G L_G [(V_{GS} - V_T - V_{DS})^3 - (V_{GS} - V_T)^3]}{3[(V_{GS} - V_T)V_{DS} - V_{DS}^2/2]}. \quad (38)$$

Then, by definition

$$C_{gs} = \frac{\partial Q_g}{\partial V_{GS}} = -\frac{\partial Q_n}{\partial V_{GS}}$$

$$= C_0 W_G L_G \left(1 - \frac{1}{12[(V_{GS} - V_T)/V_{DS} - \frac{1}{2}]^2}\right). \quad (39)$$

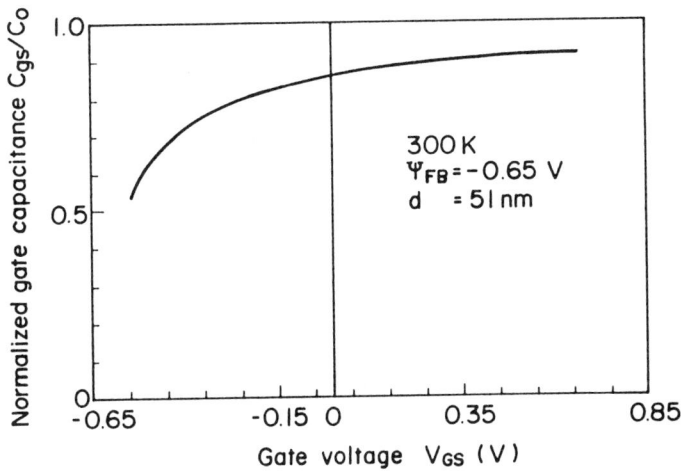

FIG. 8. Normalized gate–source capacitance C_{gs} as a function of gate voltage V_{GS}.

Thus, at the pinchoff voltage $V_{DS} = V_{GS} - V_T$, the total gate-source capacitance C_{gs} reduces to two-thirds of the capacitance of the gate AlGaAs layer, $C_0 W_G L_G$.

3. High-Frequency Characteristics

It is important to access the current-gain cutoff frequency f_T of the HEMT, since high-speed capability depends on the f_T value of the devices, as will be discussed in the next section.

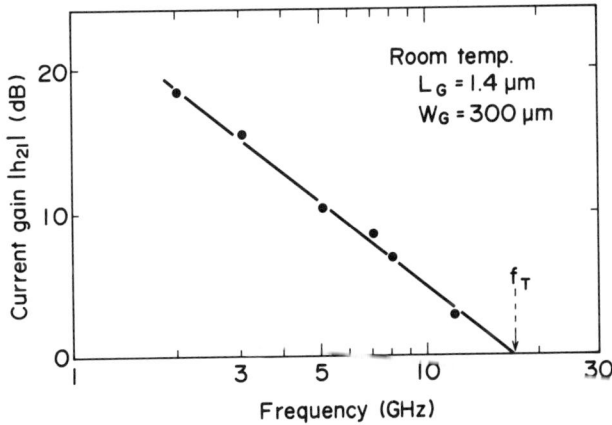

FIG. 9. Current gain $|h_{21}|$ for the HEMT, derived from the measured s-parameters, plotted as a function of frequency.

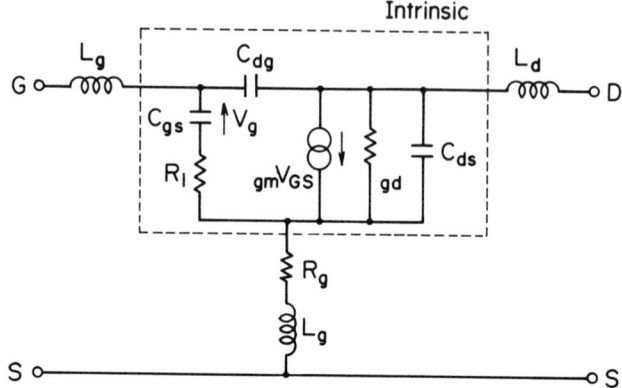

FIG. 10. Equivalent circuit model describing the small-signal characteristics of the HEMT.

The s-parameters were measured to derive the current gain $|h_{21}|$. The measurement current gain $|h_{21}|$ for the 1.4-μm-gate HEMT is plotted in Fig. 9 as a function of frequency. The current-gain cutoff frequency at which $|h_{21}| = 0$ dB is 17 GHz.

The equivalent circuit model used to describe the small-signal characteristics of the HEMT is shown in Fig. 10. The equivalent circuit includes intrinsic elements as well as extrinsic elements. The intrinsic device consists of low-frequency transconductance g_m, drain conductance g_d, channel resistance R_i, gate–source capacitance C_{gs}, and gate–drain feedback capacitance C_{gd}. The extrinsic elements are parasitic source resistance R_s, drain–source capacitance C_{ds}, and source, drain, and gate lead inductance L_s, L_d, and L_g, respectively.

The values of equivalent circuit elements are determined by fitting the frequency dependence of s-parameters simulated from the equivalent circuit to the measured one. Typical values are listed in Table I.

TABLE I

EQUIVALENT CIRCUIT PARAMETERS DEFINED IN FIG. 10[a]

Intrinsic elements	Extrinsic elements
$g_m = 37.9$ mS	$C_{ds} = 0.01$ F
$C_{gs} = 0.3$ pF	$R_s = 20.0$ Ω
$C_{dg} = 0.021$ pF	$L_s = 0.012$ nH
$R_i = 14.1$ Ω	$L_g = 0.56$ nH
$g_d = 5$ mS	$L_d = 0.39$ nH

[a] $L_G/W_G = 1.4\,\mu\text{m}/300\,\mu\text{m}$.

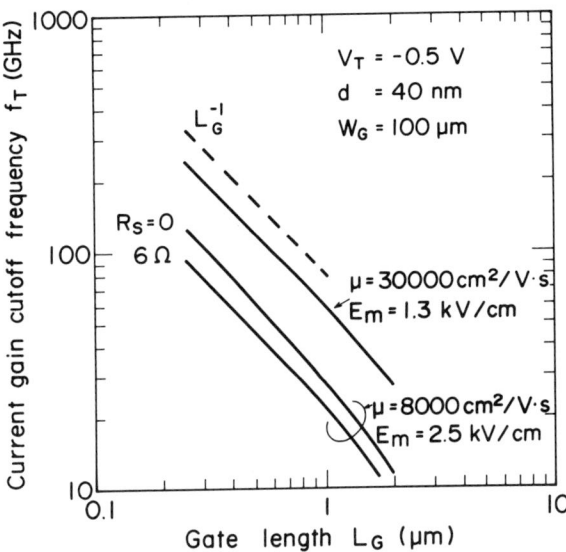

FIG. 11. Dependence of the intrinsic current gain cutoff frequency on the gate length with various values of low field mobility μ, peak field E_m and parasitic source resistance R_s.

The intrinsic f_T that is defined by $g_m/2\pi C_{gs}$ was derived to be 20 GHz using the values listed in Table I.

In HEMTs, dominant factors that affect high-frequency performance are gate length, electron mobility, effective electron saturation velocity, and parasitic source series resistance. Figure 11 shows the calculated results for f_T versus the gate length L_G with mobility μ, peak field E_m and parasitic source resistance R_S as parameters. In these calculations, g_m was derived from Eqs. (22) and (32) and C_{gs} was approximated by Eq. (39) and assumed to be two-thirds of the capacitance of the gate AlGaAs layer. In Fig. 11, it is noted that f_T depends on L_G as $f_T \propto L_G^{-1}$ for $L_G < 1\,\mu\text{m}$. The L_G^{-1}-dependence of f_T is due to the velocity saturation effect at fields higher than E_m. It is also noted that f_T is severely reduced by the effect of R_S even at $R_S = 0.6\,\Omega\text{-mm}$, which is a typical value measured for the HEMT structures.

Here we compare room temperature f_T values of a HEMT with those of a GaAs MESFET. To analyze the current–voltage characteristics of the MESFET, the device model given by Hower and Bechtel (1973) is used. The gate–source capacitance of the MESFET is approximated by

$$C_{gs} = L_G W_G (\varepsilon_s q N_D / 2 V_B)^{1/2},$$

where the notations are listed in Table II. The device and material parameters assumed for the HEMT and MESFET with the same threshold voltage are

TABLE II
Device and Material Parameters Assumed for a HEMT and a MESFET

Parameters	HEMT (I)	MESFET	HEMT (II)
Gate length L_G (μm)	1	1	1
Gate width W_G (μm)	20	20	20
Mobility μ_n (cm^2/V)	4000	4000	8000
Peak field E_m (kV/cm)	4	4	2.5
AlGaAs thickness t (Å)	510	—	510
GaAs thickness t (Å)	—	1100	—
Doping concentration N_D (10^{17} cm^{-3})			
$\quad N_D$ in AlGaAs	7.9	—	7.9
$\quad N_D$ in GaAs	—	1.5	—
Permittivity ε_S (pF/cm)			
$\quad$ AlGaAs	1.06		1.06
$\quad$ GaAs		1.15	
Built-in potential V_B (V)	—	0.65	—

also listed in Table II. In Table II, two mobility values of the HEMT are assumed; these are the same values (4000 cm^2/Vsec) as that of the MESFET, and an obtainable value (8000 cm^2/Vsec) in the HEMT channels at 300 K.

Calculated results are listed in Table III. It is noted in Table III that f_T of the HEMT having the same low-field mobility as that of the MESFET is about 30 percent higher than that of the GaAs MESFET. The HEMT structure that has an inherent advantage over the MESFET structure in high-speed capability. f_T of the HEMT with a higher mobility value of 8000 cm^2/Vsec is about twice that of the MESFET.

In Fig. 12, plots of room temperature f_T as a function of gate length summarize leading performances of experimental HEMTs and GaAs MESFETs reported so far (Joshin et al., 1983; Camnitz et al., 1984; Mishra et al., 1985; Hueschen et al., 1984; Arnold et al., 1984; Feng et al., 1984; Chye and Huang, 1982). It is noted in Fig. 12 that the values of f_T for HEMTs are much higher than those for GaAs MESFETs over gate lengths ranging from

TABLE III
Electrical Parameters Calculated for a HEMT and a MESFET

	V_T (V)	I_{DSS} (mA)	g_m (mS)	C_{gs} (fF)	$f_T{}^a$ (GHz)
HEMT (I)	−0.650	2.32	5.45	35.6	24
MESFET	−0.650	1.25	3.37	29.1	18
HEMT (II)	−0.650	3.21	8.35	35.6	37

[a] f_T (current gain cutoff frequency) = $g_m/2\pi C_{gs}$.

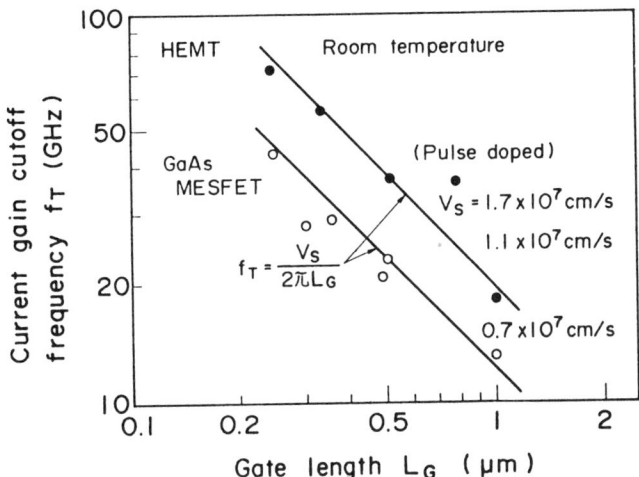

FIG. 12. Current gain cutoff frequency as a function of gate length for experimental HEMTs and GaAs MESFETs reported so far.

0.25 μm to 1.0 μm. Solid lines in Fig. 12 are calculated from $f_T = V_s/2\pi L_G$, where V_s is the effective saturated drift velocity.

The deduced V_s value for HEMTs is much higher than that for MESFETs: $V_s = 0.7 \times 10^7$ cm/sec for GaAs MESFETs and $V_s = 1.1 \times 10^7$ cm/sec for HEMTs. The V_s value of HEMTs deduced from experimental values of f_T is much lower than those expected from DC models (Su et al., 1982; Drummond et al., 1982; Takanashi and Kobayashi, 1985). This discrepancy probably comes from parasitic capacitances, such as fringing capacitance of the gate, which reduces f_T values. The pulse-doped HEMT structure (Hueschen et al., 1984) gives the highest V_s value of 1.7×10^7 cm/sec, which agrees well with that expected from DC models. In the pulse-doped structure, the donors in an AlGaAs layer are confined to a thin layer, about 10 nm thick, close to the heterointerface. The gate of the pulse-doped HEMT is formed on an undoped portion of an AlGaAs layer. Fringing capacitance in a pulse-doped structure may be smaller because of its undoped GaAs cap layer and partially undoped AlGaAs layer.

IV. LSI Technology

4. Gate Performance

Figure 13 shows the circuit diagram of a direct-coupled FET logic (DCFL). In a first-order approximation of gate delays, the pull-up time t_{pu} and the

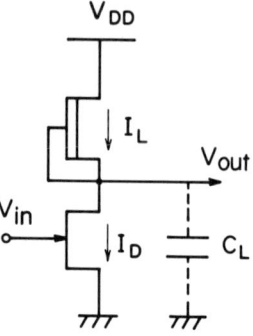

FIG. 13. Circuit diagram of a direct-coupled FET logic (DCFL).

pull-down time t_{pd} of the inverter are given by

$$t_{pu} = \frac{\Delta V C_L}{I_L}, \tag{40}$$

$$t_{pd} = \frac{\Delta V C_L}{I_D - I_L}, \tag{41}$$

where ΔV is the logic voltage swing, C_L the total load capacitance during switching, I_L the load current and I_D the driver content.

To minimize the sum of t_{pu} and t_{pd}, one should choose I_L to be about one-half of I_D, because this would achieve equal t_{pu} and t_{pd}.

Then, if $I_L = I_D/2$,

$$t_{pu} = t_{pd} = \frac{2C_L \Delta V}{I_D}. \tag{42}$$

If we assume that ΔV is small enough and C_L can be replaced by gate capacitance C_{gs} of the driver transistor, Eq. (42) can be written as

$$t_{pu} = \frac{2C_{gs}}{g_m} = \frac{1}{\pi f_T}. \tag{43}$$

Thus, once $I_L = I_D/2$, circuit switching speed is limited by f_T of the driver transistor. HEMTs have the highest f_T value as shown in the previous section, so we can expect that HEMTs would provide high-speed low-power integrated circuits.

5. Fabrication Technology

a. Selective Dry Etching Technique

DCFL inverters consist of enhancement-mode (E) driver transistors and depletion-mode (D) load transistors. To fabricate E- and DHEMTs in the

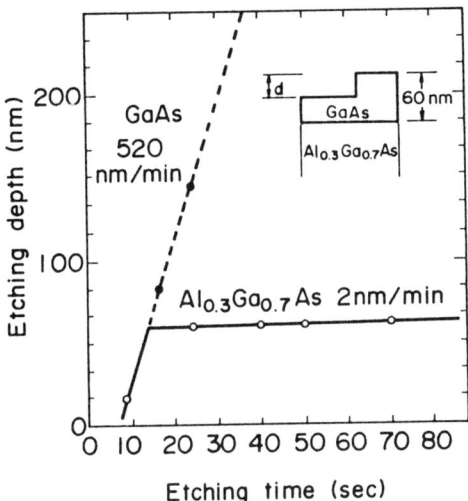

FIG. 14. Characteristics of selective dry etching performed by CCl_2F_2 and He plasma. Selectivity ratio between AlGaAs and GaAs exceeds 260.

same wafer, we use the basic epilayer structure consisting of a 600-nm undoped GaAs layer, a 30-nm $Al_{0.3}Ga_{0.7}As$ layer doped to about 2×10^{18} cm^{-3} with Si, a 70-nm GaAs cap layer, and a thin $Al_{0.3}Ga_{0.7}As$ layer embedded in the GaAs cap layer grown on a semi-insulating substrate. Some epilayers were grown with a thin undoped $Al_{0.3}Ga_{0.7}As$ spacer inserted between the undoped GaAs and n-$Al_{0.3}Ga_{0.7}As$ layers to enhance electron mobility (see Chapter 2).

The selective dry etching technique is used to achieve precise control of the gate recessing process for E- DHEMTs in the same wafer.

Figure 14 shows etching characteristics in CCl_2F_2 and He discharges by using GaAs (60 nm thick)–$Al_{0.3}Ga_{0.7}As$ heterojunction material. A high selectivity ratio of more than 260 is achieved, where the etching rate of $Al_{0.3}Ga_{0.7}As$ is as low as 2 nm/min and that of GaAs is about 520 nm/min at 140 V is self-generated bias voltage.

Figure 15 shows threshold voltage as a function of thickness between the surface and the hetero-interface at 300 K. Selective dry etching is carried out to remove the GaAs cap layer, exposing the top surface of the thin $Al_{0.3}Ga_{0.7}As$ stopper for DHEMTs. To fabricate EHEMTs, the thin $Al_{0.3}Ga_{0.7}As$ stopper is removed by nonselective wet-chemical etching followed by selective dry etching of the cap GaAs layer under the stopper. When the temperature is lowered to 77 K, threshold voltage shifts toward the positive by around 0.1 V compared with that at 300 K.

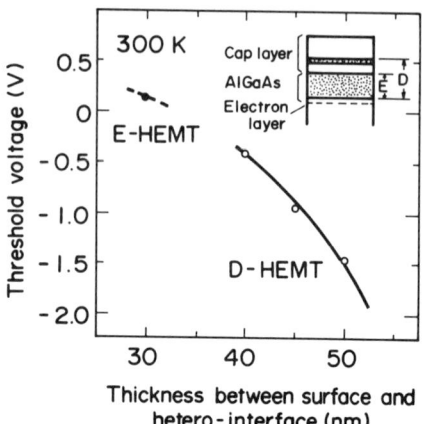

Fig. 15. Threshold voltages as a function of thickness between the surface and the heterointerface.

b. Inverter Structure and I–V Characteristics

Figure 16 indicates the sequence of the self-aligned gate process in the fabrication of HEMT LSIs, including enhancement-mode and depletion-mode HEMTs.

First of all, the active region is isolated by a shallow mesa step (180 nm), which is produced by a very simple process and can be made nearly planar. The source and drain for E- and DHEMTs are metallized with AuGe eutectic alloy and Au overlay alloying to form ohmic contacts with the electron layer. Then, fine gate patterns are formed for EHEMTs, and the top GaAs layer and thin $Al_{0.3}Ga_{0.7}As$ stopper are etched off by nonselective chemical etching. Using the same resist, after formation of gate patterns for DHEMTs, selective dry etching is performed to remove the top GaAs layer for DHEMTs and also to remove the GaAs layer under the thin $Al_{0.3}Ga_{0.7}As$ stopper for EHEMTs. Next, Schottky contacts for the E- and DHEMT gates are provided by depositing Al, the Schottky gate contacts and the GaAs cap layer for ohmic contacts being self-aligned to achieve high-speed performance. Finally, electrical connection from the interconnecting metal, composed of Ti/Pt/Au, to the device terminals are provided through contact holes etched in a crossover insulator film.

As described above, a unique epistructure in combination with self-teminating selective dry recess etching makes it possible to fabricate superuniform E- and DHEMTs simultaneously.

The uniformity of the threshold voltage is a key parameter in obtaining LSI circuits with optimum logic power–delay product. State-of-the-art results for the standard deviations of the threshold voltage were 24 mV for

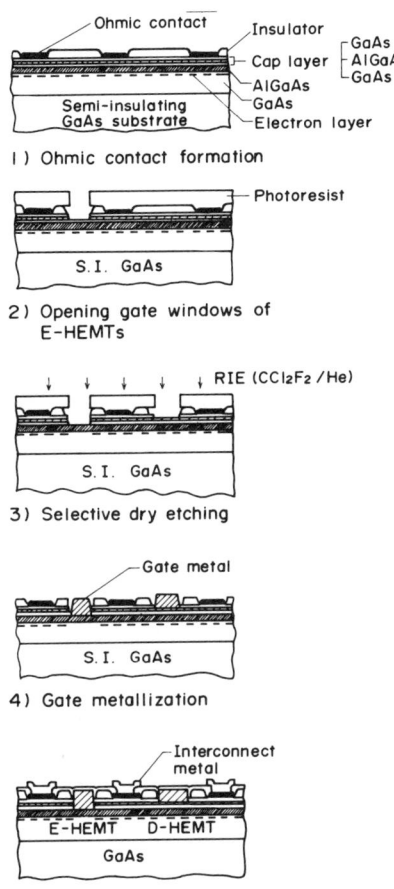

FIG. 16. Processing steps for HEMT DCFL circuits: (a) ohmic contact formation; (b) opening gate windows of EHEMTs; (c) selective dry etching; (d) gate metallization; (e) interconnect metallization.

DHEMTs and 16 mV for EHEMTs distributed over an entire 2-inch-diameter wafer. Short-range uniformity within a 1×1 cm area was also checked for E- and DHEMTs. The standard deviations were 11 mV for DHEMTs and 6 mV for EHEMTs. These are the best results ever reported in GaAs IC fields. Figure 17 shows the transfer characteristics of E/D inverters distributed over a 2-inch-diameter wafer. Variations of the output high level ΔV_H, the threshold voltage ΔV_M and the output low level ΔV_L are predominantly determined by Schottky clamp voltage variation of the next-stage inverter. ΔV_T of the EHEMTs and I_{DSS} of the DHEMTs. Values of

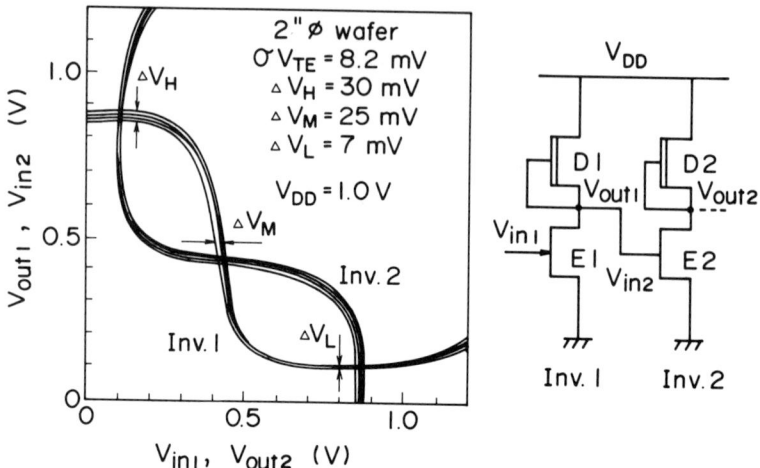

FIG. 17. State-of-the-art results for transfer characteristic uniformities of E/D inverters distributed over a 2-inch-diameter wafer.

ΔV_H, ΔV_M, and ΔV_L are very small compared with noise margins and logic voltage swing, indicating a controllability in device characteristics that is more than adequate for a high yield of HEMT LSI circuits.

In HEMT structures, the AlGaAs layer, which is heavily doped with donors such as Si, contains DX centers (Lang et al., 1979) that behave as electron traps at low temperatures. Some anomalous behaviors at low temperatures are believed to be related to these traps. These include distortion of drain I–V characteristics, an unexpected threshold voltage shift at low temperatures and highly sensitive persistent photoconduction.

We have found that the distortion of drain I–V characteristics (current collapse) is related to the type of device structures. In a conventional partial gate HEMT structure that has relatively long ($>100\,\mu$m) exposed surfaces of AlGaAs at both sides of the gate, electrons heated by the drain field have sufficient energy to transfer from a GaAs channel to the AlGaAs region with exposed surface on the drain side of the gate, where they are trapped at DX centers. As a result, space charges build up at the drain side of the gate. This eventually increases the drain output resistance in the linear region of operation, leading to drain-current collapse.

To eliminate drain-current collapse at low temperatures, we adopt a full gate structure as shown in Fig. 18a. The n-AlGaAs layer is completely covered by the n-GaAs cap layer. There are no exposed surfaces at the drain end of the gate. In the structure, high-energy electrons can easily pass through the thin n-AlGaAs layer (30 nm) without being trapped and can reach

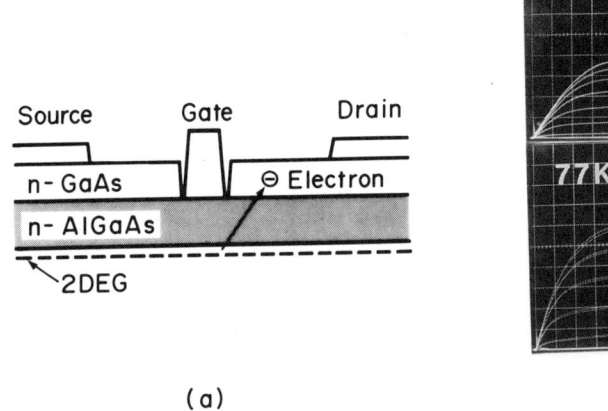

FIG. 18. (a) Schematic diagram of a full gate HEMT structure; (b) drain current–voltage characteristics. Note that there is no collapse at low temperature.

the n-GaAs cap layer, eliminating the anomalous drain I–V characteristics at low temperature, as shown in Fig. 18b (lower traces).

6. INTEGRATED CIRCUIT IMPLEMENTATIONS

Since the first demonstration of the high-speed capability of HEMT digital integrated circuits in 1981, great strides have been made at many laboratories throughout the world in improving performance and increasing circuit complexities. Figure 19 shows the first HEMT ring oscillator with 1.7 μm gate-length (Mimura *et al.*, 1981b). In 1983, Lee *et al.* fabricated 1-μm-gate HEMT ring oscillators and obtained a switching delay of 12.2 psec with 1.1 mW power dissipation per gate at room temperature. Recently, self-aligned 1 μm-gate HEMT ring oscillators were found to have a propagation

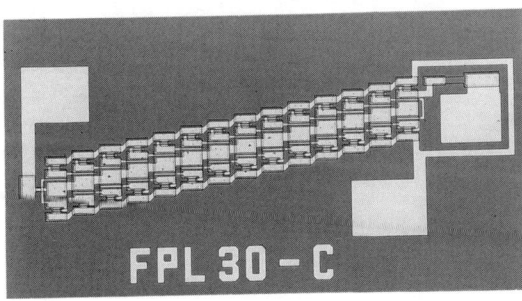

FIG. 19. The first HEMT ring oscillator.

TABLE IV
PERFORMANCE OF HEMT RING OSCILLATORS

References	Gate length (µm)	Propagation delay (psec)	Power-delay product (fJ)	Temperature (K)
Fujitsu (Mimura et al., 1981b)	1.7	17	16	77
Fujitsu (Abe et al., 1983)	0.5	15	18	300
Fujitsu (Awano et al., 1987)	0.28	9.2	38	300
Thomson CSF (Tung et al., 1982)	0.7	18	17	300
Bell Laboratories (Hendel et al., 1984)	0.7	9.4	42	77
	0.7	17	23	300
Bell Laboratories (Shah et al., 1986)	0.35	5.8	10	77
Rockwell (Lee et al., 1983)	1	12	14	300
Honeywell (Cirillo et al., 1985)		8.5	22	77
		11.6	18	300

delay of 8.5 psec at 77 K with 2.59 mW power dissipation per gate (Cirillo, Jr., and Abrokwah, 1985). In 1986, Shah et al. reported 0.35-µm-gate ring oscillators with a delay time of 5.8 psec at 77 K.

In Table IV, HEMT ring oscillator results are summarized. It is noted that propagation delays around 10–20 psec are repeatedly obtained even at room temperature.

To evaluate the high-speed capability of HEMTs in complex logic circuits, a single-clocked divide-by-two circuit based on a master–slave flip-flop consisting of 8 DCFL NOR gates, 1 inverter and 4 output buffers was fabricated (Mimura et al., 1983). The photograph of this circuit is shown in Fig. 20. The circuit has a fan-out of up to 3 and 0.5-mm-long interconnects, giving a more meaningful indication of the overall performance of HEMT ICs than that obtained with a simple ring oscillator.

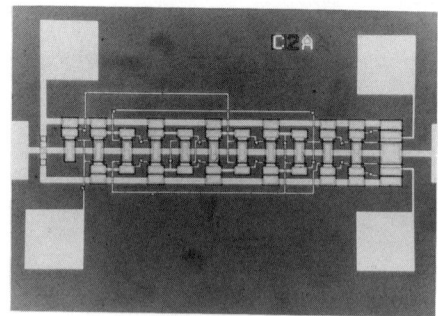

FIG. 20. The first HEMT single-clocked divide-by-two circuit fabricated with electron beam lithography.

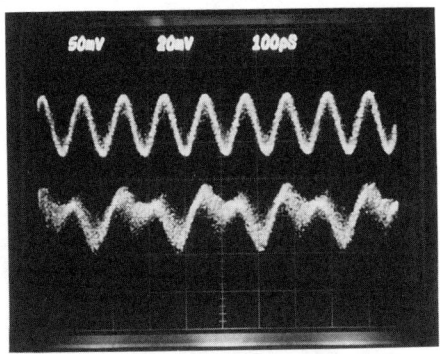

FIG. 21. Oscilloscope traces of 8.9 GHz input (upper) and divide-by-two output waveforms.

The basic gate consists of a 0.5 μm × 20 μm gate EHEMT and a saturated resistor as load. Direct writing electron-beam lithography and lift-off techniques were used throughout the fabrication process.

The maximum clock frequency achieved with these circuits at 77 K is 8.9 GHz with a supply voltage of 0.98 V and with a total power dissipation of 36.8 mW. Oscilloscope traces of 8.9 GHz input and divide-by-two output are shown in Fig. 21. At room temperature, the maximum clock frequency is 5.5 GHz with a supply voltage of 1.31 V and with 38.2 mW power dissipation. The logic delays determined from the dividing frequencies are 22 psec/gate with a 2.8 mW/gate power dissipation at 77 K, and 36 psec/gate with a 2.9 mW/gate power dissipation at room temperature. The logic delay of 36 psec/gate at room temperature is more than twice as long as that obtained from the 19-stage ring oscillators with fan-out of 1 fabricated on the same chip.

Hendel *et al.* (1984) also reported 0.7-μm-gate frequency-divider circuits with dual-clocked master–slave DCFL AND/NOR gates, achieving a maximum clock frequency of 6.3 GHz at 300 K (13 GHz at 77 K) with 35.4 mW power dissipation (33.7 mW at 77 K). The maximum clock frequency achieved with room-temperature HEMT technology is roughly two times as high as that achieved with GaAs MESFET technology with a comparable geometry.

A 1-kbit HEMT static RAM was fabricated with an E/D-type DCFL circuit configuration. Address access time of 0.87 nsec with 360 mW power dissipation at 77 K has been obtained (Nishiuchi *et al.*, 1984). This was the first demonstration of a HEMT LSI circuit.

Recently, Kuroda *et al.* (1984) succeeded in fabricating 4-kbit HEMT static RAMs. Figure 22 shows a picture of the completed 4-kbit HEMT static RAM. The RAM is organized into 4096 words × 1 bit, and arranged as a 64 × 64 matrix. Using depletion-mode HEMTs for load devices, E/D-type

Fig. 22. The first 4-kbit HEMT static RAM (Kuroda et al., © 1984).

DCFL circuits are employed as the basic circuit. The memory cell is a six-transistor cross-coupled flip-flop circuit with switching devices having gate lengths of 2 μm. For peripheral circuits, a 1.5-μm-gate switching device is chosen for performance reasons. The memory cell is 55 μm × 29 μm, the chip is 4.76 mm × 4.35 mm and 26864 HEMTs are integrated in a 4-kbit SRAM. The design rule adopted is 3 μm line width and spacing at minimum. The minimum size of contact hole is 2 μm × 2 μm.

A normal read–write operation was confirmed both at 300 K and 77 K. Figure 23 is a typical oscillograph of read–write operations at 77 K. From top to bottom, these traces are x-address input, data-input pulses, write-enable pulses and data-output signals. Figure 24 shows an oscillograph of x-address input and output waveforms at minimum memory access time at 77 K. From

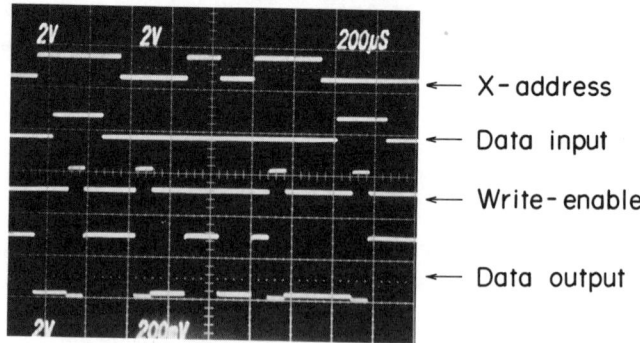

Fig. 23. Oscilloscope traces of read–write operations (Kuroda et al., © 1984).

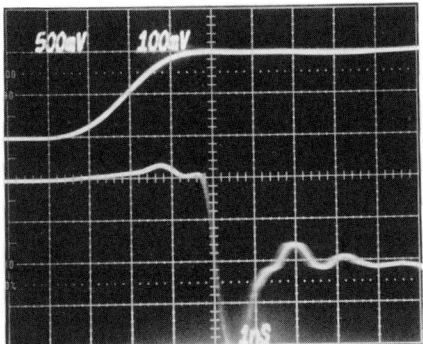

FIG. 24. Oscilloscope traces of x-address input (upper) and data output waveforms at an access time of 2.0 nsec (Kuroda et al., © 1984).

this oscillograph, the minimum address access time was estimated to be 2.0 nsec, with chip dissipation power of 1.6 W with a supply voltage of 1.54 V. At 300 K, typical address access time was 4.4 nsec with chip dissipation of 0.86 W.

Advances in HEMT integration are very rapid, fourfold each year. This is mainly because the Si technology now available is compatible with HEMT LSI technology. Another reason is the inherent large flexibility in the device structure of HEMTs, as shown in Fig. 16, with a stopper layer that is introduced to utilize selective dry etching characteristics.

V. Summary

In summary, the future of HEMT technology for high-speed digital and high-frequency analog applications looks very bright. Future developments of HEMT will follow the same history that Si devices have passed through previously. Applications of HEMTs will start with discrete devices such as low-noise devices that will be used in microwave systems operated at or near room temperature. Recent remarkable advances in low-noise performance have shown the feasibility of HEMTs in such applications, especially beyond 12 GHz.

In a digital world, the high f_T and f_{max} performance attained with HEMTs can be advantageous for SSI and MSI level circuits such as frequency counters and high-speed A/D converters. The fastest frequency divider circuits with logic delay of 22 psec per gate at 77 K (36 psec at 300 K) have been achieved. For LSI/VLSI level circuits, uniformity and controllability of device parameters are critical issues. The state-of-the-art result for threshold uniformity over a 2-inch wafer is less than 20 mV, adequate for VLSI circuits. By operating HEMT VLSI circuits at 77 K, we can reduce power without sacrificing speed. This would be advantageous in reducing the volume of processors, resulting in improved system performance. At this point, only complementary Si MOSFET technology at cryogenic temperatures will be able to compete with the HEMT technology.

However, complementary MOSFET circuits cannot work as fast as HEMT LSI circuits of a comparable geometrical size. Thus the HEMT technology at 77 K will take a main position in large-scale digital systems such as mainframes and supercomputers. HEMT technology will also open the door to high-speed microprocessors at low temperatures if small-volume cost-effective refrigerators are available. Based on the results to date, we can anticipate that the HEMT technology will be developed into one of the most important semiconductor technologies in the 20th century, and a world-wide effort is in progress.

Acknowledgments

The author wishes to thank T. Misugi, M. Kobayashi, A. Shibatomi, and M. Abe for their encouragement and support. The author also wishes to thank his colleagues, whose many contributions have made possible the results described here.

The present research effort is part of the National Research and Development Program on "Scientific Computing System", conducted under a program set by the Agency of Industrial Science and Technology, Ministry of International Trade and Industry, Japan.

Appendix A: Capacitance–Voltage Relationship in Depletion Regime and a Donor Concentration in the Undoped GaAs Layer

Consider the isotype (n-type) heterojunction shown in Fig. 25, in which the negative voltage V_G is applied to the metal electrode and causes electrons to be depleted from the surface of the undoped GaAs layer. Here, V_{bi} is the built-in potential, ΔE_c is the conduction band discontinuity, d_1 is the metallurgical thickness of the AlGaAs layer, d_2 is the thickness of the depletion layer in undoped GaAs. Hence,

$$V_{bi} - V_G = \frac{qN_D d_2^2}{2\varepsilon_{s1}} + \frac{qn_D d_1^2}{2\varepsilon_{s1}} + \frac{qN_D d_1 d_2}{\varepsilon_{s1}} + \frac{\Delta E_c}{q}. \tag{1}$$

The semiconductor charge Q_{sc} under the depletion condition is given by

$$Q_{sc} = qn_D d_1 + qN_D d_2. \tag{2}$$

Substituting Eq. (1) into Eq. (2) and noting that

$$C = \frac{dQ_{sc}}{dV_G}, \tag{3}$$

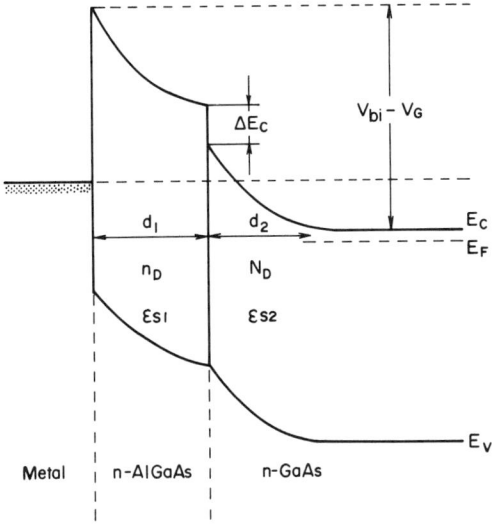

FIG. 25. Energy band diagram for the isotype heterojunction with negative metal voltage V_G.

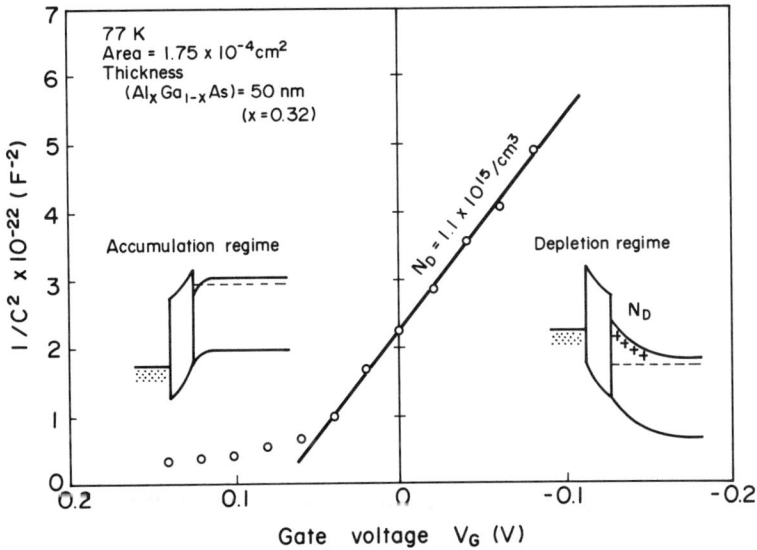

FIG. 26. Experimental plot of $1/C^2$ as a function of V_G for the n-$Al_{0.3}Ga_{0.7}As$–undoped GaAs heterojunction, indicating a donor concentration of 1.1×10^{15} cm^{-3}.

we obtain for the depletion capacitance,

$$C = \left[\frac{2}{qn_D\varepsilon_{s2}}\left\{V_{bi} - V_G - \frac{\Delta E_c}{q} + \frac{qN_D d_1^2}{2\varepsilon_{s1}}\left(\frac{\varepsilon_{s2}}{\varepsilon_{s1}} - \frac{n_D}{N_D}\right)\right\}\right]^{-1/2} \quad (4)$$

Thus, the expression for N_D from Eq. (4) becomes

$$N_D = \frac{2(-dV_G)}{q\varepsilon_{s2} d(1/C^2)}. \quad (5)$$

We see in Eq. (4) that $1/C^2$ is a linear function of the applied voltage V_G. The donor concentration in the GaAs layer can be calculated from the slope of a curve plotting $1/C^2$ against V_G, as expressed by Eq. (5). Such an experimental plot for the n-$Al_{0.3}Ga_{0.7}As$–undoped GaAs heterojunction is shown in Fig. 26, indicating a donor concentration in the GaAs layer of 1.1×10^{15} cm^{-3}.

Appendix B: Mobility Variation with Carrier Concentration

Hall measurements were performed at 77 K using the HEMT structure with voltage probes (see insert, Fig. 27) (Mimura, 1982). The variation of electron concentration n_s and mobility μ_H in the two-dimensional channel as a function of applied gate voltage above the threshold ($V_{GS} - V_T$) is shown in Fig. 27. The threshold voltage V_T of the Hall bridge was -1.05 V. The

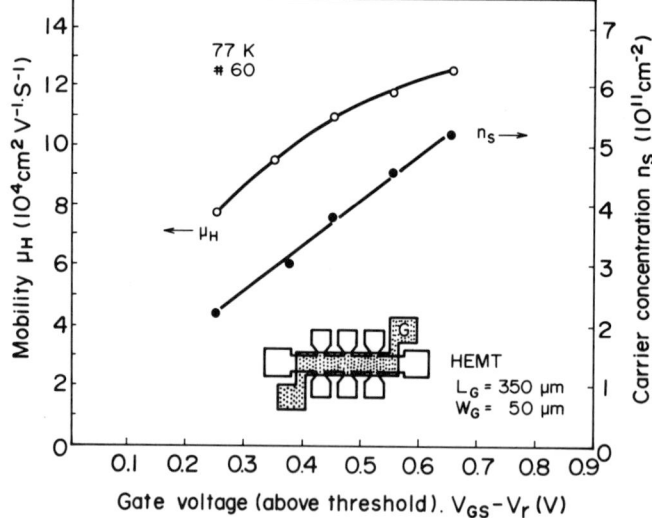

FIG. 27. Hall mobility μ_H and electron concentration n_s as a function of gate voltage above the threshold.

electron concentration variation varies linearly with gate voltage. The mobility increases when the gate voltage is increased. The mobility μ_H was found to follow approximately the law $\mu_H \propto n^k$. The value of k was experimentally found to vary from sample to sample, ranging from $\frac{1}{2}$ to 2.

It was also found that high-quality samples with higher mobility values at low electron concentrations have smaller k values approaching $k = \frac{1}{2}$. The mobility μ_H of the sample (#60) in Fig. 27 follows the law $\mu_H \propto n^{1/2}$. Thus, the mobility variation measured over the voltage range can be considered to be within about 30 percent, sufficiently small to ensure that the constant mobility assumption is valid for device modeling.

Appendix C: Maximum Thickness of AlGaAs for Complete Depletion

Let us consider the metal-semiconductor system shown in Fig. 28, in which the space charge neutral region in the n-AlGaAs layer is assumed to be infinitely thin (a complete depletion of the n-AlGaAs layer) and the Fermi level of the neutral region is denoted as ζ_1. The notations given in Fig. 28 have their usual meanings.

We define the maximum thickness of the AlGaAs layer, d_{max}, by

$$d_{max} = (d_B + d_1 + d_0)_{max}, \qquad (1)$$

where d_B is the thickness of the surface depletion region, d_1 is the thickness

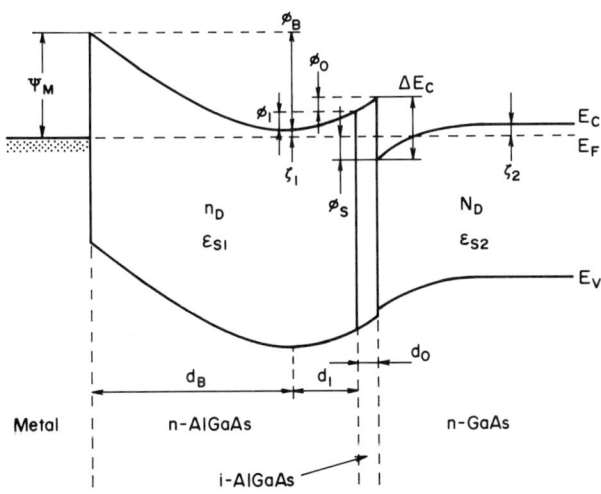

FIG. 28. Energy band diagram for metal-semiconductor heterojunction system. The space charge neutral region in the n-AlGaAs is assumed to be infinitely thin.

of the interface depletion region and d_0 is the thickness of the undoped AlGaAs layer. From the depletion approximation, Eq. (1) can be expressed by

$$d_{max} = \sqrt{\frac{2\varepsilon_{s1}\phi_B}{qn_D} - \frac{Q_s}{qn_D}} + d_0, \tag{2}$$

where Q_s is the surface charge of electrons accumulating at the undoped GaAs side of the heterojunction. For the heterojunction barrier shown in Fig. 28, we obtain

$$\phi_1 + \phi_0 + \phi_s + \frac{\zeta_1}{q} = \frac{\Delta E_c}{q}, \tag{3}$$

where

$$\phi_1 = \frac{Q_s^2}{2\varepsilon_{s1}qn_D}, \tag{4}$$

and

$$\phi_0 = \frac{Q_s d_0}{\varepsilon_{s1}}. \tag{5}$$

Substituting Eqs. (4) and (5) into Eq. (3), we have

$$\frac{Q_s}{2\varepsilon_{s1}n_D} + \frac{Q_s d_0 q}{\varepsilon_{s1}} + \zeta_1 + q\phi_s - \Delta E_c = 0. \tag{6}$$

We shall calculate the surface charge Q_s as a function of ϕ_s.

Under accumulation conditions, since the bottom of the conduction band in an undoped GaAs becomes lower than the Fermi level at the heterojunction interface, Fermi statistics must be used instead of Maxwell-Boltzmann statistics. Here, we use the Joyce–Dixon approximation (Joyce and Dixon, 1977) for the Fermi energy in a degenerate heterojunction neglecting quantum size effects. If $q\phi = E_F - E_c$, the undoped GaAs region ϕ can be written in the Joyce–Dixon approximation as

$$\frac{q\phi}{kT} = \ln\left(\frac{n}{N_c}\right) + \sum_m A_m \left(\frac{n}{N_c}\right)^m, \tag{7}$$

where n denotes the electron concentration and N_c the effective density of state of GaAs. The first four terms of A_m will be given by

$$A_1 = 1/\sqrt{8},$$
$$A_2 = -4.95009 \times 10^{-3},$$
$$A_3 = 1.48386 \times 10^{-4},$$
$$A_4 = -4.42563 \times 10^{-6}. \tag{8}$$

Differentiating in Eq. (7) with respect to z, the distance from the heterojunction interface and equating Poisson's equation $dE/dz = q(n - N_D)/\varepsilon_{s2}$ (Where N_D is the concentration of donors in GaAs and ε_{s2} is the dielectric constant of GaAs), we obtain an equation to express the relationship between the electric field E and n/N_c as

$$\frac{\varepsilon_{s2}E^2}{2N_D kT} = \frac{N_c}{N_D} \sum_m B_m \left\{ \left(\frac{n}{N_c}\right)^m - \left(\frac{N_D}{N_c}\right)^m \right\} - \ln \frac{n}{N_D} \quad (9)$$

(Krömer, 1981), where B_m is given by

$$B_1 = 1 - A_i\left(\frac{N_D}{N_c}\right); \quad B_m = A_{m-1}\left(\frac{m-1}{m}\right) - A_m\left(\frac{N_D}{N_c}\right) \quad (m > 1).$$
(10)

The surface charge Q_s can be obtained by the application of Gauss's law to the surface:

$$Q_s = -\varepsilon_{s2} E_S, \quad (11)$$

where E_s is the electric field at the surface. From Eq. (7) to (11), we obtain the relationship between Q_s and $\phi_s = \phi$ $(z = 0)$. The relationship is shown in Fig. 29.

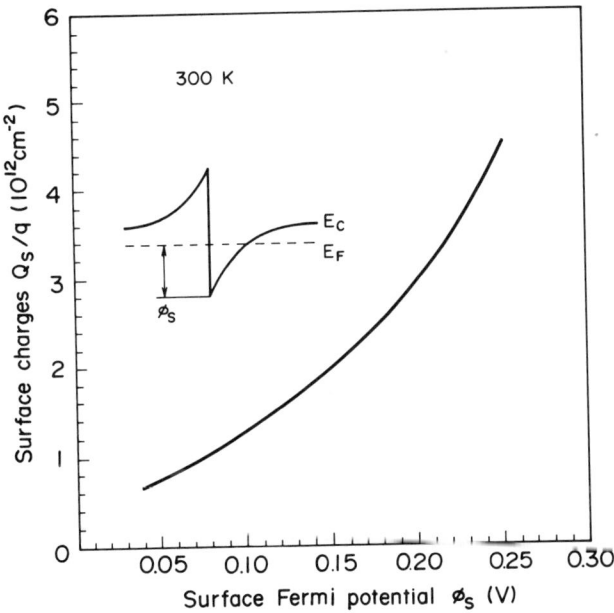

FIG. 29. Surface charge Q_s vs. ϕ_s.

Further, n_D can be expressed by assuming the degeneracy of the donor energy level to be 2:

$$n_D = n_c \left[\exp\left(-\frac{\zeta_1}{kT}\right) + 2\exp\left(\frac{2\zeta_1}{kT}\right)\exp\left(\frac{E_D}{kT}\right) \right], \quad (12)$$

where n_c is the effective state density of AlGaAs and E_D is the donor energy level.

The donor energy level E_D is assumed to be about 60 meV in the case of a Si donor impurity for the mole fraction of AlAs, $x = 0.3$. Then, using Eqs. (6) and (12), n_D can be found for each given Q_s value, which determines the ϕ_s value in Eq. (6) through the relationship between Q_s and ϕ_s, as shown in Fig. 29. The calculated electron concentration n_s as a function of donor concentration n_D for various thicknesses of the undoped AlGaAs layer as a parameter are shown in Fig. 30. In Fig. 30, the experimental result for $d_0 = 0$ is plotted, exhibiting a reasonable agreement with the theoretical result. Now d_{max} can be calculated for a given value of n_D. The results are plotted in Fig. 31. The maximum thickness d_{max} for $d_0 = 0$ decreases when n_D is increased as $d_{max} \propto n_D^{-1/2}$ over the values of the donor concentration in Fig. 31, and d_{max} increases with d_0.

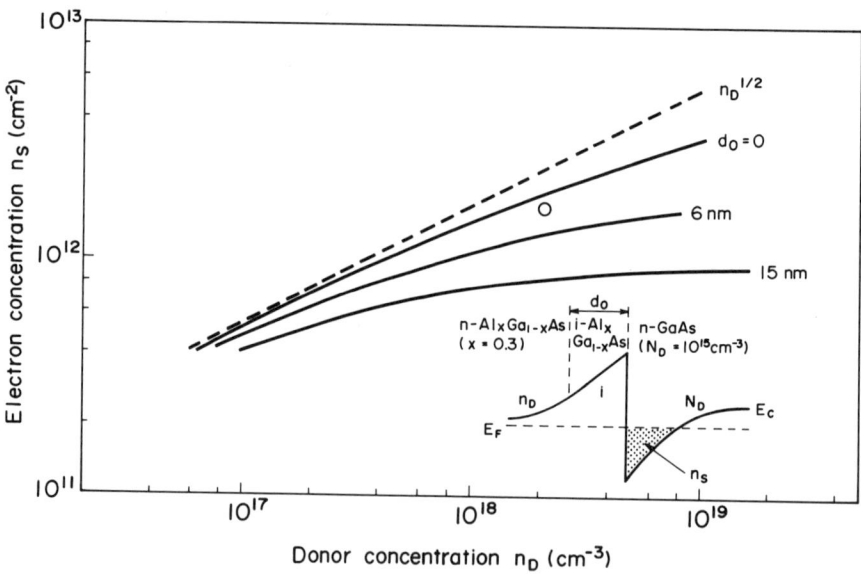

FIG. 30. Calculated electron concentration n_s as a function of donor concentration n_D for various thicknesses of undoped AlGaAs, with d_0 as a parameter. Experimental result for $d_0 = 0$ is plotted (open circle).

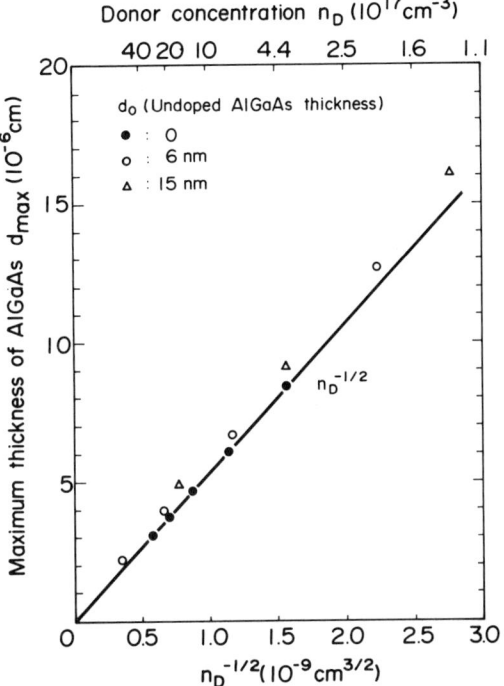

FIG. 31. Maximum thickness of AlGaAs, d_{max}, as a function of donor concentration n_D, calculated for various values of the undoped thickness d_0: 0 nm (●), 6 nm (○), and 15, nm (△).

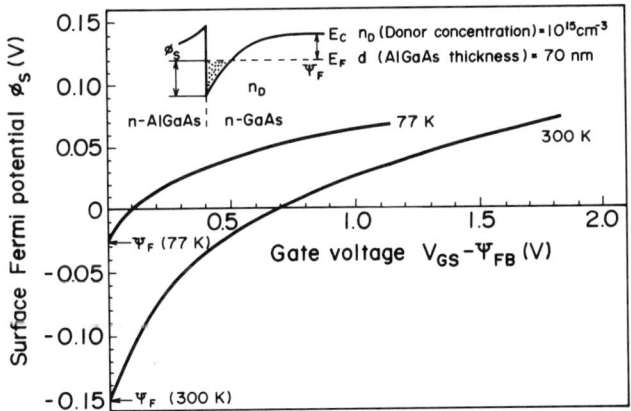

FIG. 32. Surface Fermi potential ϕ_s vs. effective gate voltage $V_{GS} - \Psi_{FB}$.

Appendix D: Surface Fermi Potential versus Effective Gate Voltage

The relationship between Q_s and ϕ_s has been derived in Appendix C (see Fig. 29). By noting $\Psi_s = \phi_s + \zeta_2$, the relationship between $(V_{GS} - \Psi_{FB})$ and ϕ_s can be obtained as shown in Fig. 32. It is noted in Fig. 32 that ϕ_s is less than one-tenth of $(V_{GS} - \Psi_{FB})$ and becomes greater than 0.07 V at 77 K. The AlGaAs layer thickness is assumed to be 0.07 μm.

References

Abe, M., Mimura, T., Nishiuchi, K., Shibatomi, A., and Kobayashi, M. (1983). *Tech. Dig.*— *1983 IEEE GaAs Ic Symposium*, Phoenix, Arizona, p. 158.

Arnold, D. F., Fisher, R., Kopp, W. F., Henderson, T. S., and Morkoç, H. (1984). *IEEE Trans. Electron Devices* **ED-31**, 1399.

Awano, Y., Kosugi, M., Mimura, T., and Abe, M. (1987). *IEEE Trans. Electron Devices Lett.* **EDL-8**, 451.

Camnitz, L. H., Tasker, P. F., Lee, H., Merwe, D. V. D., and Eastman, L. F. (1984). *1984 IEDM Tech. Dig.*, p. 360.

Chye, W., and Hung, C. (1982). *IEEE Trans. Electron Device Lett.* **EDL-3**, 401.

Cirillo, N. C., and Abrokwah, J. K. (1985). *Tech. Dig.*—43rd Device Research Conf., Boulder, Colorado, IIA-7.

Dingle, R., Störmer, H. L., Gossard, A. C., and Wiegmann, W. (1978). *Appl. Phys. Lett.* **33**, 655.

Drummond, T. J., Su, S. L., Kopp, W., Fischer, R., Thorne, R. E., and Morkoç, H. (1982). *1982 IEDM Tech. Dig.*, p. 586.

Feng, M., Kanber, H., Eu, V. K., Watkins, E., and Hackette, L. R. (1984). *Appl. Phys. Lett.* **44**, 231.

Hendel, R. H., Pei, S. S., Tu, C. W., Roman, B. J., Shah, N. J., and Dingle, R. (1984). *1984 IEDM Tech. Dig.*, p. 856.

Hikosaka, K., Mimura, T., and Joshin, K. (1981). *Jpn. J. Appl. Phys.* **20**, L847.

Hower, P. L., and Bechtel, N. G. (1973). *IEEE Trans. Electron Devices* **ED-20**, 214.

Hueschen, M., Moll, N., Gowen, E., and Miller, J. (1984). *1984 IEDM Tech. Dig.*, p. 348.

Joshin, K., Mimura, T., Niori, M., Yamashita, Y., Kosemura, K., and Saito, J. (1983). *1983 IEEE MTT-S Dig.*, p. 563.

Joyce, W. B., and Dixon, R. W. (1977). *Appl. Phys. Lett.* **31**, 354.

Krömer, H. (1981). *J. Appl. Phys.* **52**, 873.

Kuroda, S., Mimura, T., Suzuki, M., Kobayashi, N., Nishiuchi, K., Shibatomi, A., and Abe, M. (1984). *Proc. IEEE GaAs IC Symposium*, Boston MA, p. 125.

Lang, D. V., Logan, R. A., and Jaros, M. (1979). *Phys. Rev. (B)* **19**, 1015.

Lee, C. P., Hou, D., Lee, S. J., Miller, D. L., and Anderson, R. J. (1983a). *Tech. Dig.*— GaAs *IC Symposium*, Phoenix, Arizona, p. 162.

Lee, K., Shur, M., Drummond, T. J., and Morkoç, H. (1983b). *IEEE Trans. Electron Devices* **ED-30**, 207.

Mead, C. A., and Spitzer, W. G. (1964). *Phys. Rev.* **134**, A713.

Mimura, T. (1982). *Surface Science* **113**, 454.

Mimura, T., Hiyamizu, S., Fujii, T., and Nanbu, K. (1980). *Jpn. J. Appl. Phys.* **19**, L225.

Mimura, T., Hiyamizu, S., Joshin, K., and Hikosaka, K. (1981a). *Jpn. J. Appl. Phys.* **20,** L317.
Mimura, T., Joshin, K., Hiyamizu, S., Hikosaka, K., and Abe, M. (1981b). *Jpn. J. Appl. Phys.* **20,** L598.
Mimura, T., Nishiuchi, K., Abe, M., Shibatomi, A., and Kobayashi, M. (1983). *1983 IEDM Tech. Dig.*, p. 99.
Mishra, U. K., Palmateer, S. C., Chao, P. C., Smith, P. M., and Hwang, J. C. M. (1985). *IEEE Trans. Electron Device Lett.* **EDL-6,** 142.
Nishiuchi, K., Kobayashi, N., Kuroda, S., Notomi, S., Mimura, T., Abe, M., and Kobayashi, M. (1984). *1984 ISSCC Digest Tech.*, p. 48.
Shah, N. J., Pei, S. S., Tu, C. W., and Tiberio, R. C. (1986). *IEEE Trans. Electron Devices* **ED-33,** 543.
Shockley, W. (1952). *Proc. IRE* **40,** 1365.
Su, S. L., Fischer, R., Drummond, T. J., Lyons, W. G., Thorne, R. E., Kopp, W., and Morkoç, H. (1982). *Electron. Lett.* **8,** 794.
Takanashi, Y., and Kobayashi, N. (1985). *IEEE Electron Device Lett.* **EDL-6,** 154.
Tung, P. N., Delescluse, P., Delagebeadeuf, D., Laviron, M., Chaplart, J., and Linh, N. T. (1982). *Electron. Lett.* **18,** 517.
Turner, J. A., and Wilson, B. L. H. (1969). *Gallium Arsenide and Related Compounds—Inst. Phys. Conf.* **7,** 195.

CHAPTER 5

Hetero-Bipolar Transistor and Its LSI Application

Takayuki Sugeta

NTT OPTO-ELECTRONICS LABORATORIES
ATSUGI, JAPAN

Tadao Ishibashi

NTT LSI LABORATORIES
ATSUGI, JAPAN

I.	INTRODUCTION	195
II.	BASIC DEVICE CHARACTERISTICS	197
	1. HBT Structures	197
	2. Current Gain	199
	3. Electron Injection over Potential Spike	201
	4. Effect of Graded Base	204
	5. Device Modelling	205
III.	FABRICATION OF AlGaAs/GaAs HBTs	208
	6. MBE Growth of HBT Epilayers	208
	7. Ion Implantation Technique and Ohmic Contact Formation	212
	8. Self-Alignment Techniques	213
	9. Current Gain Shrinkage	213
	10. Flexibility of HBT Structure	214
IV.	HIGH-FREQUENCY CHARACTERISTICS	216
	11. Equivalent Circuit.	216
	12. Comparison between HBTs and Si-BTs	221
V.	HIGH-SPEED INTEGRATED CIRCUITS	221
	13. Basic Digital Circuits	221
	14. Ultrahigh-Speed Possibility	223
	15. IC Applications	224
VI.	CONCLUSIONS	226
	REFERENCES	227

I. Introduction

The proposal of a heterojunction bipolar transistor (HBT) was made by Shockley in 1948. Several years later, Kroemer formulated the current gain relations of HBT by a diffusion model (Kroemer, 1957a). The basic idea of

HBT is the use of a wide-gap emitter in the emitter/base junction to provide a higher emitter efficiency—that is, a higher current gain in HBTs than in homojunction bipolar transistors. This emitter efficiency enhancement is generated by the suppression of base current, which results from a confinement effect on hole flow from the base into the emitter (in the case of n–p–n type). The wide-bandgap emitter removes the restriction for doping in the emitter and base that is necessary to maintain the current gain for homojunction transistors. HBTs offer the following general advantages: first, the base resistance can be reduced by higher base doping, and second, the emitter capacitance can be minimized by the lower emitter doping.

Advantages of HBTs over homojunction (Si) bipolar transistors are also produced by the better electron transport properties of III–V semiconductor materials. Electron mobility plays an important role in the intrinsic device performance similar to that for homojunction transistors. Higher electron mobility provides a shorter base transit time, resulting in faster transistor operation. AlGaAs and InGaAsP alloy systems that are useful for both high speed and optoelectronic applications have high electron mobilities in the direct bandgap regions. Transiently high electron velocity associated with the overshoot phenomenon in these materials is also beneficial for transit time reduction. For example, electron mobility in GaAs is about ten times higher than that in Si at a doping density of -10^{18} cm^{-3}. As described later, a cutoff frequency for HBT over 100 GHz is expected, where the short electron transit time is reflected in the total reduction of charging time under a very small emitter charging time with high collector current density.

Earlier attempts at HBT fabrication were made with heterojunctions constructed of Ge and/or binary semiconductors, such as Ge/GaAs, Ge/ZnSe, ZnSe/GaAs, etc., in the 1960s (Jadus and Feucht, 1969; Hovel and Milnes, 1969; Sleger et al., 1970). The interface quality of these junctions was not satisfactory. In the 1970s, the liquid-phase epitaxy (LPE) technique for III–V semiconductors was applied to HBT fabrications, and the high current-gain feature was confirmed experimentally (Konagai and Takahashi, 1975; Konagai et al., 1977). However, epitaxial layer thickness and doping control in LPE-grown HBT was not sufficient for high-frequency operation.

More recently, the development of new epitaxial growth techniques, molecular-beam epitaxy (MBE) and metal organic chemical vapor deposition (MOCVD), has enabled fabrication of HBT epilayers suitable for high-frequency and high-speed devices. By these techniques, very fine HBT epilayer structures have been grown with a dimensional controllability below 100 Å. It is also very important that heavy p-type doping of the GaAs base, up to around 10^{20} cm^{-3}, can be done using Be, in both MBE and MOCVD.

Microwave characteristics of AlGaAs/GaAs HBTs fabricated by MBE were first reported around 1982 (Harris et al., 1982; Ankri et al., 1983).

A current gain cutoff frequency over 20 GHz already has been reported (Ito et al., 1984a; Asbeck et al., 1984a). Application of HBT to digital circuits is also very attractive, because HBTs are superior in the high cutoff frequency produced by the material properties of III–V semiconductors and in the good threshold control that is common to bipolar transistors. In 1983, the first emitter coupled logic (ECL) circuit using AlGaAs/GaAs HBTs was reported (Asbeck et al., 1983). A minimum propagation delay time t_{pd} of below 30 psec in non-threshold logic (NTL) gate also has been reported (Asbeck et al., 1984b; Chang et al., 1986). In a current mode logic (CML) gate, a t_{pd} of as fast as 40 psec (Asbeck et al., 1984c) and in an ECL gate, one of 65 psec (Nagata et al., 1985) have been achieved. Theoretically, a t_{pd} below 10 psec in an ECL gate (fan-in, fan-out = 1) has been simulated (Harris et al., 1982). There are attempts at large-scale integration of HBTs by I^2L-like gates, of which a 1K gate array has been fabricated (Yuan et al., 1984).

II. Basic Device Characteristics

1. HBT Structures

First, typical HBT structures are described briefly. Band diagrams of three basic HBT structures are shown in Fig. 1. They are abrupt emitter, graded emitter, and graded base structures. In AlGaAs/GaAs systems, all of these

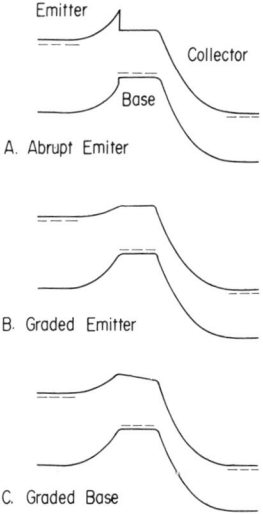

Fig. 1. Band diagrams of three basic HBT structures (a) abrupt emitter, (b) graded emitter, (c) graded base.

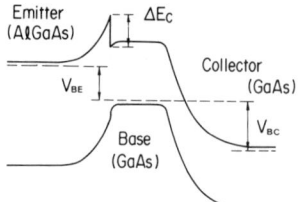

FIG. 2. Band diagram of an abrupt-emitter HBT in bias condition.

structures can be fabricated more easily, without the problem of lattice mismatching in the bandgap tailoring procedure, than in the other quarternary systems such as InGaAsP/InP.

a. Abrupt Emitter Structure

As the conduction band discontinuity ΔE_c exists, a potential spike appears at the interface of the emitter/base junction in the abrupt emitter structure (see Fig. 2 discussion later). When an electron injection is controlled by the spike, the higher built-in voltage associated with ΔE_c compensates partly for the effect of the wide-gap emitter. Thus, in the abrupt emitter structure, the bandgap energy of the emitter must be properly designed so as to reject a hole injection from the base into the emitter. When thermal equilibrium of electrons is not maintained at the emitter–base interface, the hot carrier state is expected, as described later.

b. Graded Emitter Structure

In the graded emitter structure, which is usually employed in MBE-grown HBTs, we can realize a larger emitter efficiency more easily, as compared with the abrupt emitter structure, since the barriers for each minority carrier differ with the amount of bandgap difference between the emitter and the base. For AlGaAs/GaAs HBTs, an AlAs content difference of as little as 10% gives practically high current gains.

c. Graded Base Structure

Another variation is the introduction of bandgap grading in the base (Kroemer, 1957b). This idea can also be combined with the former two structures. A very high quasi-electric field that acts for one kind of carrier in the graded base provides a drift motion of injected minority carriers, reducing the base transit time. This contributes to both a current gain enhancement and an improvement in high-frequency characteristics. When electron transport in the base is dominated by the drift motion, a thicker base layer (and/or higher base doping) can be adopted than in the conventional uniform base structure, without increasing the base transit time very much.

2. Current Gain

The classical diffusion model provides a basic understanding of HBTs (Kroemer, 1957a). In this model, the collector current is determined by minority carrier diffusion and recombinations in the base. Here, we consider HBTs in a simple one-dimensional case, without extrinsic effects such as generation recombination current through the emitter–base depletion layer and parasitic resistance. In this case, treatment similar to that for homojunction transistors (see for instance Sze,1981a) offers the resultant expressions for HBTs. Although many theoretical analyses, in which various structural and physical parameters are taken into account precisely, have been reported (Marty *et al.*, 1979; Asbeck *et al.*, 1982; Yokoyama *et al.*, 1984; Tomozawa *et al.*, 1984; Kurata and Yoshida, 1984), conventional expressions are given here for simplicity.

Figure 2 shows the band diagram of AlGaAs/GaAs HBT (n–p–n-type) when the emitter–base junction is forward-biased and the base–collector junction is reverse-biased. Each applied voltage V_{BE} and V_{BC} is provided from the base potential. The minority electron density in the neutral base layer at the boundary of the emitter–base junction side is postulated to be near thermal equilibrium with that in the emitter. The emitter current density J_E is given for neutral base layer thickness W_B by

$$J_E = q(D_e n_{p0}/L_e)\coth(W_B/L_e)((\exp(qV_{BE}/kT) - 1)$$
$$- (\cosh(W_B/L_e))^{-1}(\exp(qV_{BC}/kT) - 1))$$
$$+ q(D_h p_{n0}/L_h)(\exp(qV_{BE}/kT) - 1), \qquad (1)$$

where notations are listed in Table I. Also, the collector current density J_C is expressed as

$$J_C = q(D_e n_{p0}/L_e)(\sinh(W_B/L_e))^{-1}[(\exp(qV_{BE}/kT) - 1)$$
$$- \coth(W_B/L_e)(\exp(qV_{BC}/kT) - 1)]$$
$$- q(D'_h p'_{n0}/L'_h)(\exp(qV_{BC}/kT) - 1). \qquad (2)$$

Here, the prime in D'_h, p'_{n0}, and L'_h is used for parameters in the collector. The base current is

$$J_B = J_E - J_C. \qquad (3)$$

In the active region of transistor operations, the common emitter DC current

TABLE I
Notations Used for HBT Transistor Equations

q	electron charge
T	temperature
k	Boltzman constant
m_e	electron effective mass
V_a	applied bias voltage of p–n junction
V_{bi}	built-in potential of p–n junction
V_{an}	applied bias voltage for n-side depletion layer
V_{bin}	built-in voltage for n-side depletion layer
N_D, N_A	donor and acceptor densities
μ_e, μ_h	mobilities of electrons and holes
t_m	momentum relaxation time
D_e, D_h	diffusion coefficients of electrons and holes
L_e, L_h	diffusion lengths of electrons and holes
n_{p0}, p_{n0}	equilibrium minority carrier densities in narrow-gap p-type and wide-gap n-type semiconductors
E_{g1}, E_{g2}	narrow- and wide-bandgap energies (usually for GaAs and AlGaAs)
N_c, N_v	effective density of states of conduction and valance bands
$\varepsilon_1, \varepsilon_2$	permittivities of narrow- and wide-gap semiconductors
τ_e, τ_h	electron and hole lifetimes
ΔE_c	conduction band discontinuity
E_f	Fermi level
J_E, J_B, J_C	emitter, base, and collector current densities
V_{BE}	emitter base bias voltage
V_{BC}	base collector bias voltage
V_{CE}	collector emitter bias voltage
V_{biEB}	built-in voltage of emitter base junction
V_{biBC}	emitter and collector junction capacitances
τ_F	emitter-to-collector delay time
τ_B, τ_C	base and collector transit times
h_{FE}	DC current gain
W_B	neutral base layer thickness
X_C	collector depletion width
f_T	current gain cutoff frequency
f_{max}	maximum oscillation frequency
E	electric field
v_d	electron drift velocity
T_c	carrier temperature in the base

gain h_{fe} is

$$h_{fe} = dJ_C/dJ_B$$
$$= (\cosh(W_B/L_e) - 1 + (p_{n0}/n_{p0})(D_h/D_e)(L_e/L_h) \sinh(W_B/L_e))^{-1}$$
$$= (\cosh(W_B/L_e) - 1 + K \sinh(W_B/L_e))^{-1}, \tag{4}$$

where

$$K = (p_{n0}/n_{p0})(D_h/D_e)(L_e/L_h). \quad (5)$$

The second term in the expression of Eq. (4) represents the contribution of hole injection current from the base into the emitter. In the wide-gap emitter structure, K can be designed to be a very small value, because we can make

$$(p_{n0}/n_{p0}) = (N_A/N_D)\exp((E_{g1} - E_{g2})/kT) \ll 1 \quad (6)$$

by choosing a sufficiently large bandgap difference between the AlGaAs emitter and the GaAs base,

$$E_{g2}(\text{AlGaAs}) - E_{g1}(\text{GaAs}) \gg kT.$$

Then, the DC current gain of an HBT with appropriately designed structures is very high. When K in Eq. (5) is negligibly small, h_{fe} is approximated to be

$$h_{fe} = (\cosh(W_B/L_e) - 1)^{-1}$$
$$= (\cosh(W_B/(\tau_e kT\mu_e/q)^{0.5}) - 1)^{-1}. \quad (7)$$

For tentative values of $\tau_e = 1$ nsec, $W_B = 0.1\ \mu\text{m}$, and $\mu_e = 1200\ \text{cm}^2/\text{Vsec}$, this formula gives an h_{fe} of 620.

In the case of homojunction transistors, on the other hand, the current gain is approximated to be

$$h_{fe} = (N_D/N_A)(\mu_e/\mu_h)(L_h/W_B) \quad (\text{for } W_B \ll L_e). \quad (8)$$

3. Electron Injection over Potential Spike

A near-thermal-equilibrium condition for carriers has been assumed at the emitter–base interface in the diffusion model. This situation requires that the base charge at the interface is supplied sufficiently from the emitter. However, the extraction rate of base minority charge is so high in the HBT that it is questioned if such a condition always holds in abrupt emitter–base structures. Electron transport at an abrupt junction with a large conduction base discontinuity is similar to that in a Schottky junction. A treatment of injection over potential spike by diffusion has been reported (Ankri and Eastman, 1982). When doping of the emitter is relatively high, which provides a high electric field at the interface, electron flow is dominated by a thermionic emission mechanism rather than a diffusion one.

In the heterojunction case, a barrier height that corresponds to the Schottky barrier height ϕ_{Bn} is not constant, but changes with the applied voltage V_a depending on the doping density of each side of the semiconductor. Then $(q\phi_{Bn} - qV_a)$ in the Schottky junction is replaced by

$$qV_{bin} + (E_c - E_f) - qV_{an} \quad (9)$$

where V_{bin} and V_{an} are the built-in voltage and the applied voltage for n-side AlGaAs, respectively, and $(E_c - E_f)$ is the energy of the conduction band minimum from the Fermi level. Equation (9) is just the potential energy of the conduction band edge at the peak of the spike measured from the Fermi level. Expressing $q(V_{bin} - V_{an})$ by V_{bi} and V_a (see, e.g., Casey and Panish, 1978), we have the following equation for thermionic emission current (see, e.g., Sze, 1981b):

$$J_{TE} = A^*T^2 \exp((E_f - E_c)/kT) \exp(-q(V_{bi} - V_a)/nkT), \qquad (10)$$

where A^* is the Richardson constant, and n is an ideality factor expressed by

$$n = 1 + \varepsilon_2 N_D/\varepsilon_1 N_A, \qquad (11)$$

where ε_1 and ε_2 are the permittivity of the base and emitter materials, respectively. The depletion layer width for the n side is

$$x_n = [(V_{bi} - V_a)\varepsilon_2/nqN_D]^{0.5}. \qquad (12)$$

The maximum electric field at the interface, which determines the injection process (thermionic emission, tunneling or diffusion) becomes

$$E_{max} = 2(V_{bi} - V_a)/nx_n. \qquad (13)$$

For a typical junction structure (Al content of 30%, $N_D = 5 \times 10^{17}$ cm^{-3}, $N_A = 10^{19}$ cm^{-3}) with forward-biased condition ($V_{bi} - V_a = 0.15$ V), $x_n = 200$ Å and $E_{max} = 1.5 \times 10^5$ V/cm.

Around the field strength of E_{max} in this case, thermionic emission is the dominant mechanism, although a small correction for tunneling to A^* is necessary in discussing precise current–voltage characteristics.

On the other hand, the diffusion current through the base in the diffusion model is

$$J_{diff} = q(D_e/L_e)n_{p0}\coth(W_B/L_e)\exp(qV_a/kT) \qquad (14)$$

in the active region of HBT operations, which is obtained from Eq. (1). When J_{TE} is smaller than J_{diff} through ΔE_c in V_{bi}, the emitter current is dominated by the thermionic emission from the emitter. In this case, the "near-thermal-equilibrium" condition cannot be maintained at the interface. Further, in abrupt junctions with sufficiently large ΔE_c, such a nonequilibrium electron state in the base with high energy is expected if the energy relaxation time of carriers is long enough to maintain a "hot" state.

A "near-ballistic" motion of electrons has also been discussed (Ankri and Eastman, 1982). When an electron is injected into the base with a kinetic energy U that corresponds to the conduction band discontinuity ΔE_c, its velocity reaches

$$v_{ballistic} = (2U/m'_e)^{0.5}, \qquad (15)$$

where m'_e is the corrected effective electron mass that takes into account the nonparabolicity of the Γ valley in GaAs band structure. For $U = 0.3$ eV, Eq. (15) predicts $v_{\text{ballistic}} = 9 \times 10^7$ cm/sec. This value is far higher than the diffusion velocity of 10^6 to 10^7 cm/sec in the classical model. If we tentatively assume a momentum relaxation time of 100 fsec for such ballistic electrons, they can travel about 1000 Å without scattering. High-frequency measurements have demonstrated electron velocity enhancement in the bases of AlGaAs/GaAs HBTs, although the effect of the hot carriers was not as significant as that expected for "ballistic" motion (Ankri et al., 1983; Ito et al., 1984a).

On the other hand, the hot carrier in the base has been observed by luminescence measurement (Ishibashi et al., 1984). Figure 3 shows luminescence spectra from the base of an HBT with an abrupt emitter. The AlAs fraction in the AlGaAs emitter is 0.3 and the base layer thickness is 1000 Å in the device. Increasing collector voltage, that is, being near the active region of transistor operation, causes a high-energy tail in the spectrum to become significant, indicating carrier heating of injected electrons. Average carrier temperature deduced from the exponential tail is over 500 K in this case. Effective energy relaxation time of electrons has also been estimated from carrier temperature. The relaxation time for 5×10^{18} cm^{-3} doped base was as short as 100 fsec. This value is compared with the mean electron to optical phonon scattering time.

When a Maxwellian distribution of carriers is assumed, the motion of injected carriers can be treated in the diffusion model. Then, the base transit time is given with carrier temperature T_c,

$$t_B^{\text{hot}} = W_B^2/2D_e^{\text{hot}} = (W_B^2/2)(q/kT_c \mu_e). \tag{16}$$

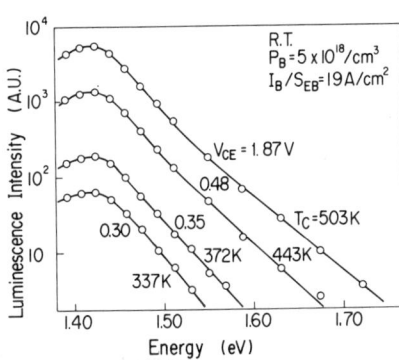

FIG. 3. Luminescence spectra observed from an abrupt-emitter HBT at room temperature. Here, the base current is kept constant, and the collector voltage V_{CE} is varied from 0.3 V to 1.87 V. No change in the spectrum is found for V_{CE} over 1.87 V. [From Ishibashi et al. (1984).] © 1985 Adam Hilger Ltd.

When the scattering process is dominated by ionized impurities, electron mobility is proportional to $T^{1.5}$. Then, we obtain the t_B reduction factor due to the hot carrier effect,

$$(t_B^{hot}/t_B) = (\mu_e/\mu_e^{hot})(T_{latt}/T_c) = (T_c/T_{latt})^{-2.5}. \tag{17}$$

From the observed T_c increase of from 330 K to 500 K, we expect a t_B reduction factor of about $\frac{1}{4}$. Microwave measurement results for the base transit time in the abrupt-emitter HBT are consistent with this estimation (Ito et al., 1984a).

4. Effect of Graded Base

The graded base structure is realized by compositional grading of an AlAs fraction over an AlGaAs base layer. A very high quasi-electric (or built-in) field that acts for one kind of carrier can be easily fabricated. For instance, 10% AlAs grading over 0.1 μm base-layer thickness results in a field of about 12 kV/cm. First, we consider the transient behavior of an injected electron. It must be noted that the static velocity–field relationship cannot be applied in every case, because the population of electrons in Γ and L valleys for a very thin graded base layer differs from that in the static case. They are rather in a "warm" state because of the restriction of maximum available energy given by the potential difference through the base.

A phenomenological kinetic equation for electrons in a field E is given as

$$d(m_e v_d)/dt = qE - m_e v_d/t_m, \tag{18}$$

where v_d and t_m are the electron drift velocity and the momentum relaxation time, respectively. When t_m is assumed to be constant, the solution for Eq. (18) becomes

$$v_d(t) = (qt_m/m_e)(1 - \exp(-t/t_m))E, \tag{19}$$

and

$$x(t) = (qt_m/m_e)[t - t_m(1 - \exp(-t/t_m))]E. \tag{20}$$

The transient region, Δx, is

$$\Delta x = x(t = t_m) = (qt_m^2/m_e)E(1/2.7). \tag{21}$$

After travelling this Δx, the electrons cross the base with almost constant drift velocity,

$$v_d = (qt_m/m_e)E = (\mu_e E). \tag{22}$$

For t_m of 80 fsec (corresponding to electron mobility of 2000 cm²/Vsec) and $E = 25$ kV/cm, Δx is calculated to be about 150 Å. This behavior of carriers is qualitatively in agreement with the results of Monte Carlo simulation

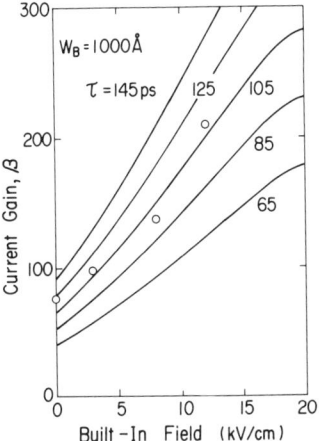

FIG. 4. Current gain dependencies on the built-in field intensity. Solid lines are calculated current gains, in which both diffusion and drift transports in the base are taken into account. [From Ito et al. (1985b).] © 1986 Adam Hilger Ltd.

(Tomizawa et al., 1984). In the simulation, a current gain cutoff frequency over 100 GHz has been predicted. Even for a small base grading and highly doped base, v_d can be made considerably higher. For example, $\mu_e = 1200$ cm^2/Vsec (in 1×10^{19} cm^{-3} doping) and 10% Al grading over 1000 Å distance give, $v_d = 1.4 \times 10^7$ cm/sec.

Reduction of base transit time also contributes to current gain enhancement (Hayes et al., 1983a; Ito et al., 1985a; Ito et al., 1985b). Figure 4 shows current gains as a function of built-in field intensity for HBTs with a constant base-layer thickness. The calculated current gain curve, in which both diffusion and drift transports are included, well explains the dependency on built-in field intensity. Temperature dependence of current gain in graded base HBTs differs from that in uniform base HBTs. A comparison of current gains in both structures is shown in Fig. 5. In the temperature region below 150 K, the current gain in the uniform base is proportional to temperature, whereas it is constant in the graded base. These results clearly demonstrate the effect of the built-in field in the graded base.

5. Device Modelling

Device modelling of HBTs for current–voltage characteristics is discussed. The Gummel–Poon model, which is typically used for the circuit simulation of bipolar transistors in the SPICE computer program (see, e.g., Sze, 1981c), is modified to express HBT characteristics here. Figure 6 shows a band diagram and various current components of an HBT. All electron and hole

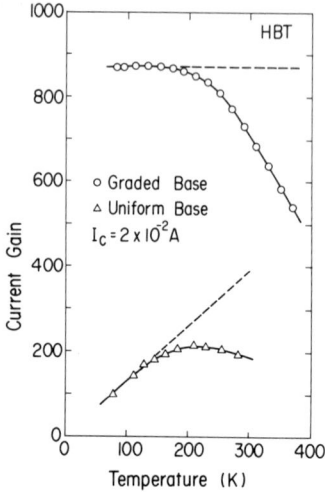

FIG. 5. Temperature dependencies of current gains in graded base (○) and uniform base (△) HBTs. [From Ito et al. (1985a).] © 1985 The Publication Board, *Jpn. J. Appl. Phys.*

currents flowing in the HBT are indicated by lines 1 to 6. These current components are defined using the following equations in the modified Gummel–Poon model:

$$I_1 = (I_{ES}/Q_B)(\exp(V_{BE}/n_{BE}V_T) - 1) \tag{23}$$

$$I_2 = (I_{CS}/Q_B)(\exp(V_{BC}/n_{BC}V_T) - 1) \tag{24}$$

$$I_3 = (I_{CS}/B_R)(\exp(V_{BC}/n_{BC}V_T) - 1) \tag{25}$$

$$I_4 = C_4 I_{CS}(\exp(V_{BC}/n_{BCL}V_T) - 1) \tag{26}$$

$$I_5 = (I_{ES}/B_F)(\exp(V_{BE}/n_{BE}V_T) - 1) \tag{27}$$

$$I_6 = C_2 I_{ES}(\exp(V_{BE}/n_{BEL}V_T) - 1), \tag{28}$$

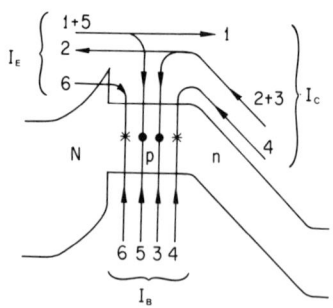

FIG. 6. Various current components of an HBT.

where I_{ES} and I_{CS} are the saturation current for the emitter and collector junctions, respectively; B_F and B_R are the forward and reverse current gains; C_2 and C_4 are the leakage current factors; and V_T is the unit thermal voltage of kT/q. Q_B is the complemental factor for an excess base charge at high injection current and is given by the following equations according to the original Gummel–Poon model:

$$Q_B = \tfrac{1}{2} Q_1 (1 + (1 + 4Q_2)^{0.5}) \tag{29}$$

$$Q_1 = (1 - V_{BC}/V_A)^{-1} \tag{30}$$

$$Q_2 = (I_{ES}/I_K)(\exp(V_{BE}/n_{BE} V_T) - 1), \tag{31}$$

where V_A is the early voltage and I_K the knee current. In this modified Gummel–Poon model, each of the ideality factors n_{BE}, n_{BC}, and n_{BEL} are newly defined for each current component to express various types of currents that depend on the HBT structures. With respect to an ideal abrupt emitter–base heterojunction, n_{BE} is given by

$$n_{BE} = 1 + \varepsilon_2 N_D / \varepsilon_1 N_A, \tag{32}$$

where ε_1 and ε_2 are the dielectric constants in the base and emitter, respectively. N_D and N_A are donor and acceptor densities in the emitter and base. The ideal graded emitter heterojunction without potential spike in conduction band energy has an n_{BE} of unity as well as does a homojunction. n_{BEL} and n_{BCL} are usually 2, since the leakage currents in emitter-base and base-collector junctions are due to the generation–recombination process.

The collector current I_C, the base current I_B and the emitter current I_E are given in terms of these parameters as follows:

$$I_B = I_3 + I_4 + I_5 + I_6, \tag{33}$$

$$I_C = I_1 - I_2 - I_3 - I_4, \tag{34}$$

$$I_E = I_C + I_B. \tag{35}$$

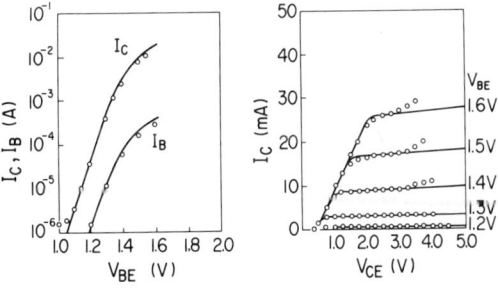

FIG. 7. Examples of I–V curve fitting for HBTs in common emitter conditions.

In practice, these parameters are determined by I–V curve fitting to the experimental data. An example of I–V fitting is shown in Fig. 7.

III. Fabrication of AlGaAs/GaAs HBTs

6. MBE Growth of HBT Epilayers

The capability of precise control in MBE growth of AlAs composition as well as of doping structures is very suitable for the preparation of HBT epilayers. A sharp interface between AlGaAs and GaAs up to the monolayer level is expected because of the low Al/Ga interdiffusion in AlGaAs materials. In addition, a gradual variation of AlAs compositions, which is very important for HBT structure, is realized simply by varying the molecular beam intensity. Although high-crystal-quality HBT epilayers can be grown by the LPE technique, the fine heterostructure required to fabricate high-speed devices is not easy to realize. Difficulties in thickness control and in high base doping are inherent in LPE.

The general requirements for HBT films are as follows:

(1) long diffusion length (long minority-carrier lifetime) in the base layer;
(2) low generation–recombination current in the emitter–base junction;
(3) well-controlled high base doping with a sharp profile; and
(4) good uniformity and reproducibility.

The deep traps are responsible both for excess generation–recombination current in the emitter–base the depletion layer and for nonradiative recombination in the neutral base. As is well known, the higher growth temperature during MBE yields better crystal quality with less incorporation of deep traps. Such a situation in growth conditions seems to be similar to that of double heterostructure laser diodes. For moderately doped base (on the order of 10^{18} cm^{-3}), the generation–recombination current tends to restrict the current gain, which fact has been generally confirmed through nonunity ideality factors in the emitter–base junction. For more heavily doped bases (greater than 1×10^{19} cm^{-3}), the electron lifetime in the base, that is, the base transport factor, comes to determine the ultimate current gain (Ito et al., 1984b).

On the other hand, the location of a p–n junction relative to a metallurgical junction significantly affects the current gain through the emitter efficiency. A redistribution of dopant during MBE, particularly for p-type dopants, has been found at higher growth temperatures, leading to a p–n junction shift. From this point of view, lower growth temperature is preferable to control the p-type base doping profile. Thus, it is necessary to optimize the growth temperature by taking into account both deep-trap incorporation and impurity redistribution.

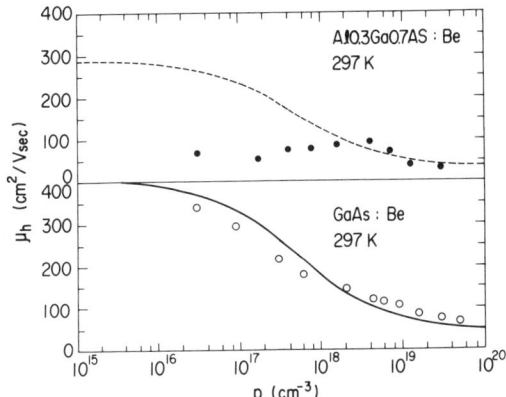

FIG. 8. Hole mobilities in Be-doped GaAs and $Al_{0.3}Ga_{0.7}As$ grown by MBE. The solid and dashed lines are highest mobilities measures in GaAs and estimated in $Al_{0.3}Ga_{0.7}As$, respectively. [From Ilegems et al. (1977).] © 1977 American Institute of Physics.

a. p-Type Doping in the Base

In MBE-grown HBT epilayers, Be is widely used as a p-type dopant in the base, since it has very high solubility in GaAs and AlGaAs. Good electrical properties in Be-doped MBE films are obtained (Ilegems, 1977). Hole mobilities in GaAs and $Al_{0.3}Ga_{0.7}As$ as a function of free carrier concentration are shown in Fig. 8. High doping by Be, over 10^{19} cm^{-3} realizes the advantage of low base resistance in the HBT structure. Recently, extremely high doping of Be, up to 2×10^{20} cm^{-3}, has been reported, where a base sheet resistance as low as 150 Ω/sheet was obtained in the HBTs grown at relatively low temperature (Lievin et al., 1985). In MOCVD, high Be doping, over 10^{20} cm^{-3}, has also been realized (Mellet et al., 1981).

At conventional growth temperatures (650–700°C), however, anomalous redistribution of Be has been reported. Be incorporation into bulk GaAs during growth has a tendency to saturate in the region over 10^{19} cm^{-3} (Enquist et al., 1985). This is due to an enhanced diffusion coefficient that is dependent on the Be doping level. Further, Be diffusion is faster in AlGaAs than it is in GaAs, as shown in Fig. 9 (Miller and Asbeck, 1985). Another anomalous behavior of Be in GaAs is a long-range "carry-forward" toward the growth direction, which is also indicated in Fig. 9.

In MBE growth of HBT epilayers, these Be redistribution behaviors must be carefully taken into account. As described before, a shift of the p–n junction position toward the wide-gap emitter side causes a reduction of the emitter efficiency. A deviation of the junction position as small as 100 Å still affects the device performance significantly, since the tolerance for the

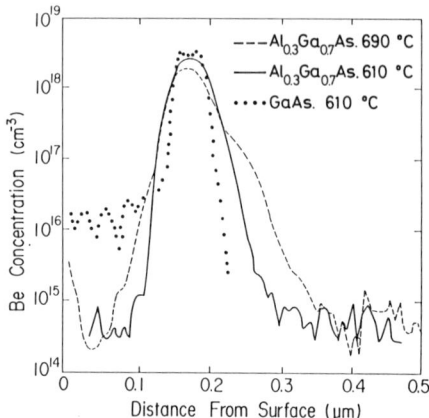

FIG. 9. SIMS profiles of Be in MBE-grown layers: $Al_{0.3}Ga_{0.7}As$ at 690°C (dashed line); $Al_{0.3}Ga_{0.3}As$ at 610°C (solid line), and GaAs at 610°C (dotted line). [From Miller et al. (1985).] © 1985 American Institute of Physics.

location of the emitter–base depletion layer is very small, a few hundred angstroms. Such influence of Be diffusion on emitter efficiency in both MBE- and MOCVD-grown HBTs have been investigated by I-V, I-L, and SIMS measurements (Enquist et al., 1984; Dubon et al., 1984). Introduction of an undoped spacer layer between the base GaAs and emitter AlGaAs has been found to be very effective. It has also been pointed out that a higher molecular beam flux ratio of As_4 over Ga during growth lowers the Be diffusion coefficient (Miller et al., 1985).

b. *Growth for Bandgap Grading in the* AlGaAs *Layer and Indium Soldering Free Growth*

Compositional bandgap grading in the AlGaAs emitter and in the AlGaAs base is performed by varying the molecular beam flux of Al and /or Ga. The Al effusion cell has a relatively low heat capacity, so that it has good time response and enables fabrication of the narrow grading layer required for HBTs at a typical growth rate of about 1 to 2 μm/h. For emitter grading, a linearly varying graded structure is conventionally used. A parabolic grading for the emitter has also been reported (Hayes et al., 1983b).

Other important aspects in MBE growth of HBT epilayers are productivity, reproducibility and reduction in defects or contamination on the wafers. In the conventional MBE growth procedure, however, wafer mounting with indium soldering to a Mo block has problems of time consumption and wafer contamination. In particular, treatment for the rough back-side of wafers is a serious problem in the IC fabrication process.

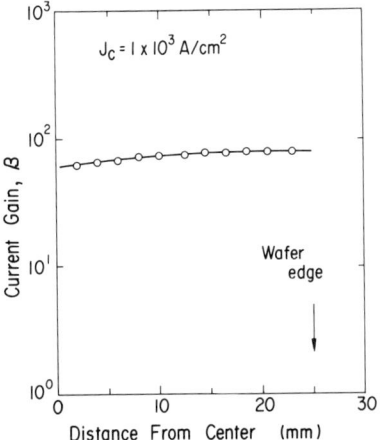

FIG. 10. Current gain uniformity in a two-inch HBT wafer grown by MBE using "indium-free" substrate heating. [From Ito et al. (1985c).] © 1985 The Publication Board, Jpn. J. Appl. Phys.

Recently, indium-free mounting of the wafer in HBT growth has been demonstrated (Shih et al., 1984; Ito et al., 1985c). Since the current gain is very sensitive to the growth temperature, inhomogeneous temperature distribution on the GaAs wafer results in a current gain distribution. As shown in Fig. 10, a good current gain uniformity with gain of over 80 has been realized by indium-free mounting growth. Here, the measured HBT with graded base had a high base doping of 4×10^{19} cm^{-3} and a base layer thickness of 1500 Å. Epilayer parameters of this HBT are listed in Table II.

TABLE II

Epitaxial Layer Parameters of an HBT Whose Current Gain Uniformity is Shown in Fig. 10.[a]

	Layer	Thickness (Å)	Doping (cm^{-3})	Al	Composition
Cap	n^+-GaAs	1500	5×10^{18}		
Emitter	n-AlGaAs	300	5×10^{17}	0–0.3	Parabolic
	n-AlGaAs	900	5×10^{17}	0.3	
	n-AlGaAs	300	5×10^{17}	0.3–0.1	Parabolic
Base	p^+-AlGaAs	1500	4×10^{19}	0.1–0	Linear
Collector	n-GaAs	3000	5×10^{16}		
Buffer	n^+-GaAs	5000	3×10^{18}		

[a] From Ito et al. (1985c). © 1985 The Publication Board, Jpn. J. Appl. Phys.

7. Ion Implantation Technique and Ohmic Contact Formation

a. Ion Implantation

In the HBT fabrication process, ion implantation is an important technique. This technique is applied to make highly doped external bases and to perform inter-device isolation. It also enables fabrication of a buried low-carrier-density layer in the external base/collector junctions for emitter-up HBTs or in external emitter/base layer junctions for collector-up HBTs, which will be described in the next subsection.

For the base contact layer, Be, Mg, or Zn ions are used. These implants exhibit fairly good activation properties with carrier density over 10^{19} cm^{-3} in GaAs, which is well known (for example, Eisen, 1984). What must be considered here is the diffusion of implants. In some cases, the dopant profile after annealing deviates largely from that calculated by LSS theory. Rapid thermal annealing is effective in reducing such undesirable diffusion of implanted ions (Asbeck et al., 1984b). Sheet resistivity of the external base layer as low as 220 Ω/sheet has been realized by rapid thermal annealing (Chang et al., 1986). On the other hand, ion implantations into AlGaAs are presenting some problems and reports are very few. Mg implantation in AlGaAs shows a rather small activation ratio and a large diffusion compared with that in GaAs (Yokota et al., 1983). More detailed study is necessary to optimize conditions for the implantation procedure and to achieve a highly doped p-AlGaAs layer.

b. Ohmic Contacts

For n-type ohmic contacts, AuGe/Ni alloyed with GaAs is widely used. Contact resistivity with these metals to n^+-GaAs doped to mid-10^{18} cm^{-2} can be made as small as 10^{-6} Ω-cm^2. A problem in AuGe/Ni contacts may be the so-called "ball-up" behavior, since this causes difficulty in the fine patterning of emitter electrodes. It has been reported that a four-layer metal structure of Au/Ti/AuGe/Ni is very effective in suppressing the ball-up and in improving contact resistivity down to 3×10^{-7} Ω-cm^2 (Ito et al., 1984c). A p-type ohmic contact is more important, as the base contact resistance determines the base resistance of HBT in most cases, affecting the high-frequency performance of devices directly. Both alloy-type and nonalloy-type contacts are employed. In addition to conventional AuZn alloyed contact, AuBe (Hayes et al., 1983a) and AgMn (Asbeck et al., 1984b) are also employed. For nonalloyed contact formation, highly doped p^+-GaAs, over about 3×10^{19} cm^{-3}, is necessary, in which the contact resistance through tunnel current is reduced down to the low range of 10^{-6} Ω-cm^2.

8. Self-Alignment Techniques

Since the intrinsic performance of the HBT is very high, the effect of parasitic elements such as base resistance, emitter resistance and extrinsic collector capacitance is still stronger than that in Si bipolar transistors. The scaling-down of device size is also an essential requirement to improve high-frequency and switching characteristics. For realizing low parasitics and device-size reduction, a self-alignment technique is very useful. However, as HBT structure is basically mesa-like, which is inherent in the starting multilayer epitaxial film, self-alignment differs largely from that of Si transistors.

Chang and his co-workers reported a self-alignment structure in which the external base area is defined by a mask for the emitter mesa (Chang et al., 1986). In this structure, the low resistive external base layer contributes to a reduction of base resistance. Another structure in which both base resistance and collector capacitance can be minimized has been proposed (Nagata et al., 1985). As shown in Fig. 11, the side surface of the emitter is surrounded by a thin ($\sim 0.2\,\mu$m) SiO_2 side-wall. This side-wall separates the base electrode from the emitter with only 0.2 μm spacing, resulting in a low base resistance and a small transistor size. An implantation of p-type dopant into the external base can also be combined with this structure.

9. Current Gain Shrinkage

In order to realize high-speed HBTs, emitter size must be minimized. This requirement is not as strong for HBTs as for homojunction bipolar transistors, but it still affects device performance significantly. It has been found in mesa-type uniform base HBTs that current gain shrinks with emitter size reduction (Nakajima et al., 1985a). Excess base leakage current by lateral flow of injected electrons has been revealed to be responsible for such current gain reduction. Since the high surface recombination rate in the external base is a dominant cause for this phenomenon, the introduction of a base contact

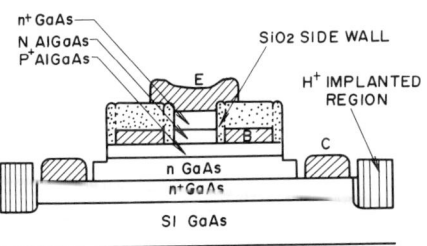

FIG. 11. A self-alignment structure HBT with a SiO_2 side-wall. [From Nagata et al. (1985).]
© 1986 Adam Hilger Ltd.

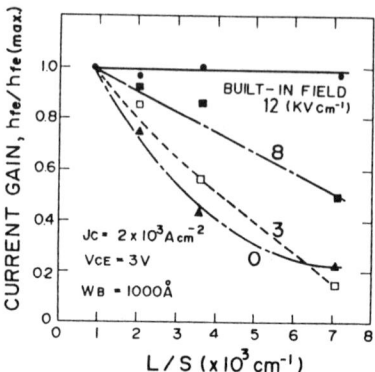

FIG. 12. Dependencies of current gain on emitter size in graded base HBTs. L/S is (emitter periphery length)/(emitter area). [From Nakajima et al. (1985b).] © 1985 The Publication Board, Jpn. J. Appl. Phys.

through the wide-gap p-AlGaAs layer seems to be effective in decreasing the excess base current. The use of the graded base structure has also been indicated to be very effective in suppressing the base leakage current (Nakajima et al., 1985b). Figure 12 shows dependencies of current gain on emitter size for various graded bases with built-in field intensity of 0 to 12 kV/cm, where L/S in the horizontal axis denotes periphery length/junction area. A built-in field of 12 kV/cm is sufficient to suppress the current gain reduction. It has been found that the generation–recombination current at the junction periphery is negligibly small in the high collector current region for practical operation.

10. FLEXIBILITY OF HBT STRUCTURE

a. Emitter-up HBT

An emitter-up HBT is most commonly used because of its structural simplicity. The collector–base junction capacitance in this structure is larger than that in the collector-up HBT described next, as it has a relatively large external base/collector area for base contacts. To decrease the extrinsic base–collector capacitance, oxygen implantations are effectively used, as shown in Fig. 13 (Asbeck et al., 1984b).

b. Collector-Up HBT

A collector-up HBT, that is, the inverted HBT, has the principal advantage of smaller collector capacitance without extrinsic base–collector junction. A higher f_{max} due to the lower collector capacitance is expected with this structure. Typical collector-up HBTs are shown in Figs. 14a and 14b

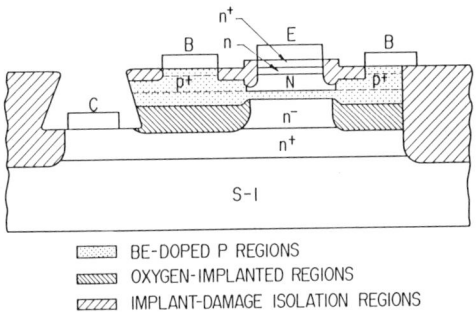

FIG. 13. Cross section of an HBT with a buried insulating layer in the external base/collector junction. [From Asbeck et al. (1984b).] © 1984 IEEE.

(Kroemer, 1982; Zhu et al., 1983). A Be ion-implanted p^+ layer is formed to produce a p-n junction inside the wide-gap emitter material in the external emitter/base junction area. This wide-gap p-n junction prevents both types of carriers from being injected over the junction. In Fig. 14b, an H^+-implanted isolation layer under the base contact is adopted to decrease the extrinsic emitter/base capacitance. A second advantage of the inverted HBT is a major reduction of the lead inductance in series with the emitter that is present in the conventional emitter-up configuration. Thus, this HBT structure is very suitable for microwave power amplification devices with high f_{max}. The inverted configuration also has more convenient possibilities for digital circuits such as multicollector I^2L application, as shown later.

c. Double Heterostructure HBT

A double heterostructure (DH) HBT with a wide-gap emitter and a collector has the following advantages (Kroemer, 1982; Beneking and Su, 1982):

(1) suppression of hole injection from the base into the collector under conditions of saturation in switching transistors;

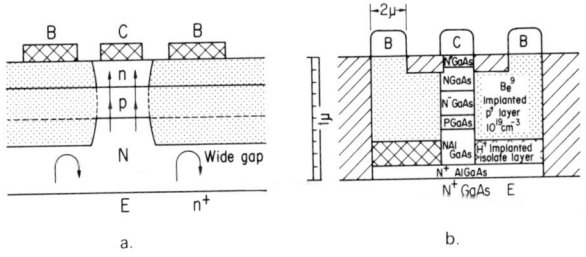

FIG. 14. Typical collector-up HBTs: [(a) from Kroemer (1982); (b) from Zhu et al. (1983).] © 1982 IEEE, © 1983 IEEE.

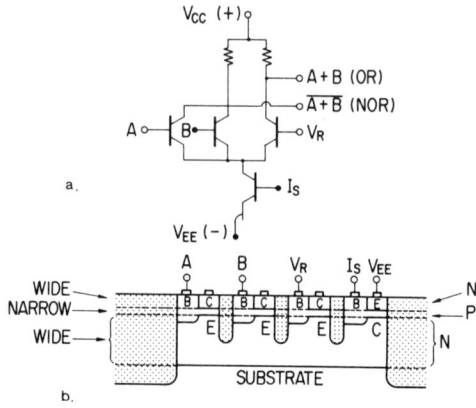

FIG. 15. (a) Basic circuit diagram of an OR/NOR gate; (b) double-hetero implementation of this ECL gate. [From Kroemer et al. (1982).] © 1982 IEEE.

(2) emitter/base interchangeability in IC applications; and
(3) separate optimization of base and collector, especially in microwave power transistors.

It is particularly noted that DH HBTs offer a major new option in the architecture of digital ICs because of the interchangeability made possible by simply changing the bias conditions while retaining a wide-gap emitter. For example, Fig. 15a gives the basic circuit diagram of the OR/NOR ECL gate. Fig. 15b shows a DH implementation of this ECL gate. In the DH design, this integration is achieved easily by implementing the top three transistors as inverted transistors and the current supply transistor as an emitter-up transistor. The emitter of the three top transistors and the collector of the bottom transistor come together in a buried n-layer on top of the substrate. All four transistors are structurally identical and differ merely in their biasing. H. Kroemer proposed this DH implementation of ECL, which he called HECL (Kroemer, 1982).

IV. High-Frequency Characteristics

11. Equivalent Circuit

The most common approach to characterizing the performance of high-frequency transistors is a combination of internal device parameters and two-port analysis. The simplified equivalent circuit is shown in Fig. 16. The current source, I_B, and I_C are defined by Eqs. (33) and (34), respectively. The small signal current gain h_{fe}, the small signal transconductance g_m, the

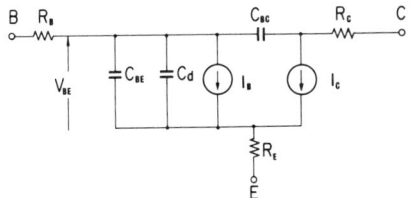

FIG. 16. A simplified equivalent circuit of an HBT.

intrinsic emitter differential resistance r_E and the emitter-base equivalent diffusion resistance r_d are defined as

$$dI_C = h_{fe} dI_B = g_m dV_{BE}, \tag{36}$$

$$r_E = (n_{BE} kT/q)/I_C = 1/g_m, \tag{37}$$

and

$$r_d = h_{fe}/g_m = h_{fe} r_E. \tag{38}$$

In Fig. 16, R_B, R_E, and R_C are the extrinsic base, emitter and collector resistances. C_{BE} and C_{BC} are the junction transition capacitance at the base/emitter and base/collector junctions. For the abrupt emitter HBTs, they are expressed by

$$C_{BE} = \frac{A_E(q\varepsilon_1\varepsilon_2 N_A N_D/(2V_{biBE}(\varepsilon_1 N_A + \varepsilon_2 N_D)))^{1/2}}{((V_{biBE} - V_{BE})/V_{biBE})^{1/2}} \tag{39}$$

and

$$C_{BC} = \frac{A_C(q\varepsilon_1 N_A N_D'/(2V_{biBC}(N_A + N_D')))^{1/2}}{((V_{biBC} - V_{BC})/V_{biBC})^{1/2}}, \tag{40}$$

where A_E and A_C are emitter and collector areas, respectively, and N_D' is collector donor density. Built-in voltages V_{biBE} and V_{biBC} are given as

$$V_{biBE} = E_{g1}/q + \Delta E_c + (kT/q)(\ln(N_D/N_c)_{emitt} + \ln(N_A/N_v)_{base}) \tag{41}$$

and

$$V_{biBC} = E_{g1}/q + (kT/q)(\ln(N_A/N_v)_{base} + \ln(N_D/N_c)). \tag{42}$$

C_d is defined as the diffusion capacitance due to the induced charge in the device and is related to carrier transit delay through the base, τ_B, and through the collector τ_C, by the equation

$$C_d = (\tau_B + \tau_C)/r_E. \tag{43}$$

The most important figures of merit for high-frequency and high-switching-speed transistors are the current gain cutoff frequency f_T, the maximum oscillation frequency f_{max}, and the circuit-limit frequency f_c, defined as follows:

$$f_T = 1/(2\pi\tau_F) = (2\pi(r_E(C_{BE} + C_{BC}) + (R_E + R_C)C_{BC} + \tau_B + \tau_C))^{-1}, \tag{44}$$

$$f_C = (2\pi R_B C_{BC})^{-1} \tag{45}$$

and

$$f_{max} = \tfrac{1}{2}(f_T f_C)^{1/2}. \tag{46}$$

The emitter-to-collector delay time τ_F consists of the emitter depletion layer charging time $r_E C_{BE}$; the collector depletion layer charging time $(r_E + R_E + R_C)C_{BC}$, including the Miller effect; the base transit time τ_B; and the collector depletion layer transit time τ_C.

a. Base Transit Time τ_B

The base transit time τ_B depends on the carrier transit mechanism in the base. Assuming carrier diffusion in the base, τ_B is given by

$$\tau_B^D = W_B^2/2D_e = qW_B^2/2kT\mu_e, \tag{47}$$

where W_B is the base width, D_e the electron diffusion constant, and μ_e the electron mobility in the base.

As described in Part II, in an abrupt-emitter HBT, hot electrons are injected into the base with a higher average energy. Thus, the effective carrier temperature T_c modulates both the electron mobility and the diffusion constant, resulting in a reduction in base transit time:

$$\tau_B^{D(hot)}/\tau_B^D = 1/(T_c/T)^{2.5}. \tag{48}$$

For instance, an increase in T_c as high as 500 K produces a reduction factor of about 1/3.6. The base transit time evaluated in abrupt-emitter HBTs has been explained by such a hot carrier effect (Ito et al., 1984a).

In the case of a graded bandgap base HBT that has an internal quasi-electric field, τ_B^G is approximately given by

$$\tau_B^G = \frac{h_{fe}^G((1 - FW_B/G^2)\sinh(GW_B) + (F/G)\cosh(GW_B))\tau_B}{(GW_B \exp(FW_B))}, \tag{49}$$

where

$$h_{fe}^G = G\exp(FW_B)/(F\sinh(GW_B) + G\cosh(GW_B)),$$
$$G = (F^2 + L_e^{-2})^{1/2},$$

and

$$F = (q/2kT)(\Delta E_G/W_B).$$

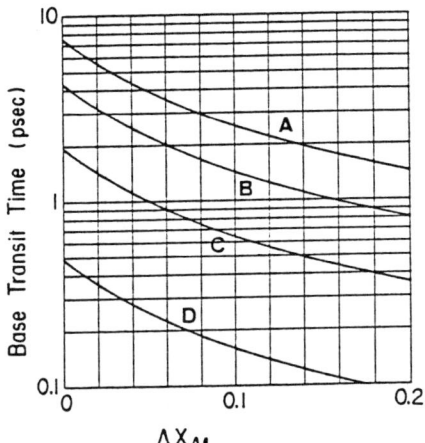

FIG. 17. Calculated base transit times of graded base HBTs as a function of Al fraction difference in the AlGaAs bases, Δx_{Al}. Base widths are 0.2, 0.15, 0.1, and 0.05 μm for curves A, B, C, and D, respectively.

ΔE_G is the bandgap difference over the base, which is related to the AlAs fraction difference Δx_{Al} by

$$\Delta E_G = 1.24 \, \Delta x_{Al}. \tag{50}$$

The graded base significantly reduces the base transit time, as shown in Fig. 17. Here, the dependence of τ_B^G on Δx_{Al} in the base with a parameter of base width is calculated. When W_B is 0.1 μm and $\Delta x_{Al} = 0.1$, the internal field in the graded base is 12.5 kV/cm and τ_B^G is about 0.7 psec, while $\tau_B^D = 2$ psec for diffusion transport in the uniform base HBT. As a result, both hot carrier injection in the abrupt-emitter structure and carrier acceleration in the internal field of the graded base structure are very effective in improving high-frequency characteristics.

b. Collector Transit Time τ_C and Collector Capacitance C_{BC}

Current flow in the collector depletion layer is the induced current so that the collector transit time delay is given by

$$\tau_C = X_C/(2v_s), \tag{51}$$

where X_C is the collector depletion width and v_s, the electron saturation velocity. Although τ_C decreases with decreasing X_C (with higher collector doping), the collector capacitance C_{BC} increases. Therefore, X_C should be designed with a compromise between τ_C and C_{BC} in each device structure or circuit.

c. Base Resistance R_B and Emitter-Base Junction Capacitance C_{BE}

Base resistance R_B can be much reduced by high base doping compared with that of homojunction transistors, which is just the advantage of the wide-gap emitter. Base doping on the order of 10^{19} cm^{-3} can be easily realized with Be by conventional MBE growth. In high doping, what must be considered is the degradation of electron lifetime, since the electron lifetime tends to decrease with increasing doping. Thicker base width also reduces R_B, although it increases the base transit time. Thus, the base must be designed by compromising among the base resistance, the base transit time and the current gain. Application of the grade base structure partly relaxes such tradeoffs, because the base transit time does not increase so much with base width. External base resistance is also very important, as the internal one is relatively small in HBTs. A reduction in C_{BE} is one more merit of the wide-gap emitter. When emitter doping is around 10^{17} cm^{-3}, the influence of C_{BE} on f_T can be made negligibly small under high collector current densities.

d. r_E, r_D, R_E, and R_C

The intrinsic emitter differential resistance r_E is the inverse of the transconductance g_m and decreases with an increase in collector current. The

TABLE III

COMPARISON BETWEEN AlGaAs/GaAs HBTs AND Si-BTs[a]

Parameters	Si	HBT
n_E (cm^{-3})	10^{20}	10^{17}
p_B (cm^{-3})	10^{18}	10^{19}
W_B (Å)	10^3	10^3
μ_B^h (cm^2/Vsec)	150	100
μ_B^e (cm^2/Vsec)	250	1000
D_B (cm^2/sec)	6.5	26
v_S (cm/sec)	10^7	10^7–10^8 [b]
ρ_B kΩ/□	4.2	0.63
R_B (extrinsic-base)	R_B^{si}	$R_B^{si}/6.6$
τ_B (diffusion)	7.68 psec	1.92 psec
$\tau_{B'}$ (drift)	—	0.5–0.05 psec
τ_C (drift)	1 psec	1–0.1 psec[b]
C_{BC}	C_{BC}^{si}	$C_{BC}^{si}/3$
f_{Tmax} (diffusion)	18 GHz	54 GHz
f_{Tmax} (drift)	—	100–1000 GHz[b]
f_C	f_C^{si}	$\sim 6 f_C^{si}$
f_{max}	f_{max}^{si}	4–$10 f_{max}^{si}$

[a] Collector depletion width ~2000 Å; $f_{Tmax} = 1/\{2\pi(\tau_B + \tau_C)\}$.
[b] Ballistic electrons.

influence of r_E on f_T can be neglected in Eq. (44) when the collector current is large enough. r_d is kept large enough to maintain the input voltage controllability if the current gain h_{fe} is significantly large. R_E and R_C are mainly determined by the emitter and collector ohmic contact resistance. They should be decreased to increase f_T according to Eq. (44).

12. COMPARISON BETWEEN HBTs AND Si-BTs

As mentioned above, HBTs are clearly noted to have great advantages as high-speed switching devices. To clarify them quantitatively, performance parameters of AlGaAs/GaAs HBTs and Si-BTs are calculated as shown in Table III, assuming typical material parameters. The high-frequency figures of merit, f_T, f_C, and f_{max}, of HBTs are several times higher than those of Si-BTs.

V. High-Speed Integrated Circuits

13. BASIC DIGITAL CIRCUITS

Three basic digital circuits elements, NTL, CML, and ECL gates, shown in Fig. 18, are the most significant ones for high-speed HBT ICs as well as for Si-BT ICs. NTL (non-threshold logic) consists of the load resistor R_L and the emitter resistor R_s with speed-up condenser C_s, which was originally developed in Si-BT ICs. DCTL (direct coupled transistor logic) consists of the transistor and the load resistor R_L without C_s and R_s in an NTL gate as shown in Fig. 18a. An ECL is a CML with an emitter-follower for improving driving capability and for a buffer to the next stage.

In order to compare the performance of these circuits, circuit simulation of a ring oscillator with basic inverters, shown in Fig. 18, has been performed by using the ASTAP simulation program. The modified Gummel–Poon model for an HBT as described in Part II is employed here. The HBT structure used is similar to that shown in Fig. 11. Spacing between the emitter

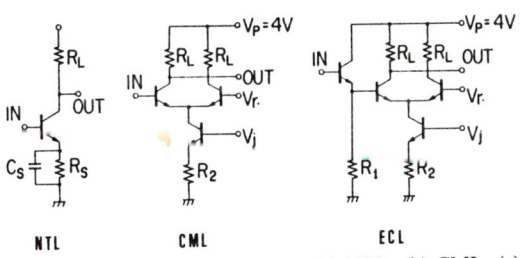

FIG. 18. Basic digital circuit elements: (a) NTL, (b) CML, (c) ECL.

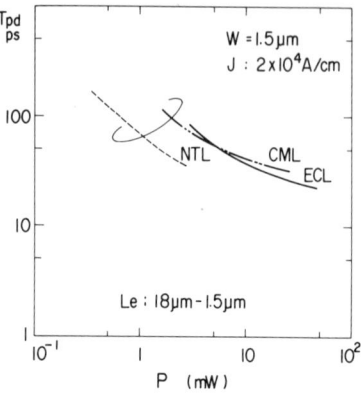

FIG. 19. Calculated propagation delay times. Widths of emitter and base electrode are 1.5 μm.

mesa and the base electrode is kept constant at 0.2 μm. Figure 19 shows typical calculated propagation delays t_{pd} as a function of power dissipation for different gates, where the emitter and base widths, W_E and W_B, are 1.5 μm. The spacing between the emitter and the base electrode is 0.2 μm. t_{pd} becomes shorter in the order NTL, CML, ECL, while power consumption increases in the same order. Figure 20 shows loci of the collector current I_C versus the voltage V_{BC} across the collector depletion layer for different inverters. The collector depletion layer capacitance C_{BC} as a function of V_{BC} is also shown in Fig. 20. Since NTL and CML inverters are directly coupled without level shift emitter followers, and base/emitter on-voltage V_{BE}^{on} of HBTs is higher than that of Si-BTs, the collector/base voltage is around zero and rather forward-biased. Thus, C_{BC} is considerably higher, as shown in Fig. 20. On the other hand, C_{BC} in the operation of ECL inverters is smaller, since the collector–base junction is fairly reverse-biased with the

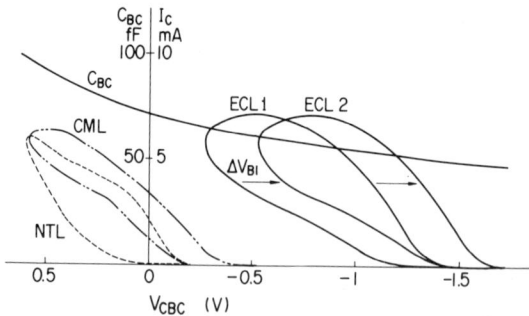

FIG. 20. Loci of collector currents and collector capacitance as a function of base/collector voltage V_{BC}.

emitter-follower. In Fig. 20, ECL1 indicates the locus of graded emitter HBT inverter operation and ECL2, that of the abrupt emitter HBT. This variation arises from a difference in level shift voltage by the emitter-follower that depends on the built-in voltage of the emitter–base junction. As a result, an ECL inverter with the higher emitter–base built-in voltage can operate at shorter propagation delay.

A few results for HBT ring oscillators with NTL, CML, and ECL gates have been reported. A propagation delay time as fast as 27.2 psec/gate in a NTL gate has been demonstrated with devices fabricated by using the self-aligned technique (emitter size, $1.5 \times 5 \mu m^2$), which is the smallest delay ever reported in bipolar transistors (Chang et al., 1986). In the ECL gate, a t_{pd} of 65 psec has been achieved with a relatively large emitter size of $3 \times 9 \mu m^2$ (Nagata et al., 1985). Reduction of both emitter and collector sizes and improvement in the fabrication technique should offer far faster switching speeds.

14. ULTRAHIGH-SPEED POSSIBILITY

The possibility of ultrahigh-speed operation in HBT ICs with smaller dimensions can be predicted by the circuit simulation described above. Table IV shows the propagation delay of HBT-ECL gates and the toggle frequency of the frequency divider for different device sizes (Ishibashi, 1985). Here, in the simulation, a self-aligned structure HBT with emitter-to-base contact spacing of $0.2 \mu m$ (shown in Fig. 11) is assumed. The widths, W, for emitter and base contact are varied from 1.5 to $0.2 \mu m$. A propagation delay of 30 psec and a toggle frequency of dividers of over 10 GHz are expected with $W = 1 \mu m$. Furthermore, a t_{pd} of less than 10 psec and a toggle frequency of higher than 30 GHz with $W = 0.2 \mu m$ are predicted.

A more detailed simulation has also been made for DCTL, CML, and ECL gate ring oscillators. Kurata and Yoshida used a method combining a realistic

TABLE IV

CALCULATED PROPAGATION DELAY TIMES OF RING OSCILLATORS AND TOGGLE FREQUENCIES OF $\frac{1}{2}$ FREQUENCY DIVIDERS WITH AlGaAs/GaAs HBT ECL GATES

Size[a] (μm)	t_{pd} for ECL (psec)	Frequency for $\frac{1}{2}$ FD (GHz)
1.5	35	10
1.0	28	13
0.5	17	22
0.2	<10	>30

[a] Emitter and base electrode widths.

physical device model that involves numerical solutions and their own circuit simulator (Kurata and Yoshida, 1984). They also predicted that sub-10-psec switching is achievable with a HBT having a $1 \times 2\ \mu m^2$ emitter pattern and two external base areas of $1 \times 2\ \mu m^2$.

15. IC Applications

Although HBTs have great potential for ultrahigh-speed ICs, there have not been many fabricated ICs as yet. Typical HBT ICs reported are frequency dividers and gate arrays. The AlGaAs/GaAs HBT divider circuit fabricated by Rockwell was based on master–slave flip-flops constructed from a D-latch circuit (Asbeck et al., 1984c). Fig. 21 shows the D-latch, which consists of series-gated CML. Divide-by-two operation was obtained by feeding the M/S flip-flop output back to the D input. Divide-by-4 operation was obtained by combining two divide-by-two sections. Emitter-follower stages were used for buffer and level shifting to interconnect feedback loops. These circuits with CMLs and emitter-followers can operate as well as ECLs. An emitter-follower output driver was also used. The divide-by-four circuit utilized 32 HBTs. The HBT used had an emitter dimension of $1.2 \times 5\ \mu m^2$, and oxygen was implanted to insert the buried insulating layer in the external base/collector junction. The divide-by-four circuit could operate up to 8.6 GHz with power dissipation of about 210 mW. A discrete device similar to that utilized in the divider yielded a current gain cutoff frequency as high as 40 GHz.

A GaAs bipolar gate array employing the combined advantages of an AlGaAs/GaAs heterojunction and I^2L-like gate structure (HI^2L) was first

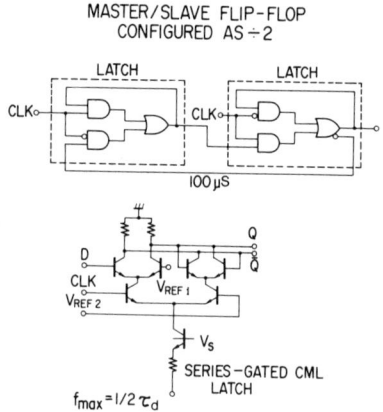

Fig. 21. HBT divider circuit based on master–slave flip-flops constructed from a D-latch circuit [From Asbeck et al. (1984d).] © 1984 IEEE.

5. HETERO-BIPOLAR TRANSISTOR AND ITS LSI APPLICATION

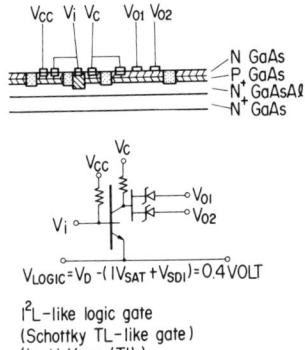

FIG. 22. An I²L-like gate structure with an inverted AlGaAs/GaAs HBT. [From Yuan *et al.* (1984).] © 1984 IEEE.

developed by the TI group (McLevige *et al.*, 1982; Yuan *et al.*, 1984). Using a nominal 3 μm design rule, a gate with a fanout of 4 occupies only 1650 μm². It also showed 0.77-nsec propagation delay and 0.3 mW power consumption per gate in the 17-stage ring oscillator. Figure 22 shows the cross-section and circuit scheme of the HI²L gate. MBE-grown GaAs/buried AlGaAs layers

TABLE V

DESIGN PARAMETERS OF 1K GATE ARRAY IMPLEMENTED WITH INVERTED AlGaAs/GaAs HBTs.[a,b]

Bar size:	3.8 × 3.56 mm²
Gate size:	72 × 24 μm²
Horizontal wire channel (first-level metal)	
Number of channels	148
Pitch	8 μm
Vertical wire channels (second-level metal)	
Number of channels	150
Pitch	9 μm
Propagation delay	
0.2 mW/gate	1.5 nsec
1.0 mW/gate	0.4 nsec
Power supply	
I/O buffers	3.5 V ± 10%
Internal gate	2 V ± 10%
I/O specifications	
Maximum output drive	2 mA
Logic level	ECL-compatible

[a] I²L-like gate structure shown in Fig. 22 is used.
[b] After Yuan *et al.*, 1984.
© 1984 IEEE.

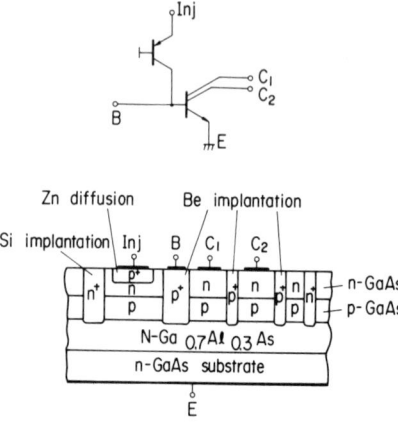

FIG. 23. An I²L inverter with a p–n–p current source. [From Narozny and Beneking (1984).] © 1985. The Institution of Electrical Engineers.

on an n^+-GaAs substrate were used as the starting material. Si and Be implants were applied to form the collector and base, respectively, in the inverted transistors. The process also used a B implant for device isolation. The design parameters of the 1K gate array are shown in Table V. The circuits were fabricated in 2 inch-diameter GaAs substrate.

An I²L inverter with the inverted p–n–p current source has also been reported in LPE-grown GaAs/AlGaAs materials using ion implantation and Zn diffusions (Narozny and Beneking, 1984, 1985). The circuit and structure of this inverter are shown in Fig. 23. The shallow p^+ emitter of the p–n–p current source was fabricated by Zn diffusion. Instead of a lateral p–n–p transistor, which is typical in Si–I²L technology, a vertical arrangement has the possibility of better current gain and better uniformity of base thickness.

VI. Conclusions

Although their development is still not matured technologically, HBTs are very promising devices for ultrahigh-speed integrated circuits because of their high cutoff frequency, large driving capability and freedom of various transistor structures. Table VI summarizes future practical applications of HBTs. In addition to these electronics applications, HBTs are also suitable devices for opto-electronic integrated circuits, since they have similar multiheterostructures to those of laser diodes and detectors, described in Chapter 6.

TABLE VI
Applications of HBTs

1. Ultrahigh-speed ICs
   Frequency dividers
   Multiplexer and demultiplexer
   Multiplier
   TD switch
   Buffer memory
   Gate array
2. Low-power ICs
   I^2L gate array
3. OE–IC
4. High-frequency and high-power amplifiers
   Collector-up HBTs
   Wide-band amplifiers

In order to realize the practical utilization of HBTs-ICs, the following technological problems must be solved:

(1) High quality, uniform and reproducible hetero-epitaxial growth technology.
(2) Good ohmic contact formations.
(3) Various ion implantation techniques.
(4) Various dry processes for device scale-down.
(5) Optimization of device structures and circuits.

HBTs definitely will play an important role in the field of ultrahigh-speed ICs in the near future as post-Si bipolar transistors.

References

Ankri, D., and Eastman, L. F. (1982). *Electron. Lett.* **18,** 750.
Ankri, D., Schaff, W. J., Smith, P., and Eastman, L. F. (1983). *Electron. Lett.* **19,** 147.
Asbeck, P. M., Miller, D. L., Asatourian, R., and Kirkpatrick, C. G. (1982). *IEEE Electron Dev. Lett.* **EDL-3,** 403.
Asbeck, P. M., Miller, D. L., Anderson, R. J., and Eisen, F. H. (1983). *Proc. GaAs IC Symp.*, p. 170.
Asbeck, P. M., Gupta, A. K., Ryan, F. J., Miller, D. L., Anderson, R. J., Liechti, C. A., and Eisen, F. H. (1984a). *Tech Dig.* **IEDM-84,** p. 864.
Asbeck, P. M., Miller, D. L., Anderson, R. J., and Eisen, F. H. (1984b). *IEEE Electron Dev. Lett.* **EDL-5,** 310.
Asbeck, P. M., Miller, D. L., Anderson, R. J., Deming, R. N., Chen, R. T., Liechti, C. A., and Eisen, F. H. (1984c). *Proc. GaAs IC Symp.*, p. 133.
Beneking, H., and Su, L. M. (1982). **18,** 25.

Casey, H. C. Jr., and Panish, M. B. (1978). *Heterostructure Lasers*, p. 217 (Academic Press, New York, 1978).

Chang, M. F., Asbeck, P. M., Miller, D. L., and Wang, K. C. (1986). *IEEE Electron Dev. Lett.* **EDL-7**, 8.

Dubon, C., Azoulay, R., Gauneau, M., Dugrand, L., Sibille, A., Dangla, J., Duchenois, A. M., and Ankri, D. (1984). *Inst. Conf. Ser.* **74**, 175.

Eisen, F. H. (1984). *Ion Implantation and Beam Processing* (Williams, J. S., and Poate, J. M. eds.), p. 327 (Academic Press, New York, 1984).

Enquist, P., Lunardi, L. M., Welch, D. F., Wicks, G. W., Shealy, J. R., Eastman, L. F., and Calawa, A. R. (1984). *Inst. Conf. Ser.* **74**, 599.

Enquist, P., Wicks, G. W., and Eastman, L. F. (1985). *J. Appl. Phys.* **58**, 4130.

Harris, J. S., Asbeck, P. M., and Miller, D. L. (1982). *Proc. 14th Conf. on Solid State Devices, Tokyo, 1982* (*Jpn. J. Appl. Phys.* **22**, Suppl. 22-1, 375).

Hayes, J. R., Capasso, F., Gossard, A. C., Malik, R. J., and Wiegman, W. (1983a). *Electron Lett.* **19**, 410.

Hayes, J. R., Capasso, F., Malik, R. J., Gossard, A. C., and Wiegman, W. (1983b). *Appl. Phys. Lett.* **43**, 949.

Hovel, H. J., and Milnes, A. G. (1969). *IEEE Trans.* **ED-16**, 766.

Ilegems, M. (1977). *J. Appl. Phys.* **48**, 1278.

Ishibashi, T. (1985). *Ohyoh-Butsuri*, **54**, 1192 (in Japanese).

Ishibashi, T., Ito, H., and Sugeta, T. (1984). *Inst. Phys. Conf. Ser.* **74**, p. 593.

Ito, H., Ishibashi, T., and Sugeta, T. (1984a). *IEEE Electron Dev. Lett.* **EDL-5**, 214.

Ito, H., Ishibashi, T., and Sugeta, T. (1984b). *Jpn. J. Appl. Phys.* **23**, L635.

Ito, H., Ishibashi, T., and Sugeta, T. (1984c). *Jpn. J. Appl. Phys.* **23**, L635.

Ito, H., Ishibashi, T., and Sugeta, T. (1985a). *Jpn. J. Appl. Phys.* **24**, L241.

Ito, H., Ishibashi, T., and Sugeta, T. (1985b). *12th Int. Symp. GaAs and Related Compounds, Inst. Phys. Conf. Ser.* **79**, 607.

Ito, H., and Ishibashi, T. (1985c). *Jpn. J. Appl. Phys.* **24**, 1567.

Jadus, D. K., and Feucht, D. L. (1969). *IEEE Trans.* **ED-16**, 102.

Konagai, M., and Takahashi, K. (1975). *J. Appl. Phys.* **46**, 2120.

Konagai, M., Katsukawa, K., and Takahashi, K. (1977). *J. Appl. Phys.* **48**, 4389.

Kroemer, H. (1957a). *Proc. IRE* **45**, 1535.

Kroemer, H. (1957b). *RCA Rev.* **18**, 332.

Kroemer, H. (1982). *Proc. IEEE* **70**, 13.

Kurata, M., and Yoshida, J. (1984). *IEEE Trans.* **ED-31**, 467.

Lievin, J. L., Dubon-Chevallier, C., Leroux, G., Dangla, J., Alexandre, F., and Ankri, D. (1985). *Presented 12th Int. Symp. GaAs and Related Compounds, Inst. Phys. Conf. Ser.* **79**, 595.

Marty, A., Rey, G., and Bailbe, J. P. (1979). *Solid State Electron.* **22**, 549.

McLevige, W. V., Yuan, H. T., Duncan, W. M., Frensley, W. R., Doerbeck, F. H., Morkoç, H., and Drummond, T. J. (1982). *IEEE Electron Dev. Lett.* **EDL-3**, 43.

Mellet, R., Azoulay, R., Dugrand, L., Rao, E. V. K., and Mircea, A. (1981). *Inst. Phys. Conf. Ser.* **63**, 583.

Miller, D. L., and Asbeck, P. M. (1985). *J. Appl. Phys.* **57**, 1816.

Miller, J. N., Collins, D. M., and Moll, N. J. (1985). *Appl. Phys. Lett.* **46**, 960.

Nagata, K., Nakajima, O., Yamauchi, Y., and Ishibashi, T. (1985). *12th Int. Symp. on GaAs and Related Compounds, Inst. Phys. Conf. Ser.* **79**, 589.

Nakajima, O., Nagata, K., Ito, H., Ishibashi, T., and Sugeta, T. (1985a). *Jpn. J. Appl. Phys.* **24**, L596.

Nakajima, O., Nagata, K., Ito, H., Ishibashi, T., and Sugeta, T. (1985b). *Jpn. J. Appl. Phys.* **24**, 1368.

Narozny, P., and Beneking, H. (1984). *Electron. Lett.* **20,** 442.
Narozny, P., and Beneking, H. (1985). *Electron. Lett.* **21,** 328.
Shih, H. D., Matteson, S. E., McLevige, W. V., and Yuan, H. T. (1984). *Proc. 3rd Int. Conf. MBE (J. Vac. Sci. Technol.* **B3,** 793).
Shockley, W. (1948). U.S. Patent 2569347.
Sleger, K., Milnes, A. G., and Feucht, D. L. (1970). *Proc. Int. Conf. on Phys. and Chem. of Semiconductors, 1970,* p. 1.
Sze, S. M. (1981a,b,c). *Physics of Semiconductor Devices, 2nd Edition,* p. 134, p. 258, p. 153 (John Wiley & Sons, New York, 1981).
Tomizawa, K., Awano, Y., and Hashizume, N. (1984). *IEEE Electron Dev. Lett.* **EDL-5,** 362.
Yokota, M., Ashigaki, S., Makita, Y., Kanayama, T., Tanoue, H., and Takayasu, I. (1983). *Nuclear Instruments and Methods* **209/210,** 711.
Yokoyama, K., Tomizawa, M., and Yoshii, A. (1984). *IEEE Trans.* **ED-31,** 1222.
Yuan, H., McLevige, M. V., Shih, H. D., and Hearn, A. S. (1984). *IEEE ISSCC-84,* p. 42.
Zhu, E. J., Ku, W. H., and Wood, C. E. C. (1983). *Digest of Papers, Monolithic Circuits Symp.* **MTT-S,** 17.

CHAPTER 6

Optoelectronic Integrated Circuits

*Hideaki Matsueda, Toshiki P. Tanaka
and Michiharu Nakamura*

CENTRAL RESEARCH LABORATORY
HITACHI LTD.
TOKYO, JAPAN

I.	INTRODUCTION	231
II.	DESIGN CONSIDERATIONS	234
	1. *Integrating Structure*	234
	2. *Circuits*	246
III.	FABRICATION PROCESSES	264
	3. *Process Flow*	264
	4. *Reflector Formation*	266
IV.	HIGH-SPEED TRANSMITTER AND RECEIVER	269
	5. *Transmitter OEIC*	269
	6. *Receiver OEIC*	275
	7. *Transmission Experiment*	279
V.	CONCLUSIONS	281
	REFERENCES	282

I. Introduction

Since the two major impacts of the realization of continuous operation of semiconductor lasers at room temperature and of the low-loss optical fiber that appeared in 1970, the technologies of optical communication have made rapid progress. Nowadays, digital optical fiber communication systems with bit rates of 400–600 Mb/sec have been widely installed for practical use. However, needs for even higher bit-rate systems are increasing. So research on systems with bit rates in multigigabits per second is proceeding.

In today's practical optical fiber communciation systems, the key components, such as optical transmitters and receivers, are solely constructed of discrete optical devices and Si electronic ICs. The interconnection between the optical devices and the associated electronic circuits is performed by conventional wire bonding and/or lead soldering techniques. The assembling technology is hybrid and is well established. However, the parasitics due to the stray capacitance of the discrete package and the inductances of lead wires and bonding wires will be a serious problem to the future multi-gigabit-per-second systems.

This problem can be solved by monolithic integration of optical devices and electronic circuits on a common substrate of III–V compound semiconductor. With the monolithic integration, design flexibility will be increased because of the reduction of parasitics. Size and cost reductions and high reliability can also be expected. The transition from the hybrid assembly to the monolithic integration is analogous to the developing history of Si devices from discrete transistors to VLSIs.

Monolithically integrated devices including photonic and electronic devices are called OEICs (optoelectronic integrated circuits) (Matsueda et al., 1983a; Hayashi, 1983), or alternatively IOECs (integrated optoelectronic circuits) (Yariv, 1983).

Fortunately, we have promising semiconductor materials that exhibit good performances in both photonics and electronics—namely, GaAs and InP. When GaAs is used as the substrate, the available optical wavelengths are in the short-wavelength regions (around 0.8 μm). In these wavelength regions, the loss and the dispersion of optical fibers are comparatively large; therefore, the distance of transmission is limited. OEICs using the GaAlAs/GaAs system are suitable for applications where a long link length is not essential. The interconnections between board packages in very high-speed computers and high-speed local area networks (LANs) are the major applications.

The loss and the dispersions of silica-based optical fibers are very small in the longer-wavelength region (1.3–1.5 μm). The emission of those wavelengths can be obtained by using the InGaAsP/InP system. InP substrates are advantageous for applications such as long-haul optical communication.

A document shows that the basic concept of monolithic integration of active optical devices and electronic circuits was proposed as early as 1972 (Somekh and Yariv). Since the first integrated device including a laser diode and a Gunn oscillator was reported (Lee et al., 1978), several kinds of OEIC devices have been fabricated. Those devices may be separated into two categories, as summarized in Tables I and II. One is the transmitter OEICs, including laser diodes and their driving electronics (Fukuzawa et al., 1980; Matsueda et al., 1980–1985) and the other is the receiver OEICs, consisting of photodetectors and pre-amplifier circuits (Leheny et al., 1980; Wada et al., 1985a,b). Both approaches use the GaAlAs/GaAs material system (Ury et al., 1979; Matsueda et al., 1980–1985; Matsueda and Nakamura, 1981, 1984; Wada et al., 1983a) and the InGaAsP/InP material systems (Koren et al., 1982b; Leheny et al., 1980).

Some devices were fabricated on conducting substrates (Katz et al., 1980; Shibata et al., 1984). However, recent devices mainly have been fabricated on semi-insulating (SI) substrates (Matsueda et al., 1983–1985; Matsueda

6. OPTOELECTRONIC INTEGRATED CIRCUITS

TABLE I
OEIC Transmitters[a]

Substrate	Structure	Photonics[b]	Electronics	Scale	Speed[c]	References
GaAs n	Vertical	Diffused stripe	MESFET	1 LD + 1 FET	0.4 nsec	Fukuzawa et al., 1980
n		MCSP	MESFET	1 LD + 2 FET	0.4 nsec	Matsueda et al., 1980; Matsueda and Nakamura, 1981
SI	Horizontal	Crowding	Gunn	1 LD + 1 Gum	1 GHz	Lee et al., 1978
SI		Crowding	MESFET	1 LD + 3 FET	—	Yust et al., 1979
SI		Mesa stripe	MESFET	1 LD + 1 FET	—	Ury et al., 1979
SI		Oxide stripe	MESFET	1 LD + 2 FET	1 GHz	Kim et al., 1984
SI		Be implantation	MESFET	1 LD + 1 FET	—	Wilt et al., 1980
n		Be implantation	HBPT	1 LD + 1 FET	—	Katz et al., 1980
n		BH	BPT	1 LD + 1 BPT	—	Bar-Chaim et al., 1982
SI		DH	MESFET	1 LED + 1 PM + 2 FET	15 MHz	Carter et al., 1982
SI		BH	MESFET	1 LD + 1 FET	4 GHz	Ury et al., 1982
SI		MQW	MESFET	1 LD + 2 FET	—	Sanada et al., 1984
SI		MQW	MESFET	1 LD + 2 FET	2 GHz	Hong et al., 1984
SI		GRIN-SCH	MESFET	1 LD + 4 FET	0.4/0.9 nsec	Sanada et al., 1985
SI		MCSP, TS	MESFET	1 LD + 1 PM + 6 FET	—	Matsueda et al., 1983a, 1983b, 1984; Matsueda and Nakamura, 1984
SI		MQW-SAS	MESFET	1 LD + 1 PM + 12 FET	2 Gb/sec	Matsueda et al., 1985
SI		TJS	MESFET	1 LD + 36 gates	160 MHz	Carney et al., 1983
SI		BH, PIN	MESFET	1 LD + 1 PD + 1 FET	178 MHz	Bar-Chaim et al., 1984
InP SI	Horizontal	TJS	MISFET	1 LD + 1 FET	—	Koren et al., 1982b
SI		BH	JFET	1 LD + 1 FET	—	Chen et al., 1983
SI		DC-PBH	MISFET	1 LD + 1 FET	2 Gb/sec	Kasahara et al., 1984b
n		DH (1.5 μm)	BPT	1 LD + 1 BPT	—	Su et al., 1985
n		BH	HBPT	1 LD + 3 BPT	—	Shibata et al., 1984
SI		BH, PIN	JFET	1 LD + 1 PD + 2 FET	1.6 GHz	Hata et al., 1985

[a] n: n-type substrate; SI: semi-insulating substrate; MCSP: modified channeled substrate planar; BH: buried heterostructure; DH: double heterostructure; GRIN-SCH: graded index waveguide separate confinement heterostructure; TS: terraced substrate; MQW-SAS: multi-quantum well self-aligned structure; TJS: transverse junction stripe; DC-PBH: double channel-planar buried heterostructure; MESFET: metal semiconductor FET; MISFET: metal insulator semiconductor FET; JFET: junction FET; BPT: bipolar transistor; HBPT: hetero BPT; LD: laser diode; PM: photomonitor; PD;: photodetector; LED: light-emitting diode.
[b] Wavelength is in 0.8 μm band for GaAs, 1.3 μm band for InP, if not otherwise specified.
[c] Rise and fall time or rise/fall times, or frequency or bit rate.

TABLE II

OEIC RECEIVERS[a]

Substrate		Photonics[b]	Electronics	Scale	Speed[c]	References
GaAs	SI	SPD	MESFET	1 PD$_t$ + 1 FET	—	Wada et al., 1983a
	SI	PIN	MESFET	1 PD$_t$ + 1FET	—	Miura et al., 1983
	SI	PIN	MESFET	1 PD$_t$ + 6 FET	40 nsec	Kolbas et al., 1983
	Si	PIN	MESFET	1 PD$_t$ + 6 FET	300 MHz	Wada et al., 1983b, 1985a
	SI	MSM	MESFET	1 PD$_t$ + 3 FET	1 GHz	Ito et al., 1984
InP	SI	PIN (1.75 μm)	JFET	1 PD$_t$ + 1 FET	—	Leheny et al., 1980
	SI	PIN	JFET	1 PD$_t$ + 1 FET	—	Inoue et al., 1983
	SI	PIN	MISFET	1 PD$_t$ + 1 FET	100 Mb/sec	Kasahara et al., 1984a
	SI	PIN (1.55 μm)	JFET	1 PD$_t$ + 1 FET	—	Hata et al., 1984
	SI	PC (<1.6 μm)	MESFET	2 PD$_t$ + 1 FET	—	Barnard et al., 1981

[a] SPD: Schottky barrier photodiode; PIN: PIN photodiode (PIN-PD); MSM: metal semiconductor metal photodiode; PC: photoconductive detector; PD$_t$: photodetector. Other abbreviations are as indicated in Table I.
[b] Wavelength is in 0.8 μm band for GaAs, 1.3 μm band for InP, if not otherwise specified.
[c] Rise and fall time, or frequency or bit rate.

and Nakamura, 1984). To realize multifunctional devices, the scale of electronic circuits must be large. Semi-insulating substrates are suitable for high-speed, large-scale integration, as is known in GaAs LSIs. So, OEIC devices on this substrate may be used more widely than devices on other substrates.

To date, the scale of the integration of reported devices is larger in the GaAlAs/GaAs system than in the InGaAsP/InP system. One of the reasons is that the fabrication technologies for both optical and electronic devices are better established for the GaAlAs/GaAs system. In this chapter, we review the present status of OEIC technologies, and then discuss the realization of high-speed transmitter and receiver OEICs, because they are the most important devices. In Part II, the design considerations for those OEIC devices are given. The fabrication process is discussed in Part III. Part IV is devoted to the operation of transmitters and receivers on GaAs semi-insulating substrates. In Part V, a summary is given and future prospects are described.

II. Design Considerations

1. INTEGRATING STRUCTURE

a. Transmitter OEIC

There are two types of integration, vertical and horizontal. The vertical may be defined as the type that uses conductive substrates to let the current flow through the substrate. In this case, a laser diode and electric circuits are integated in a two-storied manner. The horizontal, in contrast, is defined as a type with semi-insulating substrates, where the current is fed into and taken off from one face of the substrate only, and a laser diode (LD) and electric circuits are laid out. Each type could be combined with low-threshold laser diodes such as the CS (channeled substrate), terrace, BH (buried heterostructure), and QW (quantum well) types (Yamakoshi *et al.*, 1983; Hong *et al.*, 1984; Matsueda *et al.*, 1985), and others as summarized in Table I.

For the sake of electronic components, the horizontal structure has a very important technical advantage, because the semi-insulating surface can be used directly. In the vertical structure, on the other hand, active as well as

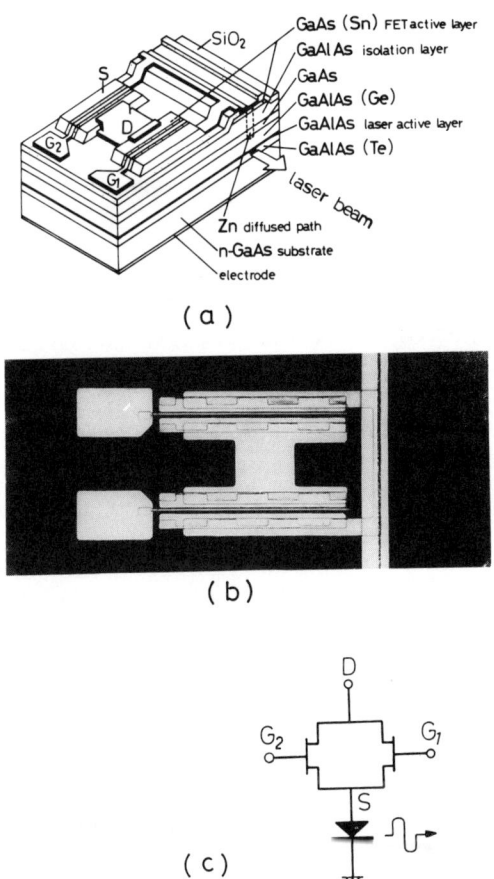

FIG. 1. A vertical GaAs OEIC transmitter with a MCSP laser and two MESFETs: (a) overall view; (b) photograph; (c) circuit diagram.

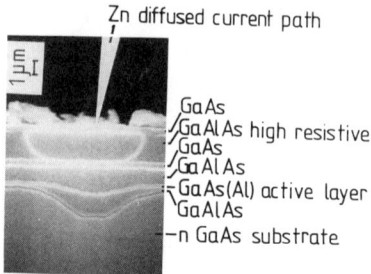

FIG. 2. Zn diffused current path.

insulating layers need to be grown for the electronics on top of the double-hetero laser layers. It is not easy to grow these, and it is also difficult to provide large areas.

The need for an additional conductive layer on the semi-insulating substrate and the resulting complexity in electrical wiring are the disadvantages of the horizontal structure.

Examples of the vertical and horizontal structures are as follows.

Vertical integration. A typical example of vertical-type integration is shown in Fig. 1a, a CS laser with two MESFETs (metal semiconductor field effect transistors) on an n-GaAs substrate; (b) is its photograph, and (c) is its circuit diagram (Matsueda *et al.*, 1980).

The MESFET's active layer and a GaAlAs high resistivity layer are grown on top of the laser's double heterostructure, as illustrated in Fig. 1a. At the laser stripe, Zn is diffused through the high-resistivity layer in order to provide a current path for the laser, as represented in Fig. 2.

Horizontal integration. A typical example of horizontal-type integration is demonstrated in Fig. 3, which shows a TS (terraced substrate) laser and a photomonitor (PM) with driving and detecting circuits of six MESFETs, along with its photograph and its circuit diagram (Matsueda *et al.*, 1983a, 1983b, 1984; Matsueda and Nakamura, 1984).

A GaAs/GaAlAs double heterostructure for the laser and monitor is grown in a counter step (a well) formed on a semi-insulating substrate, so that the overall surface is kept as planar as possible, as shown in Fig. 3a. The surface step between the laser and the FET is no more than 3 μm.

An n^+-conducting GaAs layer is necessary between the semi-insulating substrate and the laser double-hetero layers. The doping level, thickness, width and length of this conducting layer is designed to minimize the series resistance of the integrated laser diode.

6. OPTOELECTRONIC INTEGRATED CIRCUITS

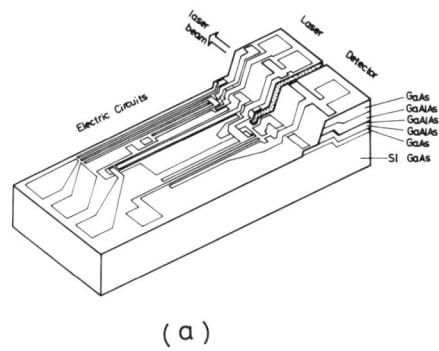

(a)

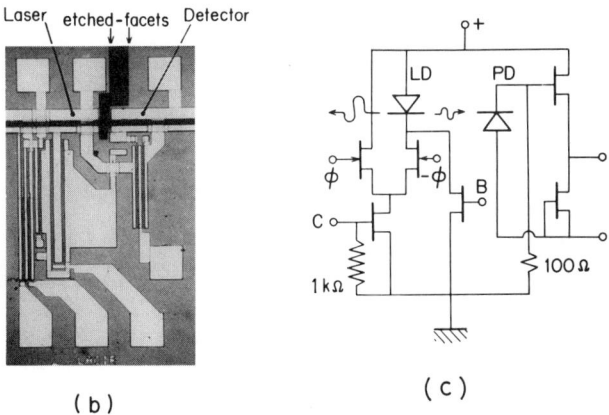

(b) (c)

FIG. 3. A horizontal GaAs OEIC transmitter with a terraced laser and six MESFETs: (a) overall view; (b) photograph; (c) circuit diagram.

The laser and the monitor are separated by a chemically (wet) etched groove, which provides the laser diode with an interior mirror facet. The separation is about 30 μm. The etched interior facet should be as smooth and vertical as possible. The depth of the etched groove is 5–10 μm. The groove is terminated in the SI substrate to secure the electrical isolation. The formation of interior mirror facets is essential for the enlargement of integration scale; otherwise, the width of the OEIC chip would be restricted to the optimum laser cavity length, which is usually 200–300 μm.

A similar type, with enlarged scale, is demonstrated in Fig. 4a, a MQW (multiquantum well) laser and a monitor with driving and detecting circuits of 12 MESFETs, (b) its photograph, (c) its schematic cross-section, and (d) its circuit diagram (Matsueda et al., 1985). The GaAs/GaAlAs double heterostructure with MQW active layer is grown in a counter step (a well)

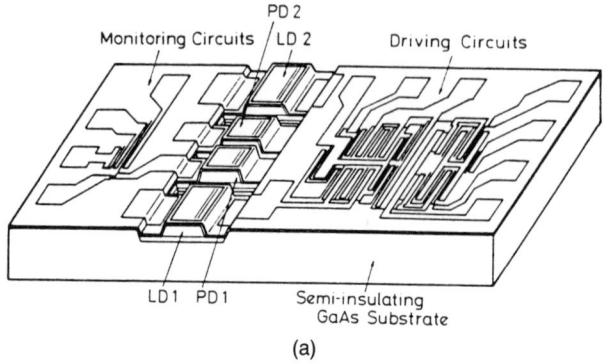

(a)

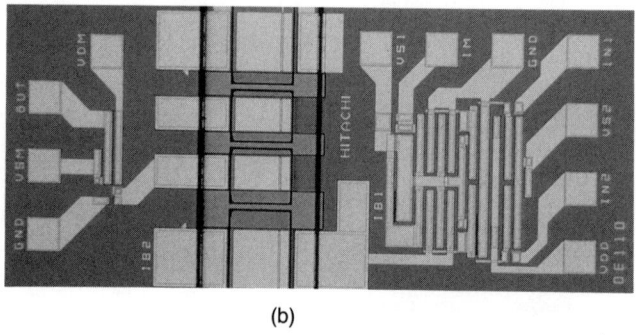

(b)

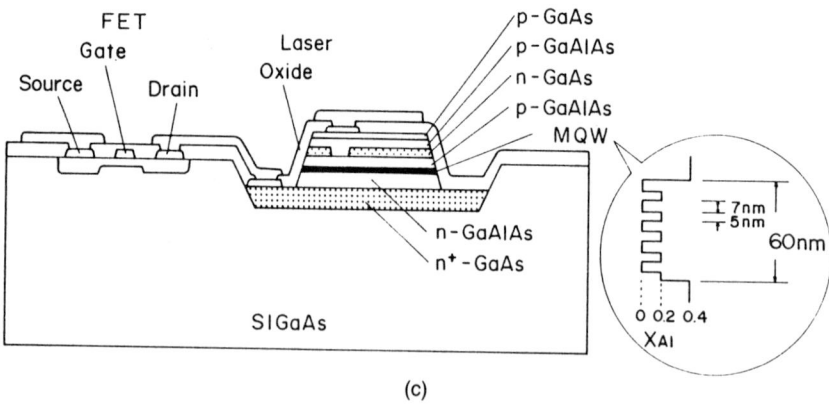

(c)

FIG. 4. A horizontal GaAs OEIC transmitter with a MQW-SAS laser and 12 MESFETs: (a) overall view; (b) photograph; (c) schematic cross-section; (d) circuit diagram.

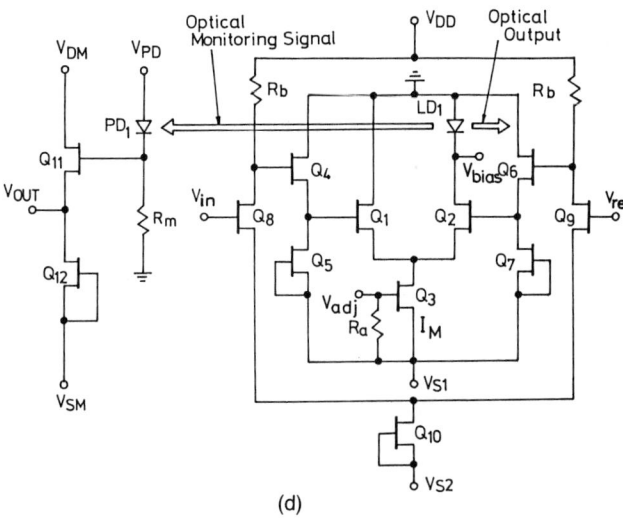

(d)

Fig. 4—continued

etched on a semi-insulating substrate, so that the overall surface is kept planar. The laser and the FET are laid out on the same level, as shown in Fig. 4c. An n^+-conducting GaAs layer lies between the semi-insulating substrate and the laser double-hetero layers.

The laser and the monitor are separated by an etched groove, by means of the dry etching technique. The separation is about 105 μm.

Detailed performances of this OEIC transmitter are explained in Part IV.

Another horizontal integration is schematically shown in Fig. 5. A laser with GRIN-SCH (graded index waveguide–separate confinement heterostructure) structure is built in a counter step well on a semi-insulating GaAs substrate (Sanada et al., 1985). MESFETs are fabricated on the

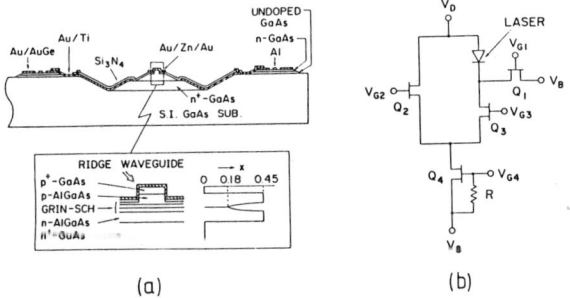

FIG. 5. A GaAs OEIC transmitter with a GRIN-SCH laser: (a) schematic cross-section; (b) circuit diagram. [From Sanada et al. (1985).]

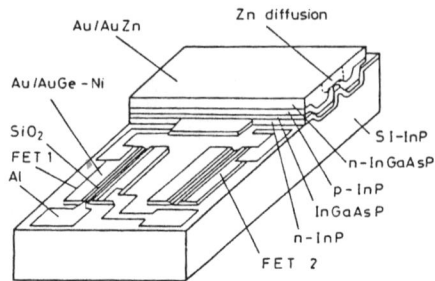

FIG. 6. An InP OEIC transmitter with a BH laser and MISFET's. [From Kasahara et al. (1984b).]

semi-insulating surface. The slopes between the laser and the FETs are flattened by an ion beam technique in order to allow smooth metallic wiring over them. A laser of this type has a threshold current as low as 5 mA and a wavelength around 0.84 μm (Wada et al., 1985b).

There are numbers of examples of the horizontal GaAs OEIC, including one with 36 gates (Carney et al., 1983), as summarized in Table I.

Attempts have been made to apply the OEIC technology to InP in order to cover the long-wavelength region. Several kinds of transistors were tried for integration with an LD.

The integration of a BH laser with MIS (metal insulator semiconductor) FETs is schematically shown in Fig. 6 (Kasahara et al., 1984b). A film of SiO_2 with thickness 0.05 μm is deposited as the insulating layer. The substrate is semi-insulating InP. The laser emits light with a wavelength of 1.3 μm.

A BH laser was also integrated with a J (junction) FET, as shown in Fig. 7 (Chen et al., 1983). The FET's jnunction is formed by diffusion of Zn. The active region of the laser is buried by vapor transport (thermal mass transport in the vapor phase). The substrate is semi-insulating InP.

A TJS (transverse junction stripe) laser was integrated with a MISFET, as shown in Fig. 8 (Koren et al., 1982b). The active layer has a crescent

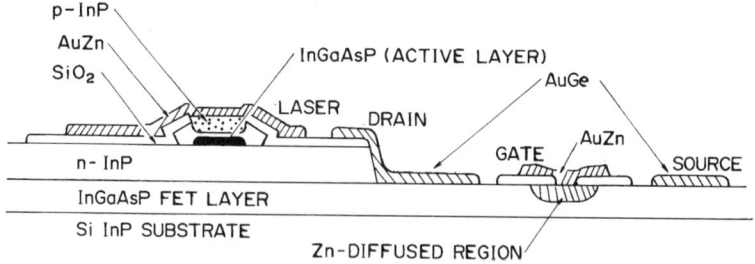

FIG. 7. An InP OEIC transmitter with a BH laser and a JFET. [From Chen et al. (1983).]

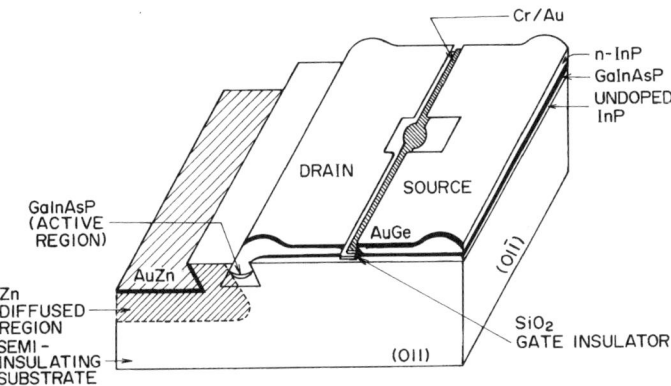

FIG. 8. An InP OEIC transmitter with a TJS laser and a MISFET. [Courtesy of U. Koren et al. (1982b).]

cross-section, and Zn is diffused on one side. The FET's insulating film is deposited externally, to a thickness of 0.06 μm. The substrate is semi-insulating InP.

One of the solutions to the difficulty in fabrication of an InP FET, is the use of bipolar transistors. Figure 9 shows the integration of a BH laser and three HBTs (hetero-bipolar transistors): (a) is the cross-section and (b) is the circuit diagram (Shibata et al., 1984). The substrate is n-type InP. The wavelength of the laser emission is 1.3 μm. Some other InP OEIC's are also listed in Table I.

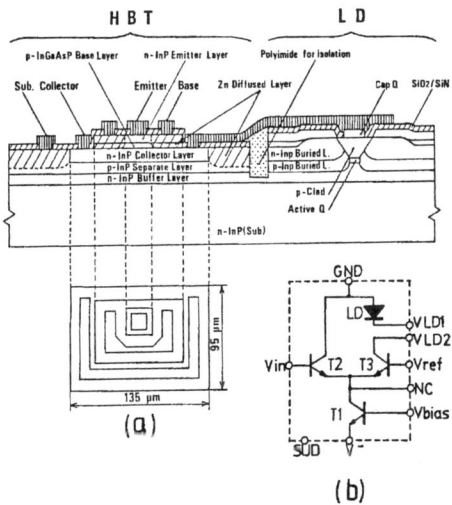

FIG. 9. An InP OEIC transmitter with a BH laser and HBTs: (a) schematic cross-section; (b) circuit diagram. [From Shibata et al. (1984).]

b. Receiver OEICs

In practical optical communication systems, PIN photodiodes (PIN-PDs) and avalanche photodiodes (APDs) are widely used. PIN-PDs have a simple *p-i-n* junction structure. The carriers excited in the *i* region by light absorption are rapidly swept out by an electric field across this region. Therefore, very high-speed operation can be realized with PIN-PDs. Further, PIN-PDs have additional merits, such as low biasing voltage and low dark current. APD has a carrier multiplicating region where the electric field is sufficiently high to cause ionization by collision. APD exhibits a higher sensitivity thanks to this carrier multiplication. However, the necessity of a higher biasing voltage is a drawback. PIN-PDs and APDs have vertical layered structures. Similar problems in integrating structure and in fabrication processes must be solved for both, in order to realize monolithic integration with electronic devices such as FETs.

Photoconductor (PC) devices are another kind of photodetector. This type of device have not been applied in practical optical communication systems because of the comparatively low sensitivity caused by a large dark current. However, the PC devices exhibit very high-speed characteristics (Auston, 1984). Furthermore, PC devices have a simple surface structure similar to that of FETs. Therefore, PC devices are one of the promising photodetectors that are suitable for receiver OEICs in some applications.

The choice of the semiconductor materials mainly depends on the wavelength to be detected. High electron mobility is also important to realize high-speed operation. From those viewpoints, it is considered that III–V compound semiconductor materials such as GaAs and InP are suitable as substrates.

Detectors using GaAlAs/GaAs system have high sensitivity in the entire wavelength emitted from GaAlAs optical sources. A wavelength range from 0.75 μm to 0.9 μm can be covered by changing the mole fraction x in $Ga_{1-x}Al_xAs$.

When InP is used as the substrate, a wavelength range from 0.9 μm to 1.65 μm can be covered by adjusting the fractions x and y in $In_{1-x}Ga_xAs_yP_{1-y}$. Several receiver OEIC devices reported to date are listed in Table II. All of those devices have been fabricated on semi-insulating substrates. PIN-PDs have been employed as photodetectors. MESFETs were used for the GaAs substrate, whereas JFETs or MISFETs were used for the InP substrate because of the lack of good Schottky metals for InP. In the following, we review the reported devices from the viewpoint of the integrating structure.

The schematic structure and the equivalent circuit of the device by Leheny *et al.* (1980) are shown in Fig. 10. A PIN-PD and a JFET were integrated

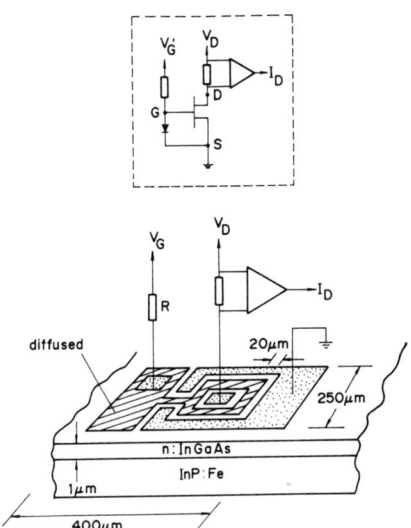

FIG. 10. An InP OEIC receiver with a PIN-PD and a JFET. [From Leheny et al. (1980).]

in an InGaAs layer grown by LPE on a semi-insulating InP substrate. The cross-hatched area indicates the Zn-diffused region forming the p-i-n junction and the gate of the JFET. The shaded regions correspond to the metallized contacts. Kasahara et al. (1984a) reported a receiver OEIC consisting of a PIN-PD and a MISFET on a semi-insulating InP substrate. The schematic view of the device is shown in Fig. 11. The junction capacitance was as low as 0.6 pF at a reverse bias of 0.6 V and the external quantum efficiency was 51% at a wavelength of 1.3 μm. Photosensitivity measurement at 100 Mb/sec (NRZ: non-return to zero) showed that a sensitivity at −34.5 dBm is required to realize a 10^{-9} bit-error rate.

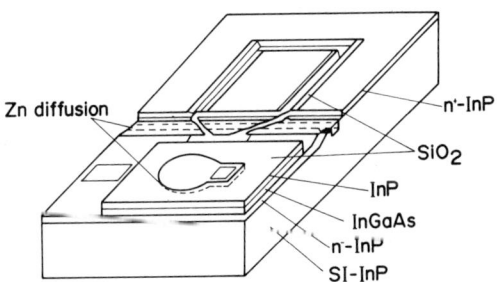

FIG. 11. An InP OEIC receiver with a PIN-PD and a MISFET. [From Kasahara et al. (1984a).]

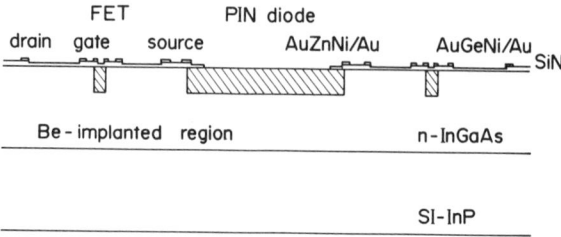

FIG. 12. A schematic cross-section of an InP OEIC with a PIN-PD and a column gate JFET. [From Hata et al. (1984).]

Another structure of the receiver OEIC device using InP substrate was proposed by Hata et al. (1984). The cross-sectional view of the device is shown in Fig. 12. In this case, p-type regions for PIN-PD and FET were formed by Be ion implantation into LPE-grown n-InGaAs on the semi-insulating InP substrate. The column gate structure surrounding the PIN-PD was utilized as a JFET. An overall effective quantum efficiency of 400% at a wavelength of 1.55 μm was reported. Receiver OEICs using GaAs substrates have been developed by several research groups. Wada et al. (1983a) have integrated a Schottky barrier photodiode (SPD) and a MESFET on a semi-insulating GaAs substrate. The cross-section of the device and the circuit diagram are shown in Fig. 13. In the circuit diagram, two resistors, R_{PD} and R_{EET}, are external elements. The isolation between the photodiode and the FET was achieved by deep chemical etching. The electrical

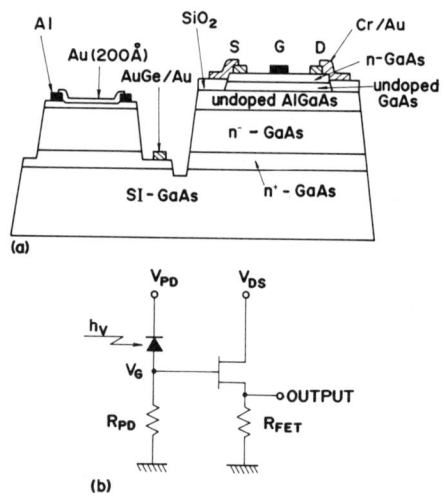

FIG. 13. An GaAs OEIC receiver with a Schottky barrier photodiode and a MESFET: (a) schematic cross-section; (b) circuit diagram. [From Wada et al. (1983a).]

connection between those elements was done by a bonding wire. The photodiode has a diameter of 200 μm, and the channel width, channel length, and gate length of the FET are 50, 8, and 2 μm respectively. A dark current as low as 80 nA at the punch-through reverse biasing condition and an amplification ratio of 19 with a 10 kΩ photodiode load resistor were obtained. An improved version of this device was reported by Miura et al. (1983). In this case, the SPD is converted to a PIN-PD by Zn diffusion, and the interconnection between the photodiode and the FET was achieved by an Al beam bridge wiring. The previous two cases (Fig. 13) are vertical integrating structures in a sense, because the FETs are on the material grown for the optical device.

Comparatively large-scale GaAlAs receiving OEIC devices including a photodetector and preamplifier circuit have been reported by Kolbas et al. (1983) and Wada et al. (1983b, 1985a). Kolbas et al. developed a quasi-planar integrating structure. A PIN-PD was fabricated using hydride VPF, in a groove etched on a semi-insulating substrate. The preamplifier was a transimpedance type of six MESFETs and five Schottky diodes. MESFETs were fabricated by selective ion implantation into the semi-insulating substrate.

The cross-sectional structure and the circuit diagram of the device developed by Wada et al. (1985a) are shown in Fig. 14. An GaAlAs/GaAs PIN-PD and a preamplifier circuit of six FETs, two Schottky barrier diodes, and a feedback resistor were integrated on a semi-insulating GaAs substrate.

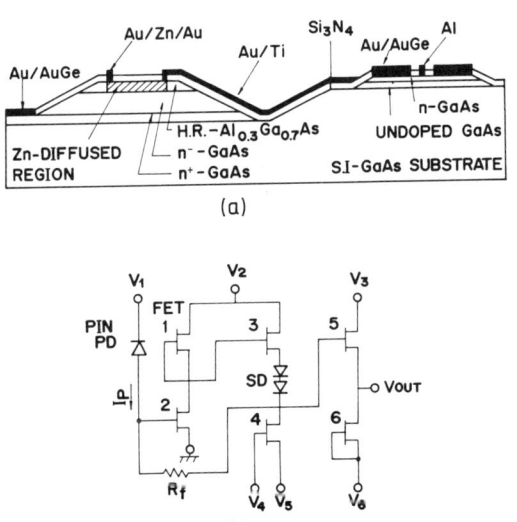

FIG. 14. An GaAs OEIC receiver with a PIN-PD and six MESFETs: (a) schematic cross-section; (b) circuit diagram. [From Wada et al. (1985a).]

The interconnection between the PIN-PD and the circuit element was realized by conventional metal wiring in this case, instead of by beam bridge wiring. A rise-and-fall time of 1.0 nsec with a feedback resistor of 1.3 kΩ were reported.

The integrating structures employed by Kolbas *et al.* and Wada *et al.* (1983b, 1985a) are recognized as horizontal integrating structures. In the horizontal integrating structure, the optical devices are embedded in a semi-insulating substrate, and the electronic circuit elements are fabricated on the upper surface of the substrate.

It is important to reduce any parasitic capacitances of the circuit elements in order to realize receiver OEICs with high-speed characteristics. The capacitances of FETs are comparatively large in the vertical structure because of the effect of the highly conductive layer under the FET layers, unless the unnecessary part of the conductive layer is removed by some means. Those capacitances can be greatly reduced by employing the horizontal structure. This is a suitable structure to develop large-scale and functional devices.

2. Circuits

One of the major merits of OEIC devices is the high-speed operation caused by the reduction of parasitics. In the hybrid-type constructions, waveforms responding to digital signals with a high bit rate suffer from damping oscillations because of parasitics such as the inductance of the bonding wire connecting optical devices, as well as the capacitance associated with the discrete packages.

An example of a theoretical comparison is shown in Fig. 15 for the case of an optical transmitter. The vertical axis indicates the penalty corresponding

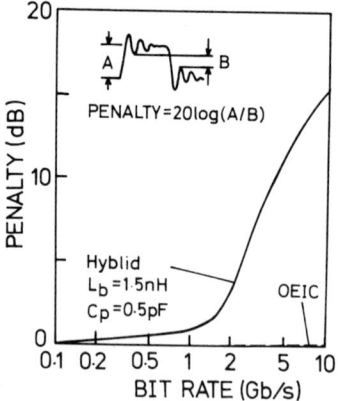

Fig. 15. Penalty in waveforms caused by parasitics in hybrid-type connection (solid curve). Penalty can be reduced by the OEIC technology (dashed curve).

to the degradation of the waveforms against waveforms without ringing (as defined in the insert in the figure). The horizontal axis is the bit rate. For simplicity, it is assumed that the time constant of the laser diode itself is very small. L_b and C_p are the inductance of the bonding wire and the capacitance of the laser package. The penalty becomes seriously larger as the bit rate goes up.

Fortunately, the parasitics can be greatly reduced by OEIC technology with which the optical devices and electronic circuit can be integrated on a substrate. The dashed curve in Fig. 15 shows the results when L_b and C_p are almost zero. It suggests that very high-speed operation without penalties is possible. This situation is also true in the optical receiver.

Although the OEIC technology shows a great advantage in reducing parasitics, there is not much difference in the design of the electronic circuit itself, compared to ICs in the hybrid approach. However, some items should be considered. At first, the active circuit elements must be selected depending on the substrate used. To date, conventional MESFETs are widely employed in the GaAlAs/GaAs system (Ury et al., 1979; Matsueda et al., 1980–1985; Matsueda and Nakamura, 1981, 1984). In the InGaAsP/InP system, on the other hand, MISFETs (Koren et al., 1982b; Kasahara et al., 1984a, 1984b) and bipolar transistors (Shibata et al., 1984; Su et al., 1985) are used, because of the lack of good Schottky gate metals.

The selection of the active circuit elements also relates to the desired functions of the OEIC devices. In a high-speed transmitter OEIC, for instance, it is required that the circuit not only operate with fast response speed, but also supply a comparatively large current that is sufficient to drive the laser diode. In the case of high-speed receiver OEICs, a fast response and low noise characteristics are important.

Although further discussion is needed concerning the selection of active elements, it seems that the approach using MESFETs is one of the promising technologies for OEIC devices, especially those employing semi-insulating GaAs substrates. In the following, we will describe the design procedures of driving and preamplifier circuits for a high-speed transmitter OEIC and a receiver OEIC, using semi-insulating GaAs substrates.

a. Transmitter OEIC

Generally, a high-speed transmitter OEIC consists of a laser diode and its driving circuit. The configuration of the laser driving circuits may be classified into two types as shown in Fig. 16. The circuit of type (a) (Goell, 1973; Filensky et al., 1977; Matsueda et al., 1980) is a simple single-ended one; the laser diode is directly driven with the drain or source current of FET Q_1. The circuit of type (b) (Gruber et al., 1978; Matsueda et al., 1983–1984;

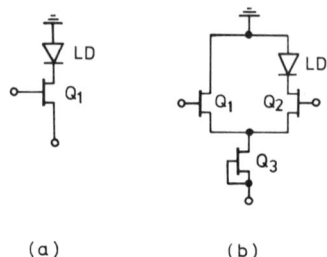

FIG. 16. Types of high-speed laser driving circuits (Q_n: FETs, LD: laser diode): (a) single-ended; (b) differential current switch.

Matsueda and Nakamura, 1984) is a differential current switching circuit. The current path is switched between FETs Q_1 and Q_2 according to the differential input signals given to the gates Q_1 and Q_2. The laser diode is inserted at the drain side of one of the FETs.

Each circuit has merits and demerits. The single-ended type is simple and the required voltage of the input driving signal is smaller than in the differential switching circuit. However, in the single-ended case, it is difficult to control the laser driving current in accordance with the fixed digital input signal voltage because of the analog nature of the circuit. The laser driving curent amplitude is strongly affected by a slight variation of the device parameters.

On the other hand, the required input voltage is larger in the differential switching circuit. This type is operated as a digital circuit, however, so the design margin for the variation of FET characteristics is larger than in the single-ended type. The larger design margin is one of the most important items for transmitter OEICs, especially for digital system applications. Therefore, the differential switching circuit is more suitable in those applications.

Figure 4d shows an example of a driving circuit based on the differential current switching principle. The basic differential switching circuit consists of FETs Q_1, Q_2, and Q_3. In this design, a buffer amplifier consisting of FETs Q_8, Q_9, and Q_{10} is employed to realize a low input driving voltage. The switching circuit and the buffer amplifier are connected through the source-follower circuit (FETs Q_4–Q_7). All FETs are of the normally-on type.

The input digital signal is given at V_{in}. V_{ref} is a constant reference voltage. The laser driving current amplitude can be adjusted by varying the voltage V_{adj}. A biasing current above the threshold level of the laser is necessary for high-speed response. The DC biasing current is fed through V_{bias} by an external circuit.

The circuit has been designed for operations up to the Gb/sec range. The initial design specifications of the driving circuit are summarized in Table III.

6. OPTOELECTRONIC INTEGRATED CIRCUITS

TABLE III

DESIGN SPECIFICATIONS OF THE LASER DRIVING CIRCUIT

Item	Specification
Input signal amplitude	0.8 V_{pp}
Rise time, fall time	less than 150 psec
Laser driving current	more than 15 mA_{pp}

The input signal voltage amplitude of 0.8 V_{pp} is compatible with the amplitude of the Si ECL logic level. The device can be operated up to 2 Gb/sec with a response time of less than 150 psec. The maximum laser driving (modulating) current is greater than 15 mA_{pp} with 0.8 V_{pp} input signal amplitude. If it is assumed that the slope efficiency of the laser diode is 0.2 mW/mA/facet, then 3 mW_{pp} optical signal amplitude will be obtained with the 15 mA_{pp} modulation.

To design the device parameters of FETs such as gate lengths and gate widths, process feasibility must be considered. Generally, in OEIC devices using semi-insulating substrates, there exist structural steps at the boundary between the two areas for optical devices and electronic circuits (Matsueda et al., 1983a). This step results from the integration of those two kinds of devices, which have very different vertical structures. The height of the step affects the dimensions and precision of fine gate patterns in photolithography processes. The height of the step must be as small as possible to obtain fine gate patterns for high-speed performances. However, it is difficult to reduce the step difference much below 1 μm. Under this condition, the achievable gate length is about 1 μm using the conventional photolithography technique. In the design, it is counted that the actual gate lengths of FETs would be 1.6 μm at maximum, using masks of 1 μm ruling.

The current–voltage characteristics of a FET are given by

$$I_{ds} = \beta W_g (V_{gs} - R_s I_{ds} - V_{th})^2, \quad (1)$$

where I_{ds}(A) and V_{gs}(V) are the drain current and gate-source voltage, respectively. The parameter β ($\mu A/V^2/\mu m$) is the transconductance coefficient per unit gate width, W_g is the gate width, R_s is the source resistance, and V_{th} is the threshold voltage. A measured example of V_{gs}–$\sqrt{I_{ds}}$ characteristics of a FET with a gate length of 1 μm (mask dimension) fabricated by the conventional processes is shown in Fig. 17. From the figure, the current versus voltage characteristics can be approximated by

$$I_{ds} = K W_g (V_{gs} - V_{th})^2, \quad (2)$$

where the parameter K ($\mu A/V^2/\mu m$) is a newly defined transconductance

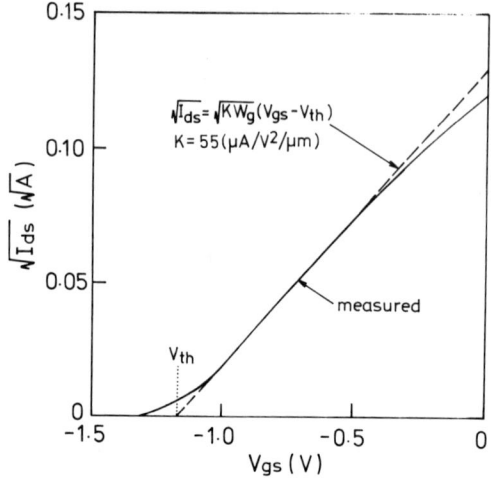

FIG. 17. Typical V_{gs}-$\sqrt{I_{ds}}$ characteristics of FETs with gate lengths of 1 μm (mask dimension).

coefficient including the influence of R_s. The relation between K and the conventional transconductance g_m is given by

$$g_m = \frac{dI_{ds}}{dV_{gs}} = 2KW_g(V_{gs} - V_{th}). \qquad (2')$$

The value of K in Fig. 17 is about 55 μA/V^2/μm. The measured device parameters of several samples and the design specifications are summarized in Table IV. The measured samples were FETs fabricated on flat semi-insulating substrates using masks in which the gate lengths were 1 μm. The

TABLE IV

MEASURED VALUES AND DESIGN SPECIFICATIONS OF FETs IN THE OEIC TRANSMITTER

Parameter	Unit	Measured		Design specification
		$\bar{x}$	σ	
β	μA/V^2/μm	—	—	60.0
K		49.7	2.0	45.0
$|V_{th}|$	V	1.19	0.09	1.20
R_s	Ω·μm	2738	794	3000
C_{gso}	fF/μm	1.15	0.04	1.2

values of K and $|V_{th}|$ were estimated with the approximation presented by Eq. (2). C_{gso} is the capacitance between the gate and the source under open drain conditions. The specification of K is set smaller than the measured average value, considering the broadening of the actual gate lengths due to the step on the substrate of the OEIC structure. When $R_s = 3\,\mathrm{k\Omega\cdot\mu m}$, $K = 45\,\mu\mathrm{A/V^2/\mu m}$ corresponds to $\beta = 60\,\mu\mathrm{A/V^2/\mu m}$.

From the specifications of device parameters shown in Table IV, the circuit parameter can be determined. At first, we consider the differential switching circuit. There is a trade-off relation between the gate width W_d of switching FETs (Q_1 and Q_2) and the required differential voltage amplitude V_{Din} at the gates of those FETs. This relation is given by

$$V_{Din} \geq \sqrt{\frac{I_M}{KW_d}}, \qquad (3)$$

where I_M is the constant drain current of Q_3. The required signal amplitude becomes smaller for wider gate widths; however, there is some dimensional limit because of the layout and fabrication. Here, $W_d = 500\,\mu\mathrm{m}$ was chosen as a reasonable value. The relation between V_{Din} and K-value in this case is shown in Fig. 18. The minimum voltage amplitude required to switch a current of 15 mA is $0.82\,V_{pp}$ for $K = 45\,\mu\mathrm{A/V^2/\mu m}$. A voltage amplitude $V_{Din} = 1.03\,V_{pp}$ was employed considering a margin of 20%. In this case, the required voltage gain of the buffer amplifier becomes 1.3 ($= V_{Din}/V_{in} = 1.03/0.8$). The gate width W_a of the current source FET Q_3 is also designed as $500\,\mu\mathrm{m}$. The saturated drain current is given by

$$I_{dss} = KW_a|V_{th}|^2; \qquad (4)$$

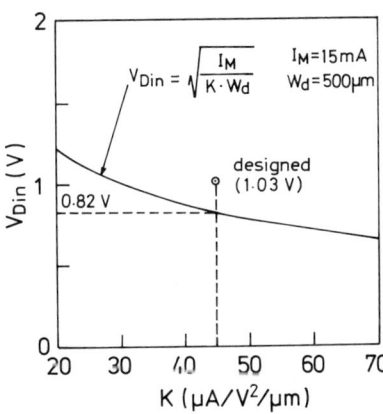

FIG. 18. Relation between K-value and the required differential input voltage amplitude for current switching.

therefore, the maximum current of Q_3 becomes about 30 mA with $W_a = 500\,\mu m$ when $K = 45\,\mu A/V^2/\mu m$. The current through Q_3 can be adjusted by varying the voltage at V_{adj} in Fig. 4d. The resistor R_a is $1\,k\Omega$.

Next, we will describe the design of the buffer amplifier. In this design, the input signal is fed to one FET (Q_8), while the input to the other FET (Q_9) is the constant reference. Let the gate widths of FETs Q_8 and Q_9 be W_b and the gate width of FET Q_{10} be W_c. The required input voltage amplitude and the maximum switching current are given by

$$V_{in} \geq 2\sqrt{\frac{I_c}{KW_b}}, \qquad (5)$$

$$I_c = I_{dss}(Q_{10}) = KW_c|V_{th}|^2. \qquad (6)$$

From the above formulae, the ratio W_c/W_b to realize digital switching operation by the buffer amplifier can be derived. When the input signal amplitude is $0.8\,V_{pp}$, W_c/W_b will be $\frac{1}{9}$, and the saturated drain current of Q_{10} becomes $\frac{1}{3}$ of the saturated drain current of Q_8 and Q_9. However, as shown in Fig. 17, the slope of the V_{gs}-$\sqrt{I_{ds}}$ curve in the small-I_{ds} region is not as steep as in the large-I_{ds} region. The effects of the smaller K-value in the small-current regions must be carefully considered, because the small K-value demands a larger input voltage amplitude for digital switching operations.

To avoid the influence of smaller K-values in the small-I_{ds} region, analog operation of the buffer amplifier by setting the operating point in a larger-current region is one of the solutions. The design model of the analog buffer amplifier is shown in Figs. 19a and 19b. Figure 19a shows the approximated V_{gs}-$\sqrt{I_{ds}}$ characteristics of FETs, and (b) is the static voltage-current transfer characteristics. In the figures, I_k is the current at the kink

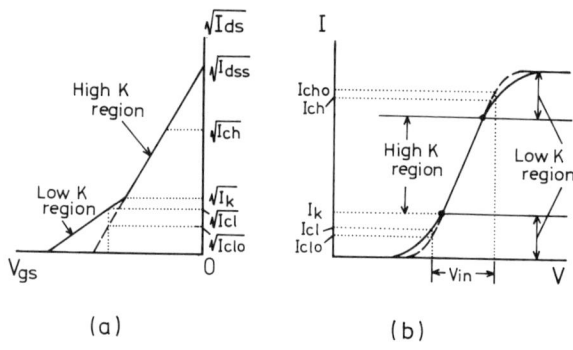

FIG. 19. Models of a buffer amplifier that functions in an analog mode: (a) simplified model fo V_{gs}-$\sqrt{I_{ds}}$ characteristics of FETs; (b) static voltage–current transfer characteristics of buffer amplifier.

in the V_{gs}-$\sqrt{I_{ds}}$ curve. I_{ch} and I_{cl} are currents corresponding to the high and low digital levels of the input digital signal, respectively. I_{ch0} and I_{cl0} are currents of high and low levels, where it is assumed that the V_{gs}-$\sqrt{I_{ds}}$ characteristics are linear as shown by a dashed line in Fig. 19a. The transfer characteristics corresponding to the constant high K-value are also shown in Fig. 19b by a dashed curve.

The relation between the gate width W_b of the buffer FETs (Q_8, Q_9) and input voltage for the analog model is given by

$$W_b = \frac{1}{KV_{in}^2}[(\sqrt{I_{ch0}} - \sqrt{I_c - I_{ch0}}) - (\sqrt{I_{cl0}} - \sqrt{I_c - I_{cl0}})]^2 \tag{7}$$

If the maximum of the constant current flowing through Q_{10} is $KW_c|V_{th}|^2$, then

$$I_{ch0} = \tfrac{1}{2}(KW_c|V_{th}|^2 + \Delta I_c)$$
$$I_{cl0} = \tfrac{1}{2}(KW_c|V_{th}|^2 - \Delta I_c), \tag{8}$$

where

$$\Delta I_c = \frac{GV_{in}}{R_b}. \tag{9}$$

G is the voltage gain of the buffer amplifier and R_b ($= R_{b1} = R_{b2}$) is the load resistance of Q_8 and Q_9. The voltage gain $G = V_{Din}/V_{in} = 1.03/0.8 \simeq 1.3$ in this case.

The gate width W_b is closely related to the allowable value of I_k. If we consider the following conditions concerning the V_{gs}-$\sqrt{I_{ds}}$ characteristics:

$$\frac{I_k}{I_{dss}} = p \leq p_0, \qquad \frac{I_k - I_{cl0}}{I_{ch0} - I_{cl0}} = c \leq c_0, \tag{10}$$

the upper limit of W_b can be derived as follows:

$$p_0 KW_b|V_{th}|^2 - I_{cl0} \leq c_0 \Delta I_c. \tag{11}$$

On the other hand, W_b has the lower limit in the relation to W_c. The switched curent flowing through Q_{10} must be smaller than the saturated drain current of Q_8 and Q_9. This condition is given by

$$\tfrac{1}{2}\left(KW_c|V_{th}|^2 + \frac{GV_{in}}{R_b}\right) \leq KW_b|V_{th}|^2. \tag{12}$$

The relation between W_b and W_c given by Eqs. (7) and (8) is shown in Fig. 20 by a solid curve. Here, $R_b = 300\,\Omega$ was used in the calculations. If we employ the condition of $p_0 = \tfrac{1}{30}$ and $c_0 = \tfrac{1}{10}$, the conditions represented by Eqs. (11) or (12) are not satisfied in the hatched areas.

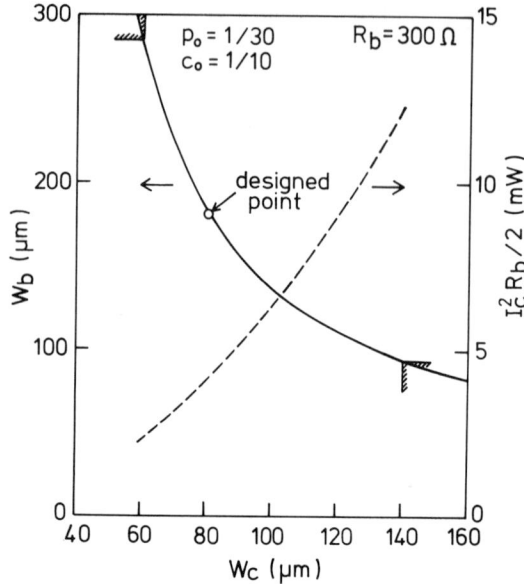

FIG. 20. Relation between the gate widths of FETs constituting the buffer amplifier and the power dissipated in the load resistance.

Although W_c has a relatively wide choice, $W_c = 80\,\mu\text{m}$ is selected, because the power dissipated in the load resistor R_b is smaller for the smaller value of W_c. In Fig. 20, the power dissipation is plotted against W_c as the dashed curve, for $R_b = 300\,\Omega$.

The value of R_b affects the response speed of the driving circuit. The response speed is limited by the time constant given by the product of R_b and gate–drain capacitance C_{gd} of FET Q_8 or Q_9. The rise time τ_r and the fall time τ_f of the laser driving circuit as a function of R_b have been evaluated by computer simulation. $W_c = 80\,\mu\text{m}$ and $C_{gd} = \frac{1}{3}C_{gso}$ were used in the calculation. Here, C_{gd} is the gate-drain capacitance. The results are shown in Fig. 21. The response is faster for smaller values of R_b, so the required W_b becomes larger. In this design, $R_b = 300\,\Omega$ was chosen as a reasonable value. The gate width W_b becomes $180\,\mu\text{m}$ from Fig. 20.

The gate widths of FETs constituting source followers have been determined from the viewpoint of response time. The rise time and the fall time of the driving circuit are plotted against the gate widths of Q_4 and Q_5 (or Q_6 and Q_7) in Fig. 22a, where it is assumed that the gate width W_e of Q_4 (Q_6) is equal to the gate width W_f of Q_5 (Q_7). The response speed becomes faster for a wider gate width; rise and fall times below 100 psec are expected for $W_e = W_f = 70\,\mu\text{m}$. The response speed can be improved by changing the

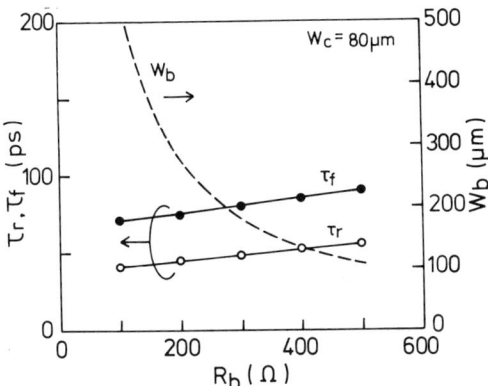

FIG. 21. Dependence of rise and fall times of the laser driving current on the value of the load resistance.

ratio W_e/W_f. The simulated response times corresponding to the different ratios are plotted in Fig. 22b. The improvement in the response time saturates for $W_e/W_f \geq 3$; therefore, $W_e/W_f = 3$ was employed in this design.

The designed driving circuit requires three power supplies and a reference voltage. Considering the threshold voltage of FETs, -3 V, -5 V, and -9 V were used for V_{DD}, V_{S1}, and V_{S2} (indicated in Fig. 4d) respectively. The reference voltage depends on the operating point of the buffer amplifier. It becomes -7.5 V under $p_0 = \frac{1}{30}$, $c_0 = \frac{1}{10}$ and the above-mentioned power supply conditions.

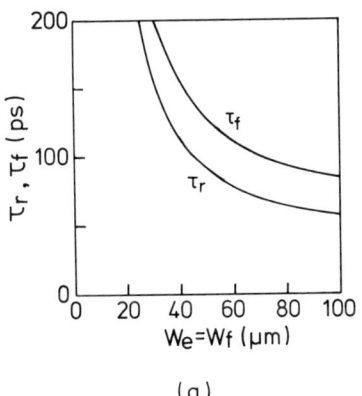
(a)

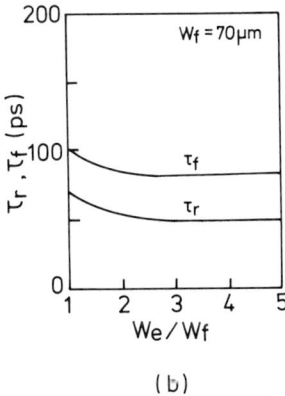
(b)

FIG. 22. Dependence of rise and fall times on the gate width of the FETs in the source follower: (a) for the case of $W_e = W_f$; (b) for the case of $W_e = nW_f$ ($n \geq 1$).

TABLE V

DESIGNED VALUES OF THE LASER DRIVING CIRCUIT

Device and circuit parameter		Design value
Gate length ℓ_g (Q_1-Q_{10})		1.6 μ max
Gate width	W_d (Q_1, Q_2)	500 μm
	W_a (Q_3)	500 μm
	W_e (Q_4, Q_6)	210 μm
	W_f (Q_5, Q_7)	70 μm
	W_b (Q_8, Q_9)	180 μm
	W_c (Q_{10})	80 μm
Load resistance $R_b = R_{b1} = R_{b2}$		300 Ω
Power supply voltage	V_{DD}	-3 V
	V_{S1}	-5 V
	V_{S2}	-9 V
Reference voltage	V_{ref}	-7.5 V

The designed dimensional parameters of the devices and other circuit parameters are summarized in Table V. The sum of the gate widths of all FETs is 2.5 mm. The total power dissipated in the circuit is 152 mW.

The region on the $|V_{th}|$-K plane where the initial design specifications shown in Table III are satisfied is shown in Fig. 23. The specifications are satisfied in the hatched area between the two curves A and B. Curve A, calculated by computer simulation, shows the limitation of the response time

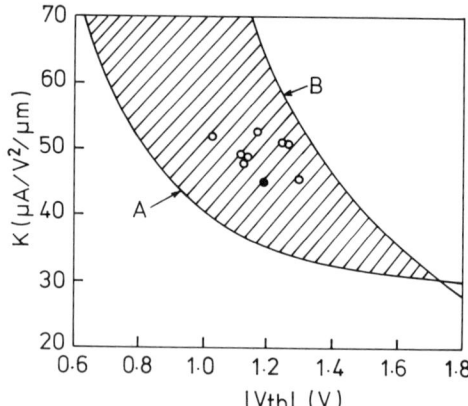

FIG. 23. The hatched area represents K and $|V_{th}|$ values that satisfy the specifications in Table III.

of 150 psec. Curve B indicates the following condition:

$$p_0 K W_d |V_{th}|^2 \leq c_0 I_M. \tag{13}$$

This is the condition for avoiding the influence of the small K-value in the small-current region of the digital switching circuit consisting of Q_1 and Q_2. The inequality of Eq.(13) can be derived by replacing $W_b \to W_d$, $I_{cl0} \to 0$, $\Delta I_c \to I_M$ in Eq. (11). Curve B was calculated for $p_0 = \frac{1}{30}$ and $c_0 = \frac{1}{10}$. The closed circle in Fig. 23 indicates the central value designed, and open circles show the measured results of FET samples fabricated by masks with 1 μm gate lengths.

The buffer amplifier is operated in an analog mode; therefore, the operating characteristics are largely influenced by the variation of threshold voltage $|V_{th}|$ of FETs Q_8 and Q_9. Figure 24 shows the dependence of the laser driving current and response times on the difference in $|V_{th}|$ of Q_8 and Q_9. The results were calculated for different $|V_{th}|$ of Q_8, while $|V_{th}|$ of Q_9 was fixed at -1.2 V. A relatively large difference in $|V_{th}|$ between Q_8 and Q_9 is allowable to realize a response time below 150 psec. But it must be smaller than ± 0.1 V to obtain the driving current, 15 mA$_{pp}$, as initially specified. The measured variation of $|V_{th}|$ is somehow larger than ± 0.1 V as shown in Fig. 23. However, those differences can be compensated for by adjusting the reference input voltage V_{ref} of Q_9.

The computer-simulated results for the waveforms of the laser driving current are shown in Fig. 25a. In the simulation, a simple parallel circuit consisting of $R = 5$ Ω and $C = 10$ pF was used as an equivalent circuit of the laser diode. The dashed curve shows the input driving signal with amplitude 0.8 V$_{pp}$, rise and fall times 50 psec.

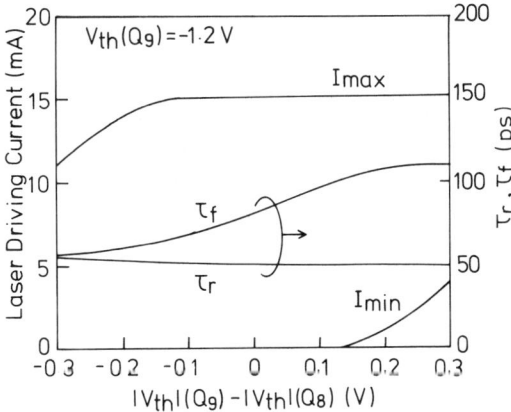

FIG. 24. Dependence of the laser driving current amplitude and response times on the difference in $|V_{th}|$ of FETs Q_8 and Q_9 in the buffer amplifier.

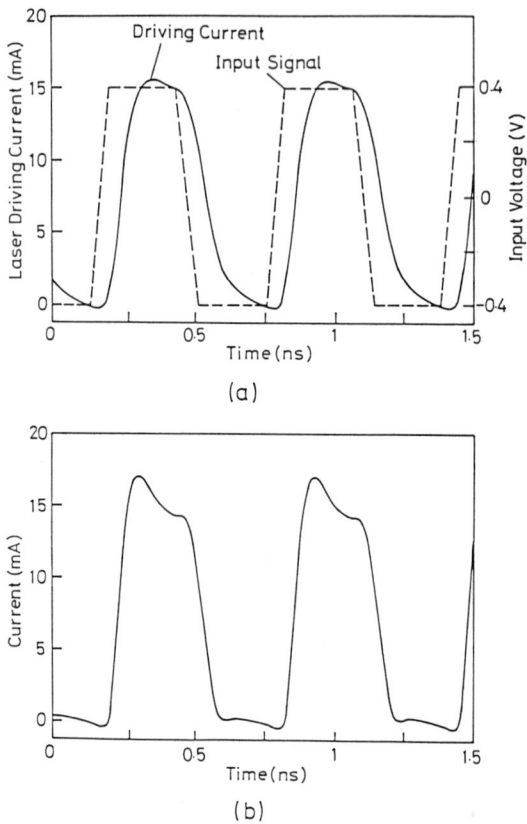

FIG. 25. Computer-simulated waveforms of the laser driving current: (a) with a laser diode (an equivalent circuit with $R = 5\,\Omega$ and $C = 10\,\mathrm{pF}$ was assumed); (b) without the laser diode.

The bit rate is 1.6 Gb/sec RZ (return to zero). A rise time and a fall time of 90 psec and 150 psec were obtained. The waveforms shown in Fig. 25b are the result with no laser diode. In this case, the rise time and the fall time are 49 psec and 81 psec, respectively. The response speed in Fig. 25a is limited by the time constant of the assumed equivalent circuit of the laser diode; therefore, it is important to reduce the time constant of the laser diode to as small a value as possible to realize very high-speed operations.

b. Receiver OEIC

In practical optical communications systems, amplifiers of high-impedance type and transimpedance type are widely used as a receiving front-end circuit. Basic configurations of a high-impedance-type and transimpedance-type

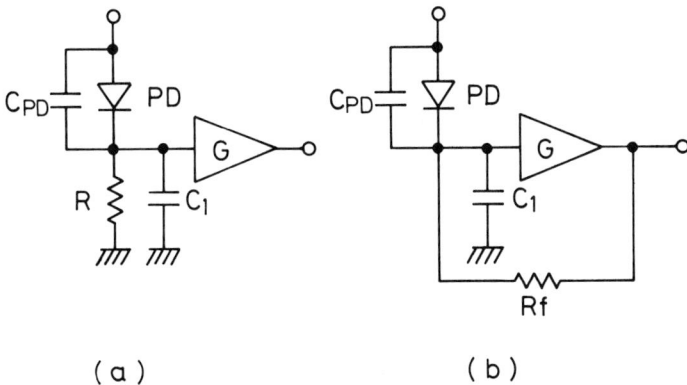

FIG. 26. Basic configurations of receiving front-end circuits: (a) high-impedance type; (b) transimpedance type.

preamplifier are shown in Figs. 26a and 26b respectively. In the figures, PD is the photodetector and C_{PD} is the capacitance of the photodetector. R and R_f are the load resistor and the feedback resistor, respectively. G is the gain of the preamplifier circuit, and C_1 is the input capacitance of the circuit. When a PIN-PD is employed as a photodetector, the sensitivity of the receiving circuit is almost proportional to $\sqrt{R}$ for the high-impedance type or to $\sqrt{R_f}$ for the transimpedance type. On the other hand, the bandwidth of the circuit of the high-impedance type is limited by a time constant given by $R(C_{PD} + C_1)$, while the bandwidth of the transimpedance type is limited by $R_f(C_{PD} + C_1)/G$. Therefore, the transimpedance type is suitable to realize a receiving front end having high-speed and high-sensitivity characteristics simultaneously.

In the following, we will describe the design procedure of a transimpedance preamplifier circuit using GaAs FETs. The circuit diagram designed is shown in Fig. 27. In the figure, PD represents a PIN-PD. A buffer circuit consisting of FETs Q_2 and Q_3 is used to realize a DC coupled feedback loop with a low output impedance. Schottky barrier diodes (SD) are used for the purpose of DC voltage level-shifting. R_1 and R_f are the load resistor and the feedback resistor, respectively.

The design specifications of the receiving circuit are shown in Table VI. The 3 dB bandwidth is considered larger than 2 GHz. The minimum receiving optical power required to obtain 10^{-9} bit error rate is -21 dBm (8 μW). In the digital optical transmission system, the output signal from the preamplifier must be amplified by the following main amplifier to an amplitude corresponding to digital logic levels, and the required output voltage amplitude V_{out} of the preamplifier depends on the gain of the main amplifier. In

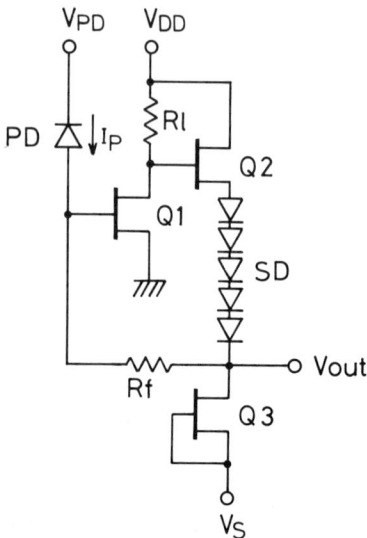

FIG. 27. Circuit diagram of the transimpedance preamplifier.

this design, V_{out} of 1.5 mV is considered. If we assume that the quantum efficiency of the detector is 70%, then the photocurrent i_p at an optical input power −21 dBm becomes 3.8 μA. In this case, the transimpedance Z_t of the preamplifier defined by V_{out}/i_p becomes about 400 Ω.

From the specifications shown in Table VI, the device and circuit parameters must be determined. The cutoff frequency f_t of a FET is given by

$$f_t = \frac{1}{\pi} \frac{v_d}{\ell_g} \qquad (14)$$

where v_d is the drift velocity of the carrier and ℓ_g is the gate length of the FET. A shorter gate length is required to realize wide-band characteristics; however, there exist some limitations in achievable gate lengths. Especially in OEIC devices, there exists an inevitable broadening in the fine gate patterns during the photolithography process. This is because of the surface

TABLE VI

DESIGN SPECIFICATIONS OF THE RECEIVING CIRCUIT

Item	Symbol	Specification
3 dB Bandwidth	B	≥ 2 GHz
Minimum receiving optical power	P_{rmin}	≤ −21 dBm
Transimpedance	Z_t	≥ 400 Ω

TABLE VII
Measured Values and Design Specifications of FETs in the OEIC Receiver

Parameter	Unit	Design specification (average of measured values)
g_m	mS/mm	100
C_{gs}	fF/μm	1.2
C_{gd}	fF/μm	0.15

step resulting from the integration of optical and electronic devices on a substrate. Therefore, it is considered that the actual gate lengths will be up to 1.6 μm, using masks with 1.0 μm gate-length patterns. In this case, f_t becomes about 20 GHz. The average value of the measured characteristic parameters of the FETs with 1.6 μm gate length are summarized in Table VII.

The output voltage V_{out} of the designed circuit can be derived using the equivalent circuit shown in Fig. 27. Neglecting the effect of the buffer circuit, V_{out} is given by

$$V_{\text{out}} = \frac{GR_f}{1 + G + j\omega(C_{\text{PD}} + C_1)R_f} i_p. \tag{15}$$

G is the gain of the amplifier, given by

$$G = g_m W_a R_1, \tag{16}$$

where g_m and W_a are the transconductance and the gate width of FET Q_1. The input capacitance of FET Q_1 considering the Miller effect becomes

$$C_1 = \{C_{gs} + (1 + G)C_{gd}\}W_a \tag{17}$$

where C_{gs} and C_{gd} are the gate–source and gate–drain capacitances of Q_1.

3 dB bandwidth and transimpedance can be derived using Eq. (14) as follows:

$$B = \frac{1 + g_m W_a R_1}{2\pi[C_{\text{PD}} + \{C_{gs} + (1 + g_m W_a R_1)C_{gd}\}W_a]R_f}, \tag{18}$$

$$Z = \frac{g_m W_a R_1}{1 + g_m W_a R_1} R_f. \tag{19}$$

On the other hand, the bit error rate P_e is given by

$$P_e = \frac{1}{\sqrt{2\pi}} \int_{\frac{1}{2}\sqrt{\frac{S_{pp}}{N_{\text{rms}}}}}^{\infty} e^{-t^2/2} \, dt. \tag{19'}$$

S_{pp}/N_{rms} is the signal-to-noise ratio of the detected signal and is given by

$$\frac{S_{pp}}{N_{rms}} = \frac{i_p^2}{2ei_p B + 4kTFB/R_f}, \tag{20}$$

where e is the charge of the electron, k is the Boltzmann constant, T is the temperature and F is the noise figure of the preamplifier (Personick, 1973). The photocurrent i_p can be expressed using photon energy hv and the quantum efficiency η of the detector, and the minimum detected optical power P_{rmin} by

$$i_p = \frac{\eta e}{hv} P_{rmin}. \tag{21}$$

The first term of the denominator in Eq. 20 represents the shot noise associated with the detection process of the optical signal, and the second term corresponds to the thermal noise of the amplifier. In the PIN-PD, the thermal noise is dominant.

Using Eqs. 18–21, the device and circuit parameters satisfying the initial specifications can be determined. An example of a calculated W_a-R_f relation is shown in Fig. 28. In this calculation, the characteristic parameters shown in Table VII were used, and it was assumed that the capacitance and the quantum efficiency of the detector are 0.6 pF and 70%, respectively, at a wavelength of 0.83 μm. A value for load resistance of 300 Ω was employed. In addition, Pucel's model (Pucel et al., 1975) was used to calculate the

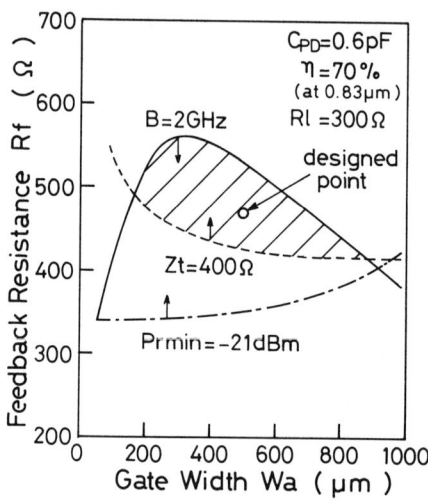

FIG. 28. Calculated relations between the gate width (W_a) and the feedback resistance (R_f) for the transimpedance-type OEIC receiver.

dependence of the noise figure (in Eq. 20) on the device parameters of the FET.

In Fig. 28, the solid, dashed, and dot-dashed curves correspond to the limitations of the bandwidth, transconductance, and minimum detected optical power, respectively. The arrows indicate the region where each specification is satisfied. The hatched area shows the region where all of the initial specifications are satisfied. The designed value of W_a and R_f is shown by an open circle ($W_a = 500\,\mu m$, $R_f = 470\,\Omega$).

The output impedance of a FET is approximately given by the inverse of the operating g_m value. Therefore, the wider the gate, the lower the output impedance. However, the input capacitance becomes larger with wider gate widths, and the large capacitance affects the achievable bandwidth. In this design, the output impedance was considered to be 50 Ω, and the gate widths W_b and W_c of FETs Q_2 and Q_3 were designed as 600 μm and 300 μm, respectively, to get rid of degradations in the bandwidth characteristics.

The voltages of the power supplies V_{DD} and V_S were designed as 6 V and -2.5 V, respectively. The gate voltage V_{gs} of Q_1 must be slightly negative, and the drain voltage V_{ds} must be sufficiently large, to obtain a high-g_m operation. When $V_{gs} = -0.5$ V and $V_{ds} = 2$ V, the voltage level of Q_2 becomes 2.5 V considering the voltage drop due to the current flowing through Q_3. Therefore, about 3 V is required as the voltage level shifting between the source of Q_2 and the output terminal. The level shift was achieved by inserting five Schottky barrier diodes having a voltage drop of 0.6 V each.

The designed device and circuit parameters are summarized in Table VIII. The calculated results of the frequency characteristics using the designed

TABLE VIII

DESIGNED VALUES OF THE RECEIVING CIRCUIT

Device and circuit parameter	Design value
Gate length ℓ_g (Q_1-Q_3)	1.6 μm max
Gate width W_a (Q_1)	500 μm
Gate width W_b (Q_2)	600 μm
Gate width W_c (Q_3)	300 μm
Load resistance R_l	300 Ω
Feedback resistance R_f	470 Ω
Power supply voltage V_{DD}	6 V
Power supply voltage V_s	-2.5 V

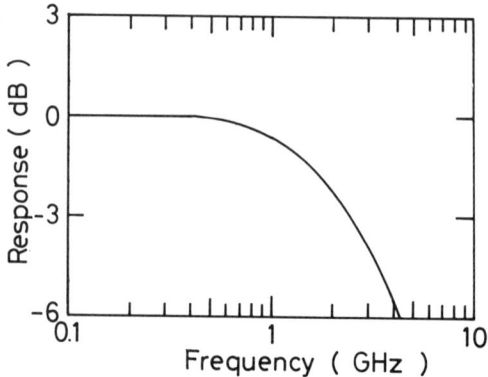

Fig. 29. Computer-simulated frequency characteristics for the transimpedance-type OEIC receiver.

parameters are shown in Fig. 29. The 3 dB bandwidth, transimpedance, and minimum receiving optical power were 2.5 GHz, 440 Ω, and -21.6 dBm, respectively.

III. Fabrication Processes

OEICs that have been fabricated so far have followed similar processes, so in Part III the process flow for the OEIC shown in Fig. 4 is first explained. Then, the process of the laser reflector is detailed because this is the most distinctive process of OEICs.

3. Process Flow

The fabrication starts with epitaxial processes, as represented in Fig. 30. (a) A counter step is chemically etched on a substrate. (b) n^+-GaAs, n-GaAlAs, MQW, p-GaAlAs, and n-GaAs layers are grown successively by the MOCVD (metal organic chemical vapor deposition) method. (c) A channel for the laser active region is chemically etched. (d) p-GaAlAs and p-GaAs layers are successively grown by MOCVD. MBE (molecular-beam epitaxy), LPE (liquid-phase epitaxy) or other methods also may be employed for these crystal growths. (e) The unnecessary part of the epilayers is chemically etched down to the n^+-GaAs conducting layer. Thus the side slopes of the LD are formed. (f) Part of the n^+-GaAs conducting layer is chemically etched to reveal the original semi-insulating surface. These etchings, especially in step (c), may better be combined with dry etching, as the reactive ion etching technique, in order to clean the interface completely.

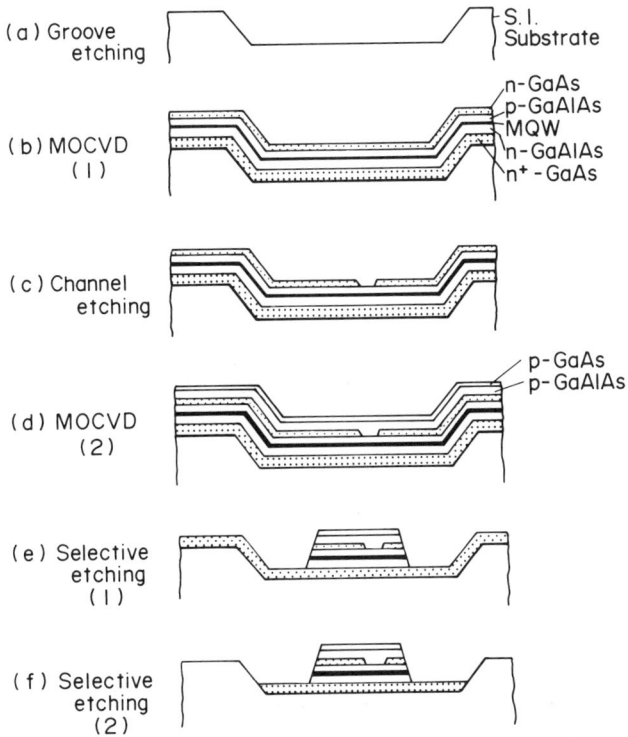

FIG. 30. Epitaxial process for the horizontal OEIC: (a) counter step (well); (b) epitaxial growth 1; (c) laser channel etching; (d) epitaxial growth 2; (e) selective etching 1; (f) selective etching 2.

The electric circuits then are fabricated on the revealed semi-insulating surface, following the processes represented in Fig. 31. (a), (b) Si^+ ions are implanted for the n^+ contact region and for the n region. The implantation is followed by annealing at around 800°C for about 30 minutes. (c) Sequential evaporation of AuGe alloy, Ni, and Au is done for the ohmic electrode. The patterning is done by the lift-off technique. The metal layers are alloyed at 400°C for several minutes. (d) The Schottky electrode is formed by sequential evaporation of Ti, Pt, and Au, followed by a lift-off procedure. (e) The wiring involving through holes is done by sequential evaporation of Mo and Au, followed by an ion-milling procedure. The ion beam direction is adjusted so that the wires at the laser side slopes are patterned correctly.

Finally, the separation between the LD and the PM is done by the RIBE (reactive ion beam etching) technique. Cl_2 is the major constituent of the etching gas. Etching power, pressure, and acceleration are important factors to obtain smooth and vertical interior facets. Wet chemical etching is an

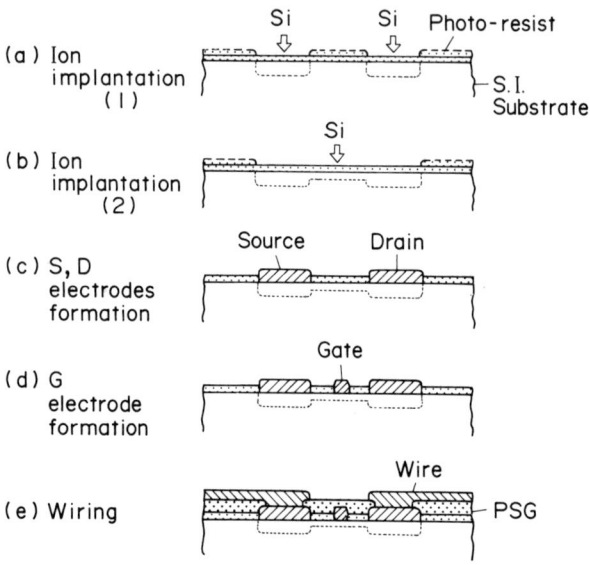

FIG. 31. IC process for the horizontal OEIC: (a) ion implantation 1; (b) ion implantation 2; (c) ohmic electrodes; (d) Schottky gates; (e) metallic wiring.

alternative method (Matsueda et al., 1983a, 1983b, 1984; Matsueda and Nakamura, 1984), although some care in crystal anisotropy is needed.

4. Reflector Formation

Conventional low-threshold lasers are all of the Fabry–Perot type, having two cleaved mirror facets 200–300 μm apart. The important factors are reflectivity, light absorption loss, and heat dissipation at the facet. The cleavage is believed to give the best facets for low-current and long-life operation. However, in the case of OEICs with considerable scale, at least one reflector should be inside the chip, for example, an interior mirror facet.

Three different methods are proposed for the reflector formation. The first is the micro-cleavage technique: i.e., to cleave the active area locally. The unnecessary portion around the laser active cavity is first etched off, to yield a beam or a cantilever structure as indicated in Fig. 32a. Then the beams are cleaved by means of ultrasonic vibration (Blauvelt et al., 1982; Koren et al., 1982a), micro-structured jigs or an adhesive tape, as in Fig. 32b. The facet is as smooth and vertical as the conventional cleavage face. However, the chemical compositions in the surrounding part are restricted, in order to enhance the preferential etching to make the beams.

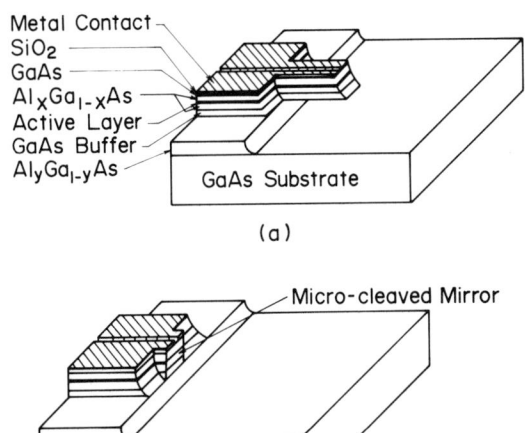

FIG. 32. Micro-cleavage: (a) under etching; (b) cleaved mirror. [From Blauvelt et al. (1982).]

The second is to make at least one interior facet by wet chemical etching or dry etching. In the case of wet chemical etching, the solution, temperature, material, and sharpness of the mask determine the etched morphology. In the case of GaAs/GaAlAs multilayers, the 1:1:3 solution of H_2O_2: H_3PO_4: $HOCH_2CH_2OH$ (ethylene glycol) is found to give good room-temperature etching results. The mask was photoresist (AZ1350J). This method was applied to the OEIC shown in Fig. 3. The etching degrades the threshold and efficiency to some extent (Matsueda et al., 1983a, 1983b, 1984, 1985; Matsueda and Nakamura, 1984).

The etching is also done by dry processes, for example, RIBE. The RIBE is done by the apparatus shown in Fig. 33. In the GaAs/GaAlAs case, Cl_2 gas of about 1 mtorr pressure was used, and the ion beam was accelerated at several hundred volts, to fabricate the OEIC shown in Fig. 4. It is better that the base vacuum level of the etching chamber is as high as possible, for morphology control. The mask was a photoresist film. The dry etching yields smooth and vertical facets free from crystal anisotropy. The current threshold was as low as 31 mA, with a slope efficiency of 0.2 mW/mA, as indicated in Fig. 34 (Matsueda et al., 1985). The degradation due to the etching is not significant as shown in Fig. 34. In future, the interior mirror facets may be buried in the epitaxial structure to form a waveguide circuit that is completely built-in.

The third possibility is the employment of a facetless laser, such as the DFB (distributed feedback) or DBR (distributed Bragg reflector) types.

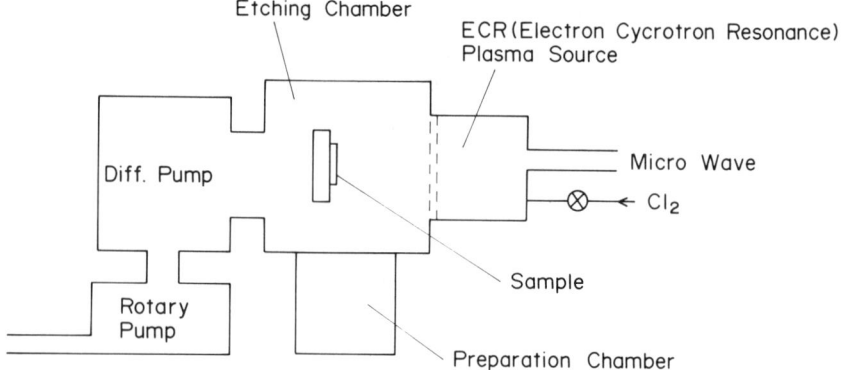

FIG. 33. Diagram of the RIBE apparatus.

Optical feedback is realized by microcorrugations etched on the substrate. This greatly increases the layout freedom. For example, a parallel arrangement of six DFB lasers with a curved waveguide manifold is shown in Fig. 35 (Aiki *et al.*, 1976). These types have also advantages in mode stability and temperature insensitivity. However, at present they consume more current and are difficult to fabricate as well.

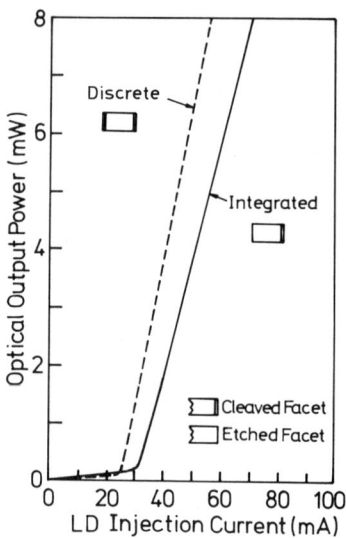

FIG. 34. Current–optical output characteristics of the etched laser compared with those of the cleaved laser.

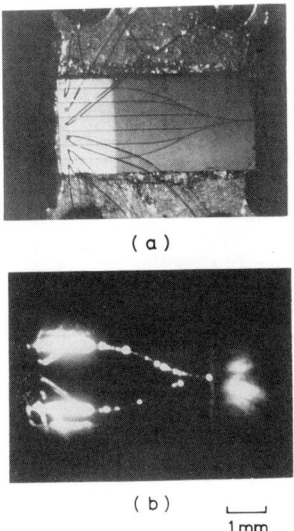

FIG. 35. Six integrated DFB lasers with curved wave guides: (a) overall photograph; (b) with two channels activated.

Furthermore, developments in surface-emitting lasers (Motegi et al., 1982; Iga et al., 1985) may in future be combined with the OEIC technology to add a fourth choice.

IV. High-Speed Transmitter and Receiver

In this section, we will describe the recently developed transmitter and receiver OEIC devices, considering applications for high bit-rate optical local area networks. Both OEIC devices have been fabricated on semi-insulating GaAs substrates (Matsueda et al., 1985).

5. Transmitter OEIC

The block diagram of the fabricated transmitter OEIC device is shown in Fig. 36. On a semi-insulating GaAs substrate, a laser, a driving circuit, a photomonitor (PM) and a monitoring circuit are integrated. The biasing current of the laser is given by an external circuit. A photodetector and a monitoring circuit are integrated for a purpose of automatic power control (APC). The photodetector detects the output power from the laser diode; the photocurrent, which is proportional to the optical output, is amplified by the monitoring circuit. The electrical output from the monitoring circuit is fed back to the biasing of the laser diode, and the optical output can be kept constant against the variation due to temperature changes.

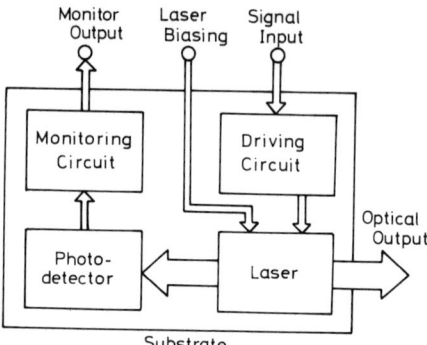

Fig. 36. Block diagram of a transmitter OEIC.

A bird's-eye view and a surface view of the chip are shown in Figs. 4a and 4b. In the figures, the optical devices are fabricated on the middle part of the substrate. The driving circuit and the monitoring circuit are integrated on the right part and the left part of the substrate, respectively. Although the OEIC device includes two laser diodes (LD_1 and LD_2) and two photodetectors (PD_1 and PD_2), only one pair (LD_1, PD_1) is connected to the circuits. Another pair is integrated to check independently the characteristics of the discrete devices. The chip size is 0.85 mm × 2.0 mm.

The optical devices have been fabricated in an etched well (groove) by the MOCVD technique, to reduce the height difference between the top of the optical device and the surface where the electronic circuits are integrated. The MESFETs are formed by the conventional ion-implanting technique. The details of the fabrication processes are as described in Part III.

The circuit diagram of the device is shown in Fig. 4d. The driving circuit is identical to the circuit described in Part II, and its specifications are summarized in Table III. The circuit and device parameters are shown in Table V. The monitoring circuit is a simple source-follower circuit that consists of two FETs (Q_{11}, Q_{12}). R_m is the load resistor.

The static switching characteristics of the driving circuit are shown in Fig. 37. In these measurements, a chip in which the laser diode was replaced by a 50 Ω dummy resistor was used to examine the electrical characteristics independently. The horizontal axis indicates the input voltage V_{in} given to the gate of FET Q_8. The input voltage V_{ref} at the gate of Q_9 was kept constant (-7.5 V) as a reference voltage. The vertical axis V_{DUMMY} indicates the voltage across the dummy resistor, and the driving current given by $V_{DUMMY}/50$ is also shown in the vertical axis. The input voltage at the gate of Q_3 is zero in these experiments. Therefore, the switching current is equal to the saturated drain current of Q_3. The gate width of Q_3 is 500 μm and the

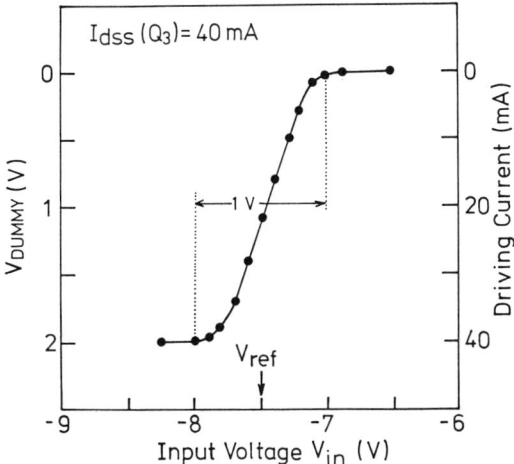

FIG. 37. Static switching characteristics of the driving circuit (the laser diode was replaced by a 50 Ω dummy resistor).

transconductance coefficient K and threshold voltage $|V_{th}|$ of the sample are 56 μA/V^2/μm and 1.2 V, respectively, so the saturated drain current I_{dss} given by $KW|V_{th}|^2$ is about 40 mA.

The results in Fig. 37 show that the required input voltage amplitude for digital switching with 40 mA$_{pp}$ current is about 1.0 V$_{pp}$. As shown by Eq. (3) in Part II, the required input voltage is proportional to the square root of the switching current. Therefore, the required voltage amplitude for the initially specified 15 mA$_{pp}$ driving is as low as 0.61 V$_{pp}$. Therefore, the input voltage specification, 0.8 V$_{pp}$, is sufficiently large for 15 mA$_{pp}$ driving.

The waveforms of the dynamic response are shown in Fig. 38. The waveforms indicate the response of the driving current flowing through the dummy resistor when the circuit is driven with pulsed input signals. Rise and fall times of 130 psec and 120 psec were obtained, respectively. The measured response is slower than the predicted one shown in Fig. 25b in Part II. The measured results include the influence of the stray capacitance and the

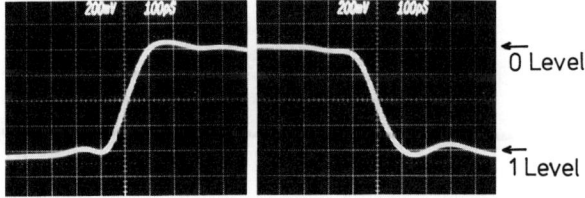

FIG. 38. Dynamic response of the driving circuit (by a 50 Ω dummy resistor).

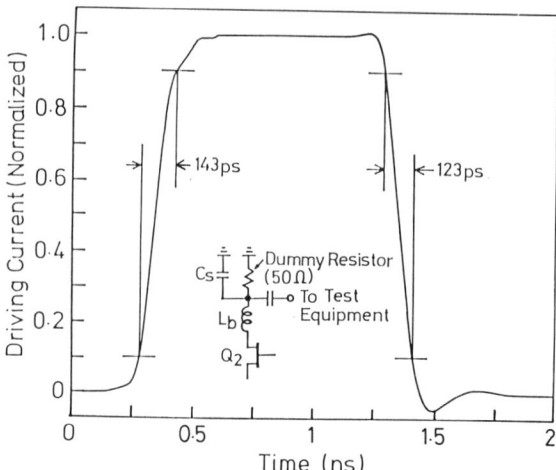

FIG. 39. Simulated response of the driving circuit considering the parasitics associated with the measuring setup. ($C_s = 2$ pF, $L_b = 4$ nH).

inductance associated with the measuring setup. The computer-simulated results which consider the influence of the parasitics associated with the measuring setup are shown in Fig. 39. In this case, the rise and fall times are 143 psec and 123 psec, respectively, which are comparable to the measured data. Figure 40 shows the eye pattern when the circuit is driven by a pseudorandom pulse train with a bit rate of 2 Gb/sec (NRZ). It shows an excellent eye-opening.

The structure of the integrated laser is shown in Fig. 4c. An active layer is formed between the lower and upper cladding layers ($Ga_{0.6}Al_{0.4}As$), and a narrow current path with 3 μm width is formed by the blocking layers. An n^+-conductive layer is embedded to realize the electrical connection between the n-side of the laser and the electronic circuit. A multiple quantum-well (MQW) structure is employed as an active layer to obtain a low threshold current and a high slope efficiency. Figures 41a and 41b show the photographs of the cross-section of the laser facet and the fine structure of the

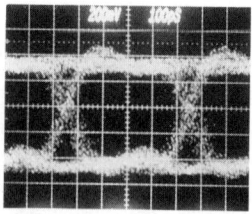

FIG. 40. Eye pattern at 2 Gb/sec (NRZ) operation of the laser driver.

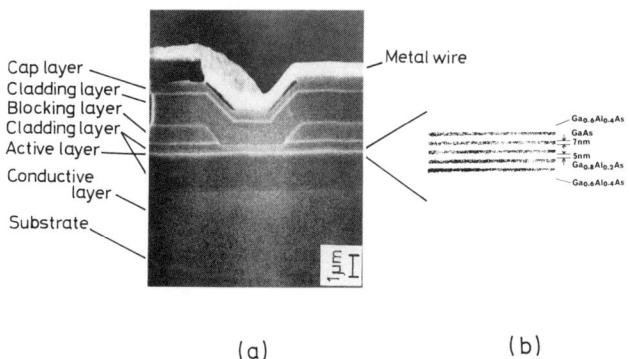

FIG. 41. Active region of the OEIC transmitter: (a) photograph of the integrated laser facet; (b) fine structure of the MQW active layer.

MQW active layer. The MQW active layer consists of 7-nm-thick GaAs well layers and 5-nm-thick $Ga_{0.8}Al_{0.2}As$ barrier layers. The total thickness of the active layer is 60 nm. The optical wavelength with this structure is about 0.83 μm (830 nm).

One facet of the integrated laser was formed using dry etching, that is, RIBE, while the other facet was formed by the conventional cleaving technique. The cavity length of the laser diode is 200 μm.

The typical lasing characteristics of the integrated laser have been shown already in Fig. 34. In the figure, the solid curve indicates the characteristics of an integrated laser having etched and cleaved facets. The dashed curve corresponds to a discrete laser having the same structure as the integrated laser, except that both facets are cleaved mirrors. Threshold current as low as 31 mA and slope efficiency of 0.20 mW/mA/facet were obtained in the integrated laser. The characteristics of the integrated laser are slightly inferior to those of the discrete laser, whose threshold current and slope efficiency are 25 mA and 0.26 mW/mA/facet, respectively. It is considered that the inferiority is mainly due to imperfections of the etched facet, such as surface roughness.

The far field patterns have single-peak characteristics at various injection current levels. This result indicates that the laser operates in single transverse mode condition.

The observed optical output waveforms from the transmitter OEIC device are shown in Fig. 42. The input signal is NRZ pulse train with a bit rate of 2 Gb/sec. In the photograph, the upper and lower levels correspond to on and off of the optical signal. The biasing level of the laser ($I_{bias}/I_{threshold}$) is 1.2; the amplitude of the laser driving current (modulation) is 15 mA_{pp}, and the amplitude of the optical signal is 3 mW_{pp}.

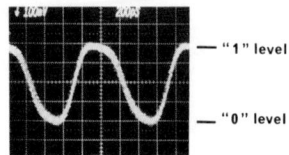

Fig. 42. Dynamic optical output waveforms of the transmitter OEIC (operating bit rate is 2 Gb/sec (NRZ)).

The rise time and the fall time are 200 psec and 250 psec, respectively. The response speed is limited by the time constant of the integrated laser in the OEIC sample. To realize very high-speed transmitter OEIC devices, it is important to reduce the time constant of the integrated laser diode itself. The laser structure employed is basically similar to that of the well-known CSP (channeled substrate planar) laser, and the capacitance is comparatively larger than other structures such as the BH (buried hetero) structure. Unfortunately, there exist some difficulties in fabricating BH-type integrated lasers at this time. It is desirable to develop the fabrication processes of integrating lasers with very small time constants.

As is well known, the frequency of the relaxation oscillation will be another limiting factor of high-speed operation. The frequency of the relaxation oscillation of a laser diode is given by

$$f_r = \frac{1}{2\pi} \sqrt{\frac{P_0 G}{\tau_p}} \qquad (22)$$

(Lau et al., 1983), where P_0 is photon density and G and τ_p are differential gain and lifetime of the photons, respectively. The frequency range of the conventional laser diode is 1–5 GHz. Lau and Yariv have shown that the oscillating frequency can be shifted higher by shortening the laser cavity (decreasing τ_p) (Lau et al., 1983), or by operating at low temperature (increasing G) (Lau et al., 1984). It also has been found that the frequency of relaxation oscillation of MQW lasers becomes twice as high as that of a conventional laser with simple DH (double-hetero) structure (Arakawa et al., 1984; Uomi et al., 1985). It is theoretically explained that the effect is due to the increase of the differential gain in the MQW structure. An MQW active layer has been employed in the above-mentioned OEIC mainly to obtain a low threshold current. Its influence on the dynamic response characteristics also seems to be promising. Further investigations are necessary to find the optimum laser structure for very-high-speed transmitter OEIC devices.

The photodetector used in the monitoring circuit have a structure identical to the integrated laser diode and are operated in reverse biasing conditions. As has been described, the facets of the optical devices were formed by a dry

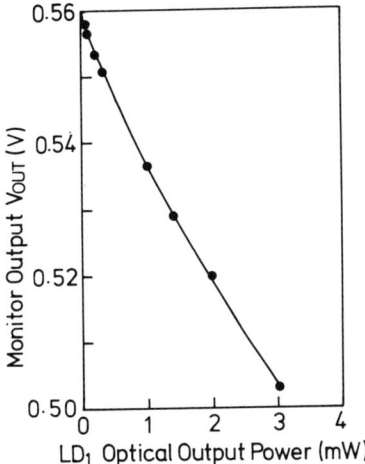

FIG. 43. Detection characteristics of the monitoring circuit.

etching process. The separation between the laser diode (LD_1) and the photo monitor (PD_1) is 105 μm.

The gate widths of FETs Q_{11} and Q_{12} in the monitoring circuit are 210 μm and 70 μm, respectively. The load resistor R_m is about 3 kΩ. The monitoring circuit requires three power supplies, +3 V and −2 V for V_{DM} and V_{SM}, respectively, and −3 V for the reverse biasing of the photodetector V_{PD}. The detection characteristics of the monitoring circuit have been examined. Figure 43 shows the results. The horizontal axis indicates the output optical power from the cleaved facet of the integrated laser LD_1. The vertical axis is the electrical monitor output voltage. Almost linear characteristics were observed. The responsivity of the photodetector estimated from the output voltage is 6×10^{-3} A/W. It is considered that the low responsivity is mainly due to the low coupling efficiency across the 105 μm separation.

6. Receiver OEIC

A bird's-eye view and a photograph of a receiver OEIC chip employing semi-insulating GaAs substrate are shown in Figs. 44a and 44b. On the substrate, two PIN-PD (PD1 and PD2) and a preamplifier are integrated. PD1 is connected to the preamplifier, while PD2 is a dummy to check the characteristics as a discrete device. The circuit of the preamplifier is identical to the circuit described in Part II. The diameter of the receiving area of the detector is 58 μm$^\phi$. The chip size is 1.2 mm × 1.3 mm.

The OEIC has the horizontal integrating structure. The detectors are embedded in an etched groove in order to reduce the height difference

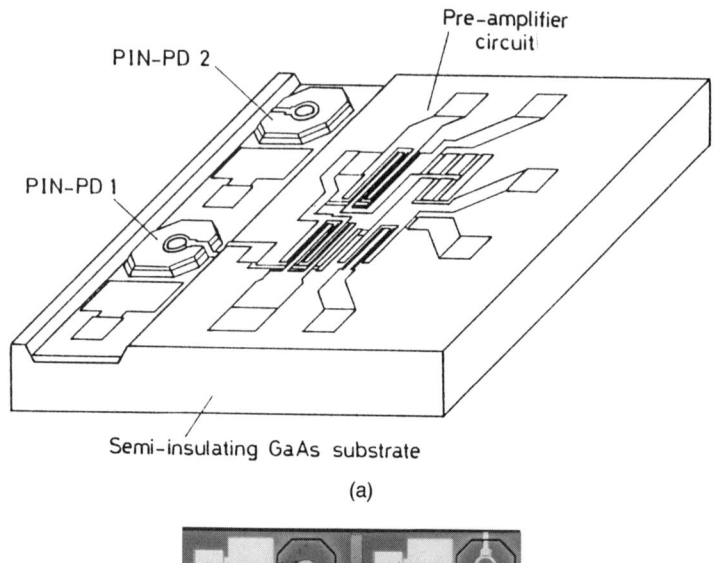

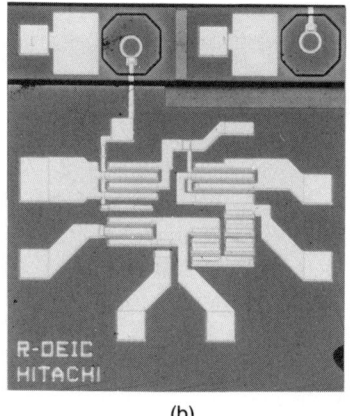

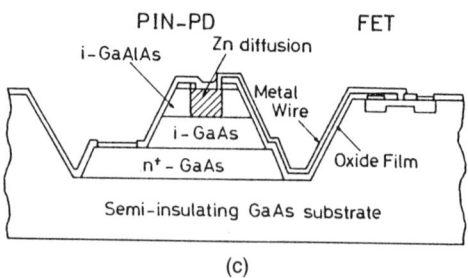

FIG. 44. A GaAs OEIC receiver with a PIN-PD and three MESFETs: (a) overall view; (b) photograph; (c) schematic cross-section.

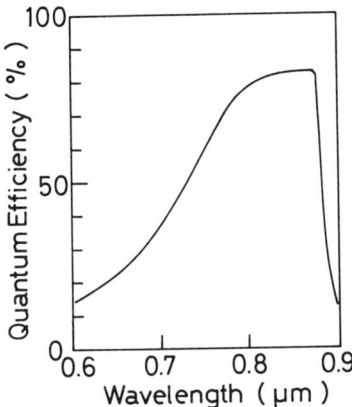

FIG. 45. Wavelength sensitivity of the integrated PIN-PD.

between the top of the detector and the surface for the preamplifier circuit. The detector is fabricated by the LPE crystal growth technique. The FETs are formed by the conventional ion-implanting technique.

The cross-sectional view around the integrated detector is shown in Fig. 44c. A p^+ region is formed by Zn diffusion in an i-GaAlAs layer. The n^+-conductive layer is used as the n-side electrode. The thickness of the i-GaAs layer is 2 μm, and the carrier concentration is about 1×10^{15} cm^{-3}.

The sensitivity of the fabricated PIN-PD is shown in Fig. 45. A quantum efficiency of up to 80% was obtained at a wavelength of 0.83 μm.

The capacitance of the PIN-PD greatly affects the bandwidth of the integrated receiver. Figure 46 shows the dependence of the capacitance (C_{PD})

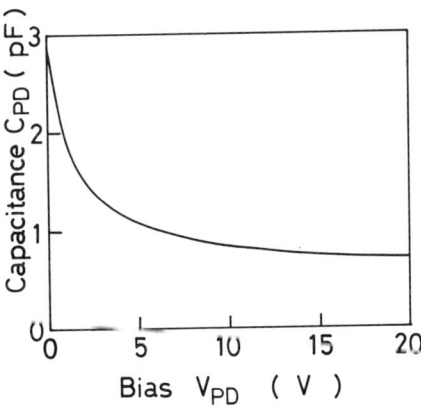

FIG. 46. Dependence of C_{PD} on the reverse biasing voltage, for the integrated PIN-PD.

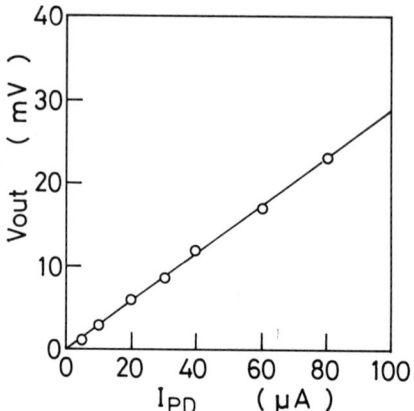

FIG. 47. Static input and output characteristics of the integrated PIN-PD.

on the reverse biasing voltage (V_{PD}). The measured values of C_{PD} include the stray capacitance associated with the metal wiring in the electronic circuit. The capacitance at the punch-through voltage of about 10 V was 1 pF. The calculated capacitance given by the sum of the junction capacitance and the stray capacitance of the wiring is about 0.5 pF. The measured value is larger than the calculated result, mainly because of the lateral diffusion of Zn in the fabrication of the p^+ region.

The typical static input and output characteristics are shown in Fig. 47. The transimpedance is estimated to be about 300 Ω from the figure. The deviation of the transconductance from the designed value in Part II is mainly due to the lower sheet resistance.

The 3 dB bandwidth characteristics measured in small-input-signal conditions are shown in Fig. 48. A bandwidth of 1.2 GHz was obtained. The

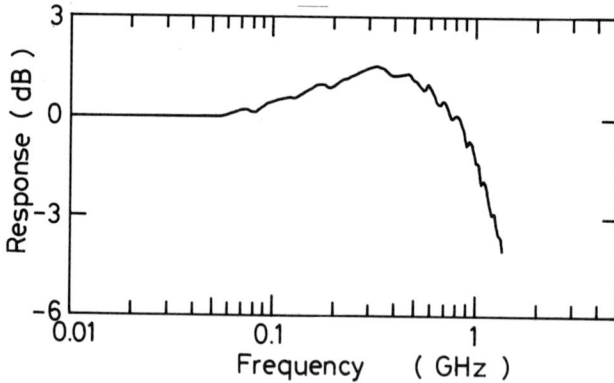

FIG. 48. Bandwidth characteristics of the OEIC receiver.

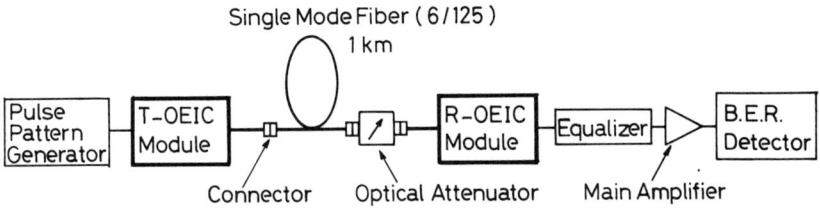

FIG. 49. Block diagram of the transmission experiments.

bandwidth is narrower than designed because of two reasons: the detector has a larger capacitance, and the load resistance in the preamplifier is smaller, so that gain is suppressed.

7. Transmission Experiment

Transmission experiments using transmitter and receiver OEIC devices were carried out by Horimatsu *et al.* (1985). They reported 4 km transmission at 400 Mb/sec NRZ signal, employing GaAs OEIC devices and multimode graded index fibers. A minimum receiving power of −18 dBm was reported.

The transmission experiments using OEIC devices shown in Fig. 44 were also carried out. The block diagram of the measuring setup is shown in Fig. 49. An equalizer was employed to improve the system bandwidth. The optical fiber used was single-mode fiber designed for the 0.83 μm wavelength; loss was 3 dB/km, and transmission length was 1 km.

Figure 50 shows the level diagram of the transmission experiments. The optical output from the transmitter OEIC module with pig-tailed single-mode fiber was −8 dBm, and the minimum receiving power at 1 Gb/sec (NRZ) for 10^{-9} bit error rate (BER) was −19.5 dBm at the OEIC receiver.

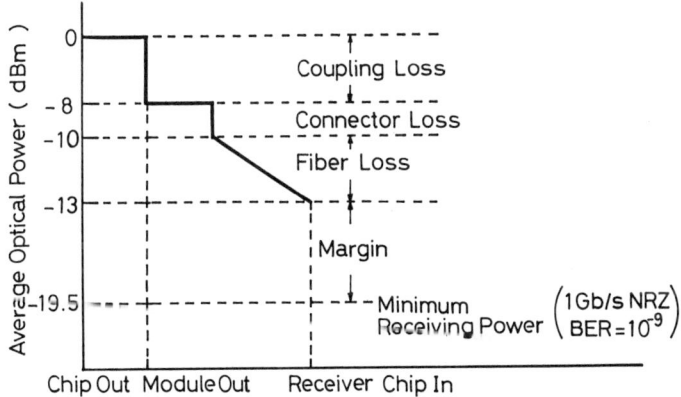

FIG. 50. Level diagram of the transmission experiments.

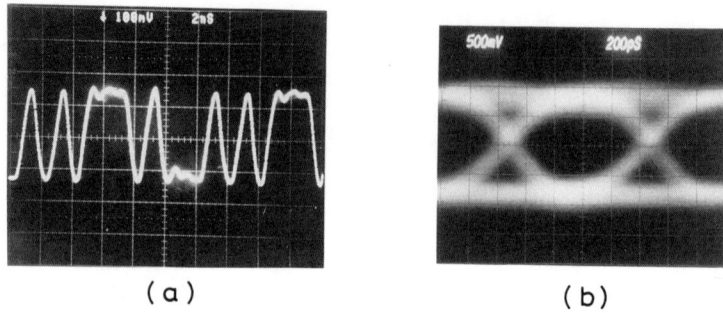

FIG. 51. Output waveforms of the receiving OEIC: (a) fixed-pulse pattern; (b) eye pattern.

The output waveforms for fixed-pulse patterns and the eye pattern for pseudorandom pulse trains of 1 Gb/sec NRZ signal are shown in Fig. 51. The relation between the receiving optical power and the bit error rate is shown in Fig. 52. The lower two curves indicate the characteristics for fiber lengths of 1 km and 1 m, respectively. A very small degradation by transmission through long fiber was observed. The characteristics at 1.6 Gb/sec were also measured, and the results are also shown in Fig. 52. A minimum receiving power of -18.5 dBm was obtained for 1.6 Gb/sec.

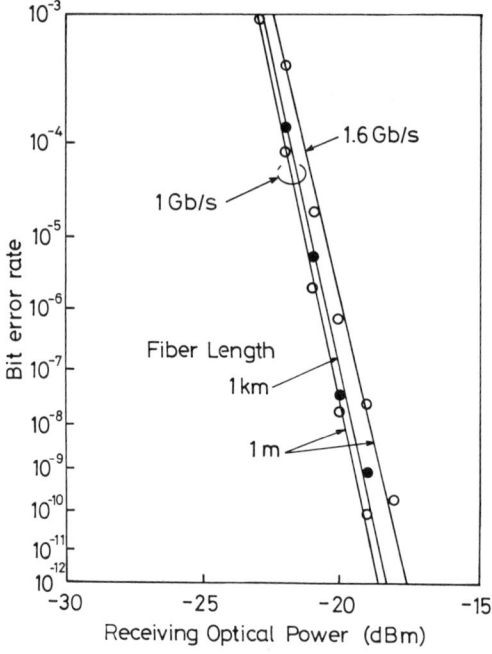

FIG. 52. Bit error rate characteristics of the transmission experiment involving the OEICs.

Transmission experiments through further distances, such as over 50 km, are in progress using different OEICs (Suzuki *et al.*, 1988).

V. Conclusions

The state of the art of OEIC technology is described, including its history, the vertical and horizontal structures, design philosophy, and fabrication processes. The materials of major concern are III–V compounds, i.e., GaAs and InP.

The capability of OEICs in the Gb/sec region has been verified by way of an OEIC transmitter and an OEIC receiver. The transmitter is of SSI (small-scale integration) scale, having RIBE interior facets for the laser. The receiver consists of a PIN detector and 3 transistors.

The realization of OEICs depends critically on the development of technology in substrate production, epitaxy, doping, lithography, interior reflectors, structural improvements of lasers and detectors, and so on. Therefore it is understood that the last 18 years, since the proposal of the idea (for example, Somekh and Yariv, 1972), have been spent to accumulate the background technologies, together with the financial resources for the investment required. As the result of our effort to combine effectively the various technologies, the OEIC has just reached the SSI scale with Gb/sec speed. This experimental verification is the beginning of its practical use, as well as the intermediate step for further revolutionary developments. As for the receiver OEIC, integration of a MSM (metal semiconductor metal) photo detector and an LSI circuit has been reported (Rogers, 1988) as of the time of the first revision of this chapter.

The immediate application of the present OEIC is for short- and long-distance optical communications and in optical interconnections within equipment or computers. The OEICs will be one of the realistic breakthroughs for the idea of inter-chip optical connections to resolve the present communication crisis in the area of VLSI circuits and systems (Goodman *et al.*, 1984).

In the future, the OEIC technology will incorporate the optical waveguide technology, which is presently advanced for dielectric materials such as $LiNbO_3$ (Tamir, ed., 1975, 1982). At this stage, intrachip connections will also be done by optical means. Therefore, no parasitic or stray capacitances are involved. The response speed will be at least 10 times faster than any electronic interconnection (Matsueda *et al.*, 1985; Matsueda, 1987).

The spatial coherency of light will also be used as a parallel data processing function. The OEIC technology may contribute to these types of advancement, too.

The developments in material technology, such as heteroepitaxy of GaAs or InP on a Si substrate (Windhorn *et al.*, 1984; Razeghi *et al.*, 1988) and selective ion beam deposition (Hashimoto and Miyauchi, 1984), are expected to add remarkable freedom for OEIC design and application. Heteroepitaxy of this kind has been reported to realize an integration of a GaAs LED (light-emitting diode) and a Si MOSFET (metal oxide semiconductor FET) (Ghosh *et al.*, 1986). It has also been reported that the integration of GaAs MESFETs and an InGaAsP LD or InGaAs PIN-PD on a semi-insulating InP substrate was successful (Suzuki *et al.*, 1988).

Concerted developments in materials, devices, and systems will accelerate the realization of OEICs in practice.

Acknowledgments

The authors are grateful for the contributions to the experiments and discussion given by Dr. Hiroshi Kodera, Messrs. Hitoshi Sato, Yasuo Minai, Yoshihiro Motegi, Dr. Hironari Matsuda, and Mr. Kazuhide Harada of the Fiberoptics Project Div., Hitachi Ltd., and Dr. Hirobumi Ouchi, Messrs. Hiroyuki Nakano, Shigeo Yamashita, Saburou Ataka, Motohisa Hirao, and Dr. Minoru Maeda at the Central Res. Lab., Hitachi Ltd.

A part of this content has been obtained through the research in the Large Scale Project, Optical Measurement and Control System, conducted under a program set up by the Agency of Industrial Science and Technology, Ministry of International Trade and Industry of Japan.

References

Aiki, K., Nakamura, M., and Umeda, M. (1976). *Appl. Phys. Lett.* **29**, 506–508.
Arakawa, Y., Vahala, K., and Yariv, A. (1984). *Appl. Phys. Lett.* **45**, 950–952.
Auston, D. H. (1984). In "Picosecond Optoelectronic Devices" (C. H. Lee, ed.), Chap. 4. Academic Press, Orlando, Florida.
Bar-Chaim, N., Harder, C., Katz, J., Margalit, S., and Yariv, A. (1982). *Appl. Phys. Lett* **40**, 556–557.
Bar-Chaim, N., Lau, K. Y., Ury, I., and Yariv, A. (1984). *Appl. Phys. Lett.* **44**, 941–943.
Barnard, J., Ohno, H., Wood, C. E. C., and Eastman, L. F. (1981). *IEEE Electron Device Lett.* **EDL-2**, 7–9.
Blauvelt, H., Bar-Chaim, N., Fekete, D., Margalit, S., and Yariv, A. (1982). *Appl. Phys. Lett.* **40**(4), 289–290.
Carney, J. K., Helix, M. J., and Kolbas, R. M. (1983). *Tech. Dig.*, GaAs IC Symposium, Phoenix, Arizona, pp. 48–51.
Carter, A. C., Forbes, N., and Goodfellow, R. C. (1982). *Electron. Lett.* **18**, 72–74.
Chen, P. C., Law, H. D., Rezek, E. A., and Weller, J. (1983). *Tech. Dig., 4th Int. Conf. Integrated Optics and Optical Fiber Communication (IOOC), Tokyo*, pp. 190–191.
Filensky, W., Klein H.-J., and Beneking, H. (1977), *IEEE J. Solid-State Circuits* **SC-12**, 276–280.
Fukuzawa, T., Nakamura, M., Hirao, M., Kuroda, T., and Umeda, J. (1980). *Appl. Phys. Lett.* **36**, 181–183.
Ghosh, R. N., Griffing, B., and Ballantyne, J. M. (1986). *Appl. Phys. Lett.* **48**, 370–371.
Goell, J. E. (1973). *Proc. IEEE*, **61**, 1504–1505.
Goodman, J. W., Leonberger, F. J., Kung, S. Y., and Athale, R. A. (1984). *Proc. IEEE* **72**, 850–866.

Gruber, J., Marten, P., Petschacher, R., and Russer, P. (1978). *IEEE Trans. on Communications* **COM-26**, 1088-1098.

Hashimoto, H., and Miyauchi, E. (1984). *Abstracts, 16th Conf. Solid State Devices and Materials, Kobe, paper C-1-2*, pp. 121-124.

Hata, S., Ikeda, M., Amano, T., Motosugi, G., and Kurumada, K. (1984). *Electron. Lett.* **20**, 947-948.

Hata, S., Ikeda, M., Noguchi, Y., and Kondo, S. (1985). *Abstracts, 17th Conf. Solid State Devices and Materials, Tokyo, paper B-1-6*, pp. 79-82.

Hayashi, I. (1983). *Tech. Dig., 4th Int. Conf. Integrated Optics and Optical Fiber Communication (IOOC), Tokyo*, pp. 170-171.

Hong, C. S., Kasemset, D., Kim., M. E., and Milano, R. A. (1984). *Electron. Lett.* **20**, 733-735.

Horimatsu, T., Iwama, T., Oikawa, Y., Touge, T., Wada, O., and Nakagawa, T. (1985). *Electron. Lett.* **21**, 320-321.

Iga, K., Ishikawa, S., Ohkouchi, S., and Nishimura, T. (1985). *IEEE J. Quantum Electron* **QE-21**, 663-668.

Inoue, K., Ohnaka, K., Uno, T., Hase, N., and Serizawa, H. (1983). *Tech. Dig., 4th Int. Conf. Integrated Optics and Optical Fiber Communication (IOOC), Tokyo*, pp.. 186-187.

Ito, M., Wada, O., Nakai, K., and Sakurai, T. (1984). *IEEE, Electron Device Lett.* **EDL-5**, 531-532.

Kasahara, K., Hayashi, J., Makita, K., Taguchi, K., Suzuki, A., Nomura, H., and Matsushita, S. (1984a), *Electron. Lett..* **20**, 314-315.

Kasahara, K., Hayashi, J., and Nomura, H. (1984b), *Electron. Lett.* **20**, 618-619.

Katz, J., Bar-Chaim, N., Chen, P. C., Margalit, S., Ury, I., Wilt, D., Yast, M., and Yariv, A. (1980). *Appl. Phys. Lett.* **37**, 211-213.

Kim, M. E., Hong, C. S., Kasemset, D., and Milano, R. A. (1984). *IEEE Electron Device Lett.* **EDL-5**, 306-309.

Kolbas, R. M., Abrokwah, J., Carney, J. K., Bradshaw, D. H., Elmer, B. R., and Biard, J. R. (1983). *Appl. Phys. Lett.* **43**, 821-823.

Koren, U., Hasson, A., Yu, K. L., Chen, T. R., Margalit, S., and Yariv, A. (1982a). *Abstracts, 8th Int. Semiconductor Laser Cong., Ottawa-Hull, Canada, paper 33*, pp. 94, 95.

Koren, U., Yu, K. L., Chen, T. R., Bar-Chaim, N., Margalit, S., and Yariv, A. (1982b). *Appl. Phys. Lett.* **40**, 643-645.

Lau, K. Y., Bar-Chaim, N., Ury, I., Harder, Ch., and Yariv, A. (1983). *Appl. Phys. Lett.* **43**, 1-3.

Lau, K. Y., Harder, Ch., and Yariv, A. (1984). *Appl. Phys. Lett.* **44**, 273-275.

Lee, C. P., Margalit, S., Ury, I., and Yariv, A. (1978). *Appl. Phys. Lett.* **32**, 806-807.

Leheny, R. F., Nahory, R. E., Pollack, M. A., Ballman, A. A., Beebe, E. D., Dewinter, J. C., and Martin, R. J. (1980). *Electron. Lett.* **16**, 353-355.

Matsueda, H. (1987). *IEEE J. Lightwave Technol.* **LT-5**, 1382-1390.

Matsueda, H., and Nakamura, M. (1981). *Digest, Inter. Electron Devices Meeting (IEDM), Washington, D.C., paper 12.1*, pp. 272-275.

Matsueda, H., and Nakamura, M. (1984). *Appl. Optics* **23**, 779-781.

Matsueda, H., Fukuzawa, T., Kuroda, T., and Nakamura, M. (1980). *12th Conf. Solid State Devices, Tokyo (J. Appl. Phys. Sup. 20-1*, pp. 193-197 (1981)).

Matsueda, H., Sasaki, S., and Nakamura, M. (1983a). *IEEE J. Lightwave Technol.* **LT-1**, 261-269.

Matsueda, H., Sasaki, S., Kohashi, T., Tanaka, T. P., Maeda, M., and Nakamura, M. (1983b). *Digest, Conf. Lasers and Electro-Optics (CLEO), Baltimore, Maryland, paper WL2*, pp. 120-121.

Matsueda, H., Tanaka, T. P., and Nakano, H. (1984), *IEE Proceedings*, **131**, Part. H, pp. 299-303.

Matsueda, H., Hirao, M., Tanaka, T. P., Kodera, H., and Nakamura, M. (1985). *Proc. 12th Int. Symp. Gallium Arsenide and Related Compounds, Karuizawa, Japan*, pp. 655-660.
Miura, S., Wada, O., Hamaguchi, H., Ito, M., Makiuchi, M., Nakai, K., and Sakurai, T., (1983). *IEEE Electron Device Lett.* **EDL-4**, 375-376.
Motegi, Y., Soda, H., and Iga, K. (1982). *Electron. Lett.* **18**, 461-463.
Personick, S. D. (1973). *Bell System Tech. J.* **52**(6), 843-886.
Pucel, R. A., Haus, H. A., and Statz, H. (1975). In "Advances in Electronics and Electron Phys." (L. Marton, ed.), Vol. 38, pp. 195-265. Academic Press, New York.
Razeghi, M., Defour, M., Omnes, F., Maurel, Ph., and Chazelas, J. (1988). *Appl. Phys. Lett.* **53**, 725-727.
Rogers, D. L. (1988). *Tech. Dig., Optical Fiber Commun. Conf.*, New Orleans, Louisiana, paper WF2, p. 63.
Sanada, T., Yamakoshi, S., Wada, O., Fujii, T., Sakurai, T., and Sasaki, M. (1984). *Appl. Phys. Lett.* **44**, 325-327.
Sanada, T., Yamakoshi, S., Hamaguchi, H., Wada, O., Fujii, T., Horimatsu, T., and Sakurai, T. (1985). *Appl. Phys. Lett.* **46**, 226-228.
Shibata, J., Nakao, I., Sasaki, Y., Kimura, S., Hase, N., and Serizawa, H. (1984). *Appl. Phys. Lett.* **45**, 191-193.
Somekh, S., and Yariv, A. (1972). *Proc. Conf. Int. Telemetry*, pp. 407-418.
Su, L. M., Grote, N., Kaumanns, R., Katzschner, W., and Bach, H. G. (1985). *IEEE Electron Device Lett.* **EDL-6**, 14-17.
Suzuki, A., Itoh, T., and Shikada, M. (1988). *Tech. Dig., 14th European Conf. Optical Commun. (ECOC)*, Brighton, UK, Part 1, pp. 11-16.
Tamir, T., Garmire, E., Hammer, J. M., Kogelnik, H., and Zernike, F. (1975, 1982). In "Integrated Optics" (T. Tamir, ed.), 1st and 2nd Ed. Springer-Verlag, New York.
Uomi, K., Chinone, N., Ohtoshi, T., and Kajimura, T. (1985). *Proc. 12th Int. Symp. Gallium Arsenide and Related Compounds, Karuizawa, Japan*, pp. 703-708.
Ury, I., Margalit, S., Yust, M., and Yariv, A. (1979). *Appl. Phys. Lett.* **34**, 430-431.
Ury, I., Lau, K. Y., Bar-Chaim, N., and Yariv, A. (1982). *Appl. Phys. Lett.* **41**, 126-128.
Wada, O., Miura, S., Ito, M., Fujii, T., Sakurai, T., and Hiyamizu, S., (1983a). *Appl. Phys. Lett.* **42**, 380-382.
Wada, O., Hamaguchi, H., Miura, S., Makiuchi, M., Yamakoshi, S., Sakurai, T., Nakai, K., and Iguchi, K. (1983b). *Electron. Lett.* **19**, 1031-1032.
Wada, O., Hamaguchi, H., Miura, S., Makiuchi, M., Nakai, K., Horimatsu, H., and Sakurai, T. (1985a). *Appl. Phys. Lett.* **46**, 981-983.
Wada, O., Sanada, T., Nobuhara, H., Kuno, M., Makiuchi, M., and Fujii, T. (1985b). *Proc. 12th Int. Symp. Gallium Arsenide and Related Compounds, Karuizawa, Japan*, pp. 685-690.
Wilt, D., Bar-Chaim, N., Margalit, S., Ury, I., and Yariv, A. (1980). *J. Quantum Electron.* **QE-16**, 390-391.
Windhorn, T. H., Metze, G. M., Tsaur, B-Y., and Fan, J. C. C. (1984). *Appl. Phys. Lett.* **45**, 309-311.
Yamakoshi, S., Sanada, T., Wada, O., Fujii, T., and Sakurai, T. (1983). *Electron. Lett.* **19**, 1020-1021.
Yariv, A. (1983). *Tech. Dig., 4th Int. Conf. Integrated Optics and Optical Fiber Communication (IOOC)*, Tokyo, pp. 182-183.
Yust, M., Bar-Chaim, N., Izadpanah, S. H., Margalit, S., Ury, I., Witt, D., and Yariv, A. (1979). *Appl. Phys. Lett.* **35**, 795-797.

Index

A

Abrupt emitter
  band diagram, 197-198
  electron injection, 210
  hot carrier, 203
Acceptor, 108-110, 116-118, 120-123, 125-127
Activation energy, 127-128
Active layer, 235, 237, 273
Amphoteric impurity, 126-127, 131
Arsenic pressure, 21
Arsine, 106, 110-111
Atmospheric-pressure metalorganic vapor-phase epitaxy (APMOVPE), 106
Atomic-layer epitaxy
  absorption equation, 42
  cycle speed, 44
  elementary process, 40
  growth apparatus, 33
  growth sequence, 33
  principle, 32
  self-limiting mechanism, 37
  surface coverage, 40
  temperature dependence, 38
  thickness uniformity, 46
Atomic planar doping of Si in GaAs/AlAs QW, 89
Auger electron spectroscopy (AES), 146
Autodoping, 149

B

Band diagram
  abrupt emitter, 197
  AlGaAs/GaAs HBT, 198
  graded base, 197
  graded emitter, 197
Band gap grading, 210
Bandwidth, 261
Base current, 199, 206-207
  excess, 213
Base resistance, 220
Base transit time, 203
  in graded base, 218
Basic digital circuit element, 221
Beryllium (Be)
  diffusion, 209
  redistribution, 209
Bipolar transistor, 241
Bit error rate, 259, 261, 279
Breakdown voltage, 141, 149
Brooks-Herring formula, 108, 123
Buffer amplifier, 248, 252, 257
Built-in voltage, 149

C

Capacitance
  collector, 217, 219
  diffusion, 217
  junction
    of base/emitter, 217
    of emitter/base, 217
  of photodiode, 277
Capacitance-voltage characteristic of HEMT, 166
CBE, 100
Chemical potential, 9
Chemical potential difference, 9
  in LPE, 10
  in MBE, 15
  in MOCVD, 14
  in VPE, 11
Circuit, 246
  differential current switching, 248
  electric, 235
  equivalent, of HBT, 216
  high-impedance-type, 258
  integrated (IC), 131
  large-scale integrated (LSI), 142, 144
  optoelectronic integrated (OEIC), 232
  single-ended, 247
  transimpedance-type, 258
  very large-scale integrated (VLSI), 141

CML, 221
Collector
  capacitance, 217, 219
  current, 199
  transit time, 219
Collector-up (C-up) structure, 214
Compensation, 111, 118, 125
Compensation ratio, 120–121, 126
Compound semiconductor (III–V), 232
Conduction band offset
  of GaAs/AlGaAs, 60, 75, 78
  of GaAs/GaInP (lattice-matched to GaAs), 96
  of InGaAs/AlGaAs (pseudomorphic), 97
  of InGaAs/GaAs (pseudomorphic), 97
  of InGaAs/InAlAs (lattice-matched to InP), 91 94
  of InGaAs/InP (lattice-matched to InP), 96
Correction factor, 26
Current gain
  cutoff frequency, 217
  of graded-base HBT, 205–206
  shrinkage, 213
Cutoff frequency, 260
  of HEMT, 81–82

D

DCTL, 221
Deep level, 127–128, 131–132, 145
  transient spectroscopy (DLTS), 128–129
Device modelling of HBT, 199
Differential current switching circuit, 248
Diffusion
  capacitance, 217
  coefficient of Si impurity
    in AlGaAs, 84
    in GaAs, 84
  model of HBT, 199
  of Zn, 277–278
Diffused Zn, 236
Digital epitaxy, 44. *See also* Epitaxy
Donor, 108–110, 116–118, 121–124, 126–127, 134
  –acceptor pair, 109, 124
  –complex center (DX center), 128, 134
Doping efficiency, 131–132

Double heterostructure (DH), 110, 114
  laser diode (DH LD), 114
Drain current, 249, 251, 270
Drift velocity of 2DEG, 80
Dry etching, 239, 264, 267
DX center
  in AlGaAs, 74, 77, 86
  in AlAs/n-GaAs superlattice, 87, 89

E

ECL, 221
EL2, 127–128
Electric circuit, 235
Electroluminescence (EL), 110–111
Electron velocity enhancement, 203
Emitter/base interchangeability, 216
Emitter-up (E-up) HBT, 214
Epitaxy
  digital, 44
  liquid-phase (LPE), 2, 105, 120
  low-pressure metalorganic vapor-phase (LPMOVPE), 106, 116
  metalorganic vapor-phase (MOVPE), 105, 120–121, 127–128, 131, 140, 141–148, *see also* Metalorganic chemical vapor deposition (MOCVD)
  molecular-beam (MBE), 2, 144, 146, 148
  selective, 48
  sidewall, 49
  vapor-phase (VPE), 2, 105, 120–123, 141–142
Equivalent circuit of HBT, 216
Escape rate, 7
Etched groove, wet chemically, 237, 264, 267
Etch pit density (EPD) 143
Eye pattern, 272, 280
Excess base current, 213
Exciton, 121

F

Fabrication process of OEIC, 264
Fall time, 249, 254, 257, 271, 274
Field effect transistor (FET), 236
  characteristics of, 252

Flash lamp annealing
  for GaAs/n-AlGaAs, 84–85
  for GaAs/n-AlGaAs with stopper layer, 85
Frequency divider, 224

G

GaAs, 232
  electron mobility, 55
Gain, 140, 148
Gate
  length, 256, 260, 263
  -source voltage, 249
  width, 249, 252, 254, 256, 263
Gate array of HBT, 224
Graded base, 197–198, 204, 214
Graded emitter, 197–198
GSMBE, 100
Gummel–Poon model, 205
Gunn diode, 136

H

HBT IC, 224
HECL, 216
HEMT, 54
  capacitance–voltage characteristic, 166
  current–voltage characteristic, 160
  2DEG property, 186
  high-frequency characteristic, 169
  of AlAs/n-GaAs superlattice, threshold voltage, 89
  of GaAs/n-AlGaAs, cutoff frequency, 81
  of GaAs/n-AlGaAs, threshold voltage, 89
  of InGaAs/n-InAlAs (lattice-matched to InP) transconductance, 96
  of n-AlGaAs/GaAs/n-AlGaAs single QW, transconductance, 86
HEMT LSI
  fabrication step, 177
  inverter, 176
  ring oscillator, 180
  sRAM, 181
Heterointerface, 144–146
Heterojunction bipolar transistor (HBT), 195–229
Heterostructure, 114, 144–146
High-electron-mobility (HEMT), 114, 144, 146–148. *See also* HEMT

High-impedance-type circuit, 258
HI^{2}L, 224
Horizontal integration, 234, 236

I

Ideality factor, 149
I^{2}L inverter of HBT, 226
Impinging rate, 4
Indium solder-free MBE growth, 210
InP, 232
Integrated circuit (IC), 131
Integrating structure, 234
Interior mirror facet, 237, 266
Ion implantation
  for external base, 212
  for interdevice isolation, 212
Ionization energy, 121, 123
Ionized impurity, 108, 110, 115–118, 120–121, 127
I–V collapse of HEMT, 82, 86. *See also* HEMT

J

JFET (junction field effect transistor), 240
Junction capacitance
  of base/emitter, 217
  of emitter/base, 217

L

Langmuir-type isotherm, 42
Large-scale integrated circuit (LSI), 142, 144
Laser diode (LD), 235–236, 264, 273
Light-emitting diode (LED), 110
Liquid-phase epitaxy (LPE), 2, 105, 120. *See also* Epitaxy
Low-pressure metalorganic vapor-phase epitaxy (LPMOVPE), 106, 116. *See also* Epitaxy

M

Mass action law, 127
Maximum oscillation frequency f_{max}, 218
MBE, 53, 56, 63, 65, 71, 91, 97, 100
  growth of HBT, 196, 208

Memory effect, 131, 132
Metal insulator semiconductor field effect transistor (MISFET), 240
Metalorganic chemical vapor deposition (MOCVD), 2, 63, 96. *See also* Epitaxy; Metalorganic vapor-phase epitaxy (MOVPE)
Metalorganic vapor-phase epitaxy (MOVPE), 105, 120–121, 127–128, 131, 140, 141–148. *See also* Epitaxy
Metal semiconductor field effect transistor (MESFET), 105, 127–128, 128–144, 149, 150, 236
Microwave, 140, 144, 146
Modulation-doped GaAs/AlGaAs superlattice, 56
Modulation-doped GaAs/AlGaAs superlattice, electron mobility, 57
Molecular-beam epitaxy (MBE), 2, 144, 146, 148. *See also* Epitaxy
  controllability, 23
  flux intensity, 27
  growth parameters, 23
  growth rate, 25
  mean free path, 25
  substrate temperature, 29
MOMBE, 100
Multiquantum well (MQW)
  active layer, 273
  laser, 237

N

Noise figure (NF), 140, 147, 148
Nonequilibrium state
  degree of, 4
  parameter, 4
NTL, 221–222

O

Ohmic contact for HBT, 198
Optimized arsenic pressure, 22
Optoelectronic integrated circuit (OEIC), 232
Outdiffusion, 143
Oxygen implantation, 214

P

Parasitic resistance, 139
Parasitics, 246, 272
Persistent photoconductivity (PPC), 128
  effect for
    AlAs/n-GaAs superlattice, 89
    GaAs/n-AlGaAs, 59, 71, 74
    GaAs/n-GaInP (lattice-matched to GaAs), 97
    InGaAs/n-AlGaAs(pseudomorphic), 98
    InGaAs/n-InP (lattice-matched to InP), 97
    Si-PD GaAs/AlAs QW, 91
Photoconductivity, 127
Photodetector, 242
Photodiode (PIN-PD[PIN]), 242
Photoluminescence (PL), 109, 120–124, 126–127
Pinch-off voltage, 141
Potential spike in abrupt emitter HBT, 198
Power HEMT, 86. *See also* HEMT
Purity
  arsine, 116
  gallium aluminum arsenide (GaAlAs), 111
  gallium arsenide (GaAs), 107, 110–111, 116, 120–121
  trimethylgallium (TMG), 106, 108, 116

Q

Quasi-field in graded-base HBT, 204

R

Rate-determining species, 5
Receiver (OEIC), 234, 242, 258, 275
Relaxation oscillation, 274
Ring oscillator, 149–150
  of HEMT, 180
Rise time, 249, 254, 257, 271, 274

S

Schottky gate, 139
Secondary ion mass spectroscopy (SIMS), 143

INDEX

Selective epitaxy, 48. *See also* Epitaxy
Selectively doped (SD) heterostructure
  AlAs/n-GaAs superlattice, 87, 99
  GaAs/n-AlGaAs, 58, 64
    2DEG concentration, 62, 64
    2DEG mobility, 63–64
    drift velocity, 80
    electronic state, 65, 67
      energy level, 66–67
      Fermi level, 67
      subband energy separation, 66
      wave function, 66–67
  GaAs/n-GaInP (lattice-matched to GaAs), 96
  InGaAs/n-InAlAs (lattice-matched to InP), 91, 99
    2DEG concentration, 95–96, 99
    2DEG mobility, 95, 97, 99
    electronic state, 94
      calculation of, 94
      energy level, 94
      subband energy, 94
      wave function, 94
  InGaAs/n-AlGaAs (pseudomorphic), 97, 99
  InGaAs/n-GaAs (pseudomorphic), 97
  n-AlGaAs/GaAs/n-AlGaAs single QW, 85, 99
  Si-PD GaAs/AlAs QW, 89–90, 99
Selective epitaxy, 48
Self-alignment, 143, 149
  of HBT structure, 213
Semiconductor laser, 127
Semi-insulating (SI) substrate, 232, 235–236
Sensitivity, wavelength, 277
Series resistance, 138
Sidewall epitaxy, 49. *See also* Epitaxy
Si impurity
  in AlGaAs
    ionization energy, 76. *See also* DX center
Signal-to-noise ratio, 262
Single-ended circuit, 247
SiO_2 sidewall, 213
Short-channel effect, 144
Shubnikov-de Hass (SdH) oscillation
  in GaAs/AlAs QW, 89–90, 99
  in InGaAs/n-InAlAs (lattice-matched to InP), 93–94

T

Thermionic emission current, 210–212
Threshold current, 271–274
Threshold voltage, 142–144, 249, 257
Toggle frequency of AlGaAs/GaAs HBT ECL gate, 223
Transconductance, 140–141, 149
  coefficient, 249
Transimpedance, 260–278
  -type circuit, 258
Transmitter OEIC, 233–234, 247, 269
Trap, 127–128
Two-dimensional electron gas (2DEG), 106, 144, 146
  AlAs/n-GaAs superlattice concentration, 88, 99
  GaAs/n-AlGaAs concentration, 64
    AlAs mole fraction dependence, 78
    anneal temperature dependence, 72
    doping concentration dependence, 72
    spacer-layer thickness dependence, 71–72
    temperature dependence, 61–62
  GaAs/n-GaInP (lattice-matched to GaAs), 97, 99
  InGaAs/n-AlGaAs (pseudomorphic), 98–99
  InGaAs/n-InAlAs (lattice-matched to InP), 99
    doping concentration dependence, 95–96
    temperature dependence, 92
  InGaAs/n-InP (lattice-matched to InP), 96, 99
  n-AlGaAs/GaAs/n-AlGaAs single QW, 86, 99
  Si-PD GaAs/AlAs QW, 90, 99
  mobility of
    AlAs/n-GaAs superlattice, 88, 99
    GaAs/n-AlGaAs, 62–64
      AlAs mole fraction dependence, 75, 78
      anneal temperature dependence, 72
      calculation of, 68–69, 74–75
      electric field dependence, 80
      electron concentration dependence, 72–73, 75
      hot electron effect, 79
      spacer-layer thickness dependence, 70

temperature dependence, 60–61, 67–68
under high electric field, 79
GaAs/n-GaInP (lattice-matched to GaAs), 97, 99
InGaAs/n-AlGaAs
  pseudomorphic, 98–99
  lattice-matched to InP, electron concentration dependence, 95, 97
InGaAs/n-InP (lattice-matched to InP), 96, 99
n-AlGaAs/GaAs/n-AlGaAs single QW, 86, 99
Si-PD GaAs/AlAs QW, 99

U

Undoped spacer layer in HBT structure, 210
Uniform base HBT, 205

V

Vacancy, 131
Vapor-phase epitaxy (VPE) 2, 105, 120–123, 141–142. *See also* Epitaxy
Vertical integration, 234, 236
Very large-scale integrated circuit (VLSI), 141
VPE, 96

W

Wet etching, 267
  chemically, 264
Wide-gap emitter, 196

Contents of Previous Volumes

Volume 1 Physics of III–V Compounds

C. Hilsum, Some Key Features of III–V Compounds
Franco Bassani, Methods of Band Calculations Applicable to III–V Compounds
E.O. Kane, The $k \cdot p$ Method
V.L. Bonch-Bruevich, Effect of Heavy Doping on the Semiconductor Band Structure
Donald Long, Energy Band Structures of Mixed Crystals of III–V Compounds
Laura M. Roth and Petros N. Argyres, Magnetic Quantum Effects
S.M. Puri and T.H. Geballe, Thermomagnetic Effects in the Quantum Region
W.M. Becker, Band Characteristics near Principal Minima from Magnetoresistance
E.H. Putley, Freeze-Out Effects, Hot Electron Effects, and Submillimeter Photoconductivity in InSb
H. Weiss, Magnetoresistance
Betsy Ancker-Johnson, Plasmas in Semiconductors and Semimetals

Volume 2 Physics of III–V Compounds

M.G. Holland, Thermal Conductivity
S.I. Novkova, Thermal Expansion
U. Piesbergen, Heat Capacity and Debye Temperatures
G. Giesecke, Lattice Constants
J.R. Drabble, Elastic Properties
A.U. Mac Rae and G.W. Gobeli, Low Energy Electron Diffraction Studies
Robert Lee Mieher, Nuclear Magnetic Resonance
Bernard Goldstein, Electron Paramagnetic Resonance
T.S. Moss, Photoconduction in III–V Compounds
E. Antončik and J. Tauc, Quantum Efficiency of the Internal Photoelectric Effect in InSb
G.W. Gobeli and F.G. Allen, Photoelectric Threshold and Work Function
P.S. Pershan, Nonlinear Optics in III–V Compounds
M. Gershenzon, Radiative Recombination in the III–V Compounds
Frank Stern, Stimulated Emission in Semiconductors

Volume 3 Optical of Properties III–V Compounds

Marvin Hass, Lattice Reflection
William G. Spitzer, Multiphonon Lattice Absorption
D.L. Stierwalt and R.F. Potter, Emittance Studies
H.R. Philipp and H. Ehrenreich, Ultraviolet Optical Properties
Manuel Cardona, Optical Absorption above the Fundamental Edge
Earnest J. Johnson, Absorption near the Fundamental Edge
John O. Dimmock, Introduction to the Theory of Exciton States in Semiconductors
B. Lax and J.G. Mavroides, Interband Magnetooptical Effects

CONTENTS OF PREVIOUS VOLUMES

H.Y. Fan, Effects of Free Carries on Optical Properties
Edward D. Palik and George B. Wright, Free-Carrier Magnetooptical Effects
Richard H. Bube, Photoelectronic Analysis
B.O. Seraphin and H.E. Bennett, Optical Constants

Volume 4 Physics of III–V Compounds

N.A. Goryunova, A.S. Borschevskii, and D.N. Tretiakov, Hardness
N.N. Sirota, Heats of Formation and Temperatures and Heats of Fusion of Compounds $A^{III}B^V$
Don L. Kendall, Diffusion
A.G. Chynoweth, Charge Multiplication Phenomena
Robert W. Keyes, The Effects of Hydrostatic Pressure on the Properties of III–V Semiconductors
L.W. Aukerman, Radiation Effects
N.A. Goryunova, F.P. Kesamanly, and D.N. Nasledov, Phenomena in Solid Solutions
R.T. Bate, Electrical Properties of Nonuniform Crystals

Volume 5 Infrared Detectors

Henry Levinstein, Characterization of Infrared Detectors
Paul W. Kruse, Indium Antimonide Photoconductive and Photoelectromagnetic Detectors
M.B. Prince, Narrowband Self-Filtering Detectors
Ivars Melngailis and T.C. Harman, Single-Crystal Lead-Tin Chalcogenides
Donald Long and Joseph L. Schmit, Mercury-Cadmium Telluride and Closely Related Alloys
E.H. Putley, The Pyroelectric Detector
Norman B. Stevens, Radiation Thermopiles
R.J. Keyes and T.M. Quist, Low Level Coherent and Incoherent Detection in the Infrared
M.C. Teich, Coherent Detection in the Infrared
F.R. Arams, E.W. Sard, B.J. Peyton, and F.P. Pace, Infrared Heterodyne Detection with Gigahertz IF Response
H.S. Sommers, Jr., Macrowave-Based Photoconductive Detector
Robert Sehr and Rainer Zuleeg, Imaging and Display

Volume 6 Injection Phenomena

Murray A. Lampert and Ronald B. Schilling, Current Injection in Solids: The Regional Approximation Method
Richard Williams, Injection by Internal Photoemission
Allen M. Barnett, Current Filament Formation
R. Baron and J.W. Mayer, Double Injection in Semiconductors
W. Ruppel, The Photoconductor-Metal Contact

Volume 7 Application and Devices
PART A

John A. Copeland and Stephen Knight, Applications Utilizing Bulk Negative Resistance
F.A. Padovani, The Voltage-Current Characteristics of Metal-Semiconductor Contacts
P.L. Hower, W.W. Hooper, B.R. Cairns, R.D. Fairman, and D.A. Tremere, The GaAs Field-Effect Transistor
Marvin H. White, MOS Transistors

CONTENTS OF PREVIOUS VOLUMES

G.R. Antell, Gallium Arsenide Transistors
T.L. Tansley, Heterojunction Properties

PART B

T. Misawa, IMPATT Diodes
H.C. Okean, Tunnel Diodes
Robert B. Campbell and Hung-Chi Chang, Silicon Carbide Junction Devices
R.E. Enstrom, H. Kressel, and L. Krassner, High-Temperature Power Rectifiers of $GaAs_{1-x}P_x$

Volume 8 Transport and Optical Phenomena

Richard J. Stirn, Band Structure and Galvanomagnetic Effects in III–V Compounds with Indirect Band Gaps
Roland W. Ure, Jr., Thermoelectric Effects in III–V Compounds
Herbert Piller, Faraday Rotation
H. Barry Bebb and E.W. Williams, Photoluminescence 1: Theory
E.W. Williams and H. Barry Bebb, Photoluminescence II: Gallium Arsenide

Volume 9 Modulation Techniques

B.O. Seraphin, Electroreflectance
R.L. Aggarwal, Modulated Interband Magnetooptics
Daniel F. Blossey and Paul Handler, Electroabsorption
Bruno Batz, Thermal and Wavelength Modulation Spectroscopy
Ivar Balslev, Piezooptical Effects
D.E. Aspnes and N. Bottka, Electric-Field Effects on the Dielectric Function of Semiconductors and Insulators

Volume 10 Transport Phenomena

R.L. Rode, Low-Field Electron Transport
J.D. Wiley, Mobility of Holes in III–V Compounds
C.M. Wolfe and G.E. Stillman, Apparent Mobility Enhancement in Inhomogeneous Crystals
Robert L. Peterson, The Magnetophonon Effect

Volume 11 Solar Cells

Harold J. Hovel, Introduction; Carrier Collection, Spectral Response, and Photocurrent; Solar Cell Electrical Characteristics; Efficiency; Thickness; Other Solar Cell Devices; Radiation Effects; Temperature and Intensity; Solar Cell Technology

Volume 12 Infrared Detectors (II)

W.L. Eiseman, J.D. Merriam, and R.F. Potter, Operational Characteristics of Infrared Photodetectors
Peter R. Bratt, Impurity Germanium and Silicon Infrared Detectors
E.H. Putley, InSb Submillimeter Photoconductive Detectors
G.E. Stillman, C.M. Wolfe, and J.O. Dimmock, Far-Infrared Photoconductivity in High Purity GaAs
G.E. Stillman and C.M. Wolfe, Avalanche Photodiodes

CONTENTS OF PREVIOUS VOLUMES

P.L. Richards, The Josephson Junction as a Detector of Microwave and Far-Infrared Radiation
E.H. Putley, The Pyroelectric Detector–An Update

Volume 13 Cadmium Telluride

Kenneth Zanio, Materials Preparation; Physics; Defects; Applications

Volume 14 Lasers, Junctions, Transport

N. Holonyak, Jr. and M.H. Lee, Photopumped III–V Semiconductor Lasers
Henry Kressel and Jerome K. Butler, Heterojunction Laser Diodes
A. Van der Ziel, Space-Charge-Limited Solid-State Diodes
Peter J. Price, Monte Carlo Calculation of Electron Transport in Solids

Volume 15 Contacts, Junctions, Emitters

B.L. Sharma, Ohmic Contacts to III–V Compound Semiconductors
Allen Nussbaum, The Theory of Semiconducting Junctions
John S. Escher, NEA Semiconductor Photoemitters

Volume 16 Defects, (HgCd)Se, (HgCd)Te

Henry Kressel, The Effect of Crystal Defects on Optoelectronic Devices
C.R. Whitsett, J.G. Broerman, and C.J. Summers, Crystal Growth and Properties of $Hg_{1-x}Cd_xSe$ Alloys
M.H. Weiler, Magnetooptical Properties of $Hg_{t-x}Cd_xTe$ Alloys
Paul W. Kruse and John G. Ready, Nonlinear Optical Effects in $Hg_{t-x}Cd_xTe$

Volume 17 CW Processing of Silicon and Other Semiconductors

James F. Gibbons, Beam Processing of Silicon
Arto Lietoila, Richard B. Gold, James F. Gibbons, and Lee A. Christel, Temperature Distributions and Solid Phase Reaction Rates Produced by Scanning CW Beams
Arto Lietoila and James F. Gibbons, Applications of CW Beam Processing to Ion Implanted Crystalline Silicon
N.M. Johnson, Electronic Defects in CW Transient Thermal Processed Silicon
K.F. Lee, T.J. Stultz, and James F. Gibbons, Beam Recrystallized Polycrystalline Silicon: Properties, Applications, and Techniques
T. Shibata, A. Wakita, T.W. Sigmon, and James F. Gibbons, Metal-Silicon Reactions and Silicide
Yves I. Nissim and James F. Gibbons, CW Beam Processing of Gallium Arsenide

Volume 18 Mercury Cadmium Telluride

Paul W. Kruse, The Emergence of $(Hg_{t-x}Cd_x)Te$ as a Modern Infrared Sensitive Material
H.E. Hirsch, S.C. Liang, and A.G. White, Preparation of High-Purity Cadmium, Mercury, and Tellurium
W.F.H. Micklethwaite, The Crystal Growth of Cadmium Mercury Telluride
Paul E. Petersen, Auger Recombination in Mercury Cadmium Telluride
R.M. Broudy and V.J. Mazurczyck, (HgCd)Te Photoconductive Detectors
M.B. Reine, A.K. Sood, and T.J. Tredwell, Photovoltaic Infrared Detectors
M.A. Kinch, Metal-Insulator-Semiconductor Infrared Detectors

CONTENTS OF PREVIOUS VOLUMES

Volume 19 Deep Levels, GaAs, Alloys, Photochemistry

G.F. Neumark and K. Kosai, Deep Levels in Wide Band-Gap III–V Semiconductors
David C. Look, The Electrical and Photoelectronic Properties of Semi-Insulating GaAs
R.F. Brebrick, Ching-Hua Su, and Pok-Kai Liao, Associated Solution Model for Ga-In-Sb and Hg-Cd-Te
Yu. Ya. Gurevich and Yu. V. Pleskov, Photoelectrochemistry of Semiconductors

Volume 20 Semi-Insulating GaAs

R.N. Thomas, H.M. Hobgood, G.W. Eldridge, D.L. Barrett, T.T. Braggins, L.B. Ta, and S.K. Wang, High-Purity LEC Growth and Direct Implantation of GaAs for Monolithic Microwave Circuits
C.A. Stolte, Ion Implantation and Materials for GaAs Integrated Circuits
C.G. Kirkpatrick, R.T. Chen, D.E. Holmes, P.M. Asbeck, K.R. Elliott, R.D. Fairman, and J.R. Oliver, LEC GaAs for Integrated Circuit Applications
J.S. Blakemore and S. Rahimi, Models for Mid-Gap Centers in Gallium Arsenide

Volume 21 Hydrogenated Amorphous Silicon
Part A

Jacques I. Pankove Introduction
Masataka Hirose, Glow Discharge; Chemical Vapor Deposition
Yoshiyuki Uchida, dc Glow Discharge
T.D. Moustakas, Sputtering
Isao Yamada, Ionized-Cluster Beam Deposition
Bruce A. Scott, Homogeneous Chemical Vapor Deposition
Frank J. Kampas, Chemical Reactions in Plasma Deposition
Paul A. Longeway, Plasma Kinetics
Herbert A. Weakliem, Diagnostics of Silane Glow Discharges Using Probes and Mass Spectroscopy
Lester Guttman, Relation between the Atomic and the Electronic Structures
A. Chenevas-Paule, Experiment Determination of Structure
S. Minomura, Pressure Effects on the Local Atomic Structure
David Adler, Defects and Density of Localized States

Part B

Jacques I. Pankove, Introduction
G.D. Cody, The Optical Absorption Edge of a-Si:H
Nabil M. Amer and Warren B. Jackson, Optical Properties of Defect States in a-Si:H
P.J. Zanzucchi, The Vibrational Spectra of a-Si:H
Yoshihiro Hamakawa, Electroreflectance and Electroabsorption
Jeffrey S. Lannin, Raman Scattering of Amorphous Si, Ge, and Their Alloys
R.A. Street, Luminescence in a-Si:H
Richard S. Crandall, Photoconductivity
J. Tauc, Time-Resolved Spectroscopy of Electronic Relaxation Processes
P.E. Vanier, IR Induced Quenching and Enhancement of Photoconductivity and Photoluminescence
H. Schade, Irradiation-Induced Metastable Effects
L. Ley, Photoelectron Emission Studies

CONTENTS OF PREVIOUS VOLUMES

Part C

Jacques I. Pankove, Introduction
J. David Cohen, Density of States from Junction Measurements in Hydrogenated Amorphous Silicon
P.C. Taylor, Magnetic Resonance Measurements in a-Si:H
K. Morigaki, Optically Detected Magnetic Resonance
J. Dresner, Carrier Mobility in a-Si:H
T. Tiedje, Information about Band-Tail States from Time-of-Flight Experiments
Arnold R. Moore, Diffusion Length in Undoped a-Si:H
W. Beyer and J. Overhof, Doping Effects in a-Si:H
H. Fritzche, Electronic Properties of Surfaces in a-Si:H
C.R. Wronski, The Staebler-Wronski Effect
R.J. Nemanich, Schottky Barriers on a-Si:H
B. Abeles and T. Tiedje, Amorphous Semiconductor Superlattices

Part D

Jacques I. Pankove, Introduction
D.E. Carlson, Solar Cells
G.A. Swartz, Closed-Form Solution of I-V Characteristic for a-Si:H Solar Cells
Isamu Shimizu, Electrophotography
Sachio Ishioka, Image Pickup Tubes
P.G. LeComber and W.E. Spear, The Development of the a-Si:H Field-Effect Transitor and Its Possible Applications
D.G. Ast, a-Si:H FET-Addressed LCD Panel
S. Kaneko, Solid-State Image Sensor
Masakiyo Matsumura, Charge-Coupled Devices
M.A. Bosch, Optical Recording
A. D'Amico and G. Fortunato, Ambient Sensors
Hiroshi Kukimoto, Amorphous Light-Emitting Devices
Robert J. Phelan, Jr., Fast Detectors and Modulators
Jacques I. Pankove, Hybrid Structures
P.G. LeComber, A.E. Owen, W.E. Spear, J. Hajto, and W.K. Choi, Electronic Switching in Amorphous Silicon Junction Devices

Volume 22 Lightwave Communications Technology
Part A

Kazuo Nakajima, The Liquid-Phase Epitaxial Growth of InGaAsP
W.T. Tsang, Molecular Beam Epitaxy for III-V Compound Semiconductors
G.B. Stringfellow, Organometallic Vapor-Phase Epitaxial Growth of III--V Semiconductors
G. Beuchet, Halide and Chloride Transport Vapor-Phase Deposition of InGaAsP and GaAs
Manijeh Razeghi, Low-Pressure Metallo-Organic Chemical Vapor Deposition of $Ga_xIn_{t-x}As_yP_{t-y}$ Alloys
P.M. Petroff, Defects in III–V Compound Semiconductors

Part B

J.P. van der Ziel, Mode Locking of Semiconductor Lasers
Kam Y. Lau and Amnon Yariv, High-Frequency Current Modulation of Semiconductor Injection Lasers
Charles H. Henry, Spectral Properties of Semiconductor Lasers

CONTENTS OF PREVIOUS VOLUMES

Yasuharu Suematsu, Katsumi Kishino, Shigehisa Arai, and Fumio Koyama, Dynamic Single-Mode Semiconductor Lasers with a Distributed Reflector
W.T. Tsang, The Cleaved-Coupled-Cavity (C^3) Laser

Part C

R.J. Nelson and N.K. Dutta, Review of InGaAsP/InP Laser Structures and Comparison of Their Performance
N. Chinone and M. Nakamura, Mode-Stabilized Semiconductor Lasers for 0.7–0.8- and 1.1–1.6-μm Regions
Yoshiji Horikoshi, Semiconductor Lasers with Wavelengths Exceeding 2 μm
B.A. Dean and M. Dixon, The Functional Reliability of Semiconductor Lasers as Optical Transmitters
R.H. Saul, T.P. Lee, and C.A. Burus, Light-Emitting Device Design
C.L. Zipfel, Light-Emitting Diode Reliability
Tien Pei Lee and Tingye Li, LED-Based Multimode Lightwave Systems
Kinichiro Ogawa, Semiconductor Noise-Mode Partition Noise

Part D

Federico Capasso, The Physics of Avalanche Photodiodes
T.P. Pearsall and M.A. Pollack, Compound Semiconductor Photodiodes
Takao Kaneda, Silicon and Germanium Avalanche Photodiodes
S.R. Forrest, Sensitivity of Avalanche Photodetector Receivers for High-Bit-Rate Long-Wavelength Optical Communication Systems
J.C. Campbell, Phototransistors for Lightwave Communications

Part E

Shyh Wang, Principles and Characteristics of Integratable Active and Passive Optical Devices
Shlomo Margalit and Amnon Yariv, Integrated Electronic and Photonic Devices
Takaaki Mukai, Yoshihisa Yamamoto, and Tatsuya Kimura, Optical Amplification by Semiconductor Lasers

Volume 23 Pulsed Laser Processing of Semiconductors

R.F. Wood, C.W. White, and R.T. Young, Laser Processing of Semiconductors: An Overview
C.W. White, Segregation, Solute Trapping, and Supersaturated Alloys
G.E. Jellison, Jr., Optical and Electrical Properties of Pulsed Laser-Annealed Silicon
R.F. Wood and G.E. Jellison, Jr., Melting Model of Pulsed Laser Processing
R.F. Wood and F.W. Young, Jr., Nonequilibrium Solidification Following Pulsed Laser Melting
D.H. Lowndes and G.E. Jellison, Jr., Time-Resolved Measurements During Pulsed Laser Irradiation of Silicon
D.M. Zehner, Surface Studies of Pulsed Laser Irradiated Semiconductors
D.H. Lowndes, Pulsed Beam Processing of Gallium Arsenide
R.B. James, Pulsed CO_2 Laser Annealing of Semiconductors
R.T. Young and R.F. Wood, Applications of Pulsed Laser Processing

Volume 24 Applications of Multiquantum Wells, Selective Doping, and Superlattices

C. Weisbuch, Fundamental Properties of III–V Semiconductor Two-Dimensional Quantized Structures: The Basis for Optical and Electronic Device Applications

CONTENTS OF PREVIOUS VOLUMES

H. Morkoc and H. Unlu, Factors Affecting the Performance of (Al, Ga)As/GaAs and (Al, Ga)As/InGaAs Modulation-Doped Field-Effect Transistors: Microwave and Digital Applications
N.T. Linh, Two-Dimensional Electron Gas FETs: Microwave Applications
M. Abe et al., Ultra-High-Speed HEMT Integrated Circuits
D.S. Chemla, D.A.B. Miller, and P.W. Smith, Nonlinear Optical Properties of Multiple Quantum Well Structures for Optical Signal Processing
F. Capasso, Graded-Gap and Superlattice Devices by Band-gap Engineering
W.T. Tsang, Quantum Confinement Heterostructure Semiconductor Lasers
G.C. Osbourn et al., Principles and Applications of Semiconductor Strained-Layer Superlattices

Volume 25 Diluted Magnetic Semiconductors

W. Giriat and J.K. Furdyna, Crystal Structure, Composition, and Materials Preparation of Diluted Magnetic Semiconductors
W.M. Becker, Band Structure and Optical Properties of Wide-Gap $A^{II}_{1-x}Mn_xB^{VI}$ Alloys at Zero Magnetic Field
Saul Oseroff and Pieter H. Keesom, Magnetic Properties: Macroscopic Studies
T. Giebultowicz and T.M. Holden, Neutron Scattering Studies of the Magnetic Structure and Dynamics of Diluted Magnetic Semiconductors
J. Kossut, Band Structure and Quantum Transport Phenomena in Narrow-Gap Diluted Magnetic Semiconductors
C. Riqaux, Magnetooptics in Narrow Gap Diluted Magnetic Semiconductors
J.A. Gaj, Magnetooptical Properties of Large-Gap Diluted Magnetic Semiconductors
J. Mycielski, Shallow Acceptors in Diluted Magnetic Semiconductors: Splitting, Boil-off, Giant Negative Magnetoresistance
A.K. Ramdas and S. Rodriquez, Raman Scattering in Diluted Magnetic Semiconductors
P.A. Wolff, Theory of Bound Magnetic Polarons in Semimagnetic Semiconductors

Volume 26 III–V Compound Semiconductors and Semiconductor Properties of Superionic Materials

Zou Yuanxi, III–V Compounds
H.V. Winston, A.T. Hunter, H. Kimura, and R.E. Lee, InAs-Alloyed GaAs Substrates for Direct Implantation
P.K. Bhattacharya and S. Dhar, Deep Levels in III–V Compound Semiconductors Grown by MBE
Yu. Ya. Gurevich and A.K. Ivanov-Shits, Semiconductor Properties of Superionic Materials

Volume 27 High Conducting Quasi-One-Dimensional Organic Crystals

E.M. Conwell, Introduction to Highly Conducting Quasi-One-Dimensional Organic Crystals
I.A. Howard, A Reference Guide to the Conducting Quasi-One-Dimensional Organic Molecular Crystals
J.P. Pouget, Structural Instabilities
E.M. Conwell, Transport Properties
C.S. Jacobsen, Optical Properties
J.C. Scott, Magnetic Properties
L. Zuppiroli, Irradiation Effects: Perfect Crystals and Real Crystals

CONTENTS OF PREVIOUS VOLUMES

Volume 28 Measurement of High-Speed Signals in Solid State Devices

J. Frey and D. Ioannou, Materials and Devices for High-Speed and Optoelectronic Applications
H. Schumacher and E. Strid, Electronic Wafer Probing Techniques
D. H. Auston, Picosecond Photoconductivity: High-Speed Measurements of Devices and Materials
J. A. Valdmanis, Electro-Optic Measurement Techniques for Picosecond Materials, Devices, and Integrated Circuits
J. M. Wiesenfeld and R. K. Jain, Direct Optical Probing of Integrated Circuits and High-Speed Devices
G. Plows, Electron-Beam Probing
A. M. Weiner and R. B. Marcus, Photoemissive Probing

Volume 29 Very High Speed Integrated Circuits: Gallium Arsenide LSI

M. Kuzuhara and T. Nozaki, Active Layer Formation by Ion Implantation
H. Hashimoto, Focused Ion Beam Implantation Technology
T. Nozaki and A. Higashisaka, Device Fabrication Process Technology
M. Ino and T. Takada, GaAs LSI Circuit Design
M. Hirayama, M. Ohmori, and K. Yamasaki, GaAs LSI Fabrication and Performance

SEP 2 8 1990